The Research about Invariants
of Ordinary Differential Equations

The Research about Invariants of Ordinary Differential Equations

Roger Chalkley

Copyright © 2018 by Roger Chalkley
All rights reserved.

ISBN-13: 9781985381193

Chalkley, Roger 1931-
    The Research about Invariants of Ordinary Differential Equations

Printed by CreateSpace

2010 *Mathematics Subject Classification.* Primary 34M15, 34 02

Key words and phrases. relative invariant, semi-invariant, differential polynomial

ABSTRACT. Several basic relative invariants for homogeneous linear differential equations were discovered during the years shortly after 1878. Also, a basic relative invariant was found by Paul Appell in 1889 for a type of nonlinear differential equation. There was little progress during the years 1892–1988 as researchers who worked with homogeneous linear differential equations were unknowingly handicapped by the standard practice of introducing binomial coefficients in the writing of their equations. They thereby failed to develop adequate formulas for the coefficients of equations resulting from a change of the independent variable. Consequently, for relative invariants as the most important kind of invariant, progress was stymied.

The notation was simplified in 1989, adequate transformation formulas were developed, and explicit expressions were deduced in 2002 for all of the basic relative invariants of homogeneous linear differential equations. In 2007, explicit formulas were obtained for all of the basic relative invariants of a type of ordinary differential equation involving two parameters m and n that represent positive integers. When $n = 1$ and $m \geq 3$, the formulas specialize to provide all of the basic relative invariants for homogeneous linear differential equations of order m; and, when $m = n = 2$, they yield all three of the basic relative invariants for the equations of Paul Appell.

A general method developed in 2014 combines two relative invariants of weights p and q for the same type of equation to explicitly obtain a relative invariant of weight $p + q + r$, for any $r \geq 0$. With that, the principal problems about relative invariants have now been solved.

This monograph provides clear perspective about the reformulation begun after 1988 and recently completed. Chapters 15 and 18 show how the major difficulties confronting earlier researchers have been overcome.

Contents

Preface   xi

Chapter 1.   Historical Introduction   1
   1.1.   Notation to avoid   1
   1.2.   Relative invariant of Edmund Laguerre   1
   1.3.   Terminology for homogeneous linear differential equations   2
   1.4.   Relative invariants of Georges-Henri Halphen   3
   1.5.   Infinitesimal Transformations of Andrew Forsyth   4
   1.6.   Laguerre-Forsyth canonical forms   5
   1.7.   Differential equations of Paul Appell   5
   1.8.   Recent developments   6
   1.9.   Principal results not in Memoirs [19] and [20]   8
   1.10.   Subsidiary details   8
   1.11.   Instructive observations   10

Part 1.   General Perspective   11

Chapter 2.   The Importance of Suitable Notation   13
   2.1.   The representation (2.1) for the equations of interest   13
   2.2.   A transitional form (2.4) for (2.1)   13
   2.3.   A rewritten form (2.10)–(2.11) for (2.1) that is ideal   14
   2.4.   Uniqueness for (2.10) subject to (2.11)   16
   2.5.   Representations in Chapter 4 for the context about invariants   17

Chapter 3.   Coefficients of Transformed Equations   19
   3.1.   The coefficients $c^*_{i_1,\, i_2,\, \ldots,\, i_n}(z)$ for (2.32)–(2.33) and (4.4)–(4.5)   19
   3.2.   Perspective   21
   3.3.   The coefficients $c^{**}_{i_1,\, i_2,\, \ldots,\, i_n}(\zeta)$ for (2.35)–(2.36) and (4.7)–(4.8)   23
   3.4.   Observations about modifications versus abstractions   26
   3.5. $\mathcal{C}_{m,n}$ as a set of elements   27

Chapter 4.   Consistent Reformulation from 1989 Onward   29
   4.1.   Transformations for the equations of $\mathcal{C}_{m,n}$ 29
   4.2.   The ring $\mathcal{R}_{m,n}$ of differential polynomials   30
   4.3.   The basic relative invariants in $\mathcal{R}_{m,n}$ for $\mathcal{C}_{m,n}$ when $m \geq 2$ 32
   4.4.   Alternative formulas for $\mathcal{I}_{m,n;\, e_1,\, \ldots,\, e_n}$ when $m \geq 2$ 34
   4.5.   Principal technique for combining relative invariants   36
   4.6.   Several immediate applications of Theorem 4.10   37
   4.7.   Representations of relative invariants   38
   4.8.   The relative invariants in $\mathcal{R}_{3,1}$ for $\mathcal{C}_{3,1}$ of weight $s \leq 13$ 38

4.9.	The terminology *relative invariant*	39
4.10.	Subjects of other chapters	40

Chapter 5. Supplementary Results 41
5.1.	Semi-invariants of the first kind	41
5.2.	Semi-invariants of the second kind	42
5.3.	The number of basic relative invariants in $\mathcal{R}_{m,n}$ for $\mathcal{C}_{m,n}$	47
5.4.	Nonsolutions of nonzero equations	48
5.5.	Relative invariants in terms of basic ones and $\boldsymbol{a}_{m,n}$	50

Chapter 6. Use of Computer Algebra 53
6.1.	$\boldsymbol{\mathcal{I}}_{m,1;\,e_1}$ in $\mathcal{R}_{m,1}$ for $\mathcal{C}_{m,1}$ when $m \geq 3$	53
6.2.	Alternative computation for $\boldsymbol{\mathcal{I}}_{m,1;\,e_1}$ when $m \geq 3$	54
6.3.	$\boldsymbol{\mathcal{I}}_{m,2;\,e_1,e_2}$ in $\mathcal{R}_{m,2}$ for $\mathcal{C}_{m,2}$ when $m \geq 2$	56
6.4.	Alternative computation for $\boldsymbol{\mathcal{I}}_{m,2;\,e_1,e_2}$ when $m \geq 2$	58
6.5.	$\boldsymbol{C}_{p,q,r}(\boldsymbol{P}, \boldsymbol{Q})$ as a differential-polynomial combination of $\boldsymbol{P}$, $\boldsymbol{Q}$, and $\boldsymbol{a}_{m,n}$ over $\mathbb{Q}$	59
6.6.	Several identities	62
6.7.	Observations about computations	63

Chapter 7. Principal Theorems Applied to Paul Appell's Study of $\mathcal{C}_{2,2}$ 65
7.1.	Solution procedures for two special kinds of equations in $\mathcal{C}_{2,2}$	65
7.2.	Representations for $\boldsymbol{E}_6$, $\boldsymbol{E}_7$, and $\boldsymbol{D}_2$	66

Chapter 8. Separate Examination of $\boldsymbol{C}_{p,q,1}(\boldsymbol{P}, \boldsymbol{Q})$ 71
8.1.	Properties of $\boldsymbol{C}_{p,q,1}(\boldsymbol{P}, \boldsymbol{Q})$	71
8.2.	The condition that $\boldsymbol{P}^q$ and $\boldsymbol{Q}^p$ are linearly independent over $\mathbb{Q}$	73
8.3.	Several identities	74

Part 2. Proof of Theorem 4.10 75

Chapter 9. Invariant Character of $\boldsymbol{C}_{p,q,r}(\boldsymbol{P}, \boldsymbol{Q})$ when $m, r \geq 2$ 77
9.1.	Introduction of $\boldsymbol{R}$ and $\phi_{h,i,j}(z)$	77
9.2.	Formula for $R^{**}(\zeta)$ that involves $\mathfrak{A}_{p,q,r,s,t}(\zeta)$ of (9.11)	79
9.3.	Reformulation for $\mathfrak{A}_{p,q,r,s,t}(g(z))$	81
9.4.	Initial simplification for $E_{h,i,j}(z)$	82
9.5.	Properties of $E_{h,i,j}(z)$	87
9.6.	Simplification for $\mathfrak{A}_{p,q,r,s,t}(\zeta)$	92
9.7.	$\boldsymbol{C}_{p,q,r}(\boldsymbol{P}, \boldsymbol{Q})$ is a relative invariant when nonzero	95

Chapter 10. Conditions for $\boldsymbol{C}_{p,q,r}(\boldsymbol{P}, \boldsymbol{Q}) \not\equiv 0$ when $m, r \geq 2$ 97
10.1.	The dependence of $\boldsymbol{C}_{q,p,r}(\boldsymbol{Q}, \boldsymbol{P})$ on $\boldsymbol{C}_{p,q,r}(\boldsymbol{P}, \boldsymbol{Q})$	97
10.2.	Several lemmas for use in Section 10.3	99
10.3.	The situations where $\boldsymbol{C}_{p,q,r}(\boldsymbol{P}, \boldsymbol{Q}) \not\equiv 0$	101

Part 3. Independent Verification for Theorem 4.10 105

Chapter 11. Symmetry with Respect to Semi-Invariants 107
11.1.	Context employed and definition of $\boldsymbol{S}$	107
11.2.	When $\boldsymbol{S}$ is a semi-invariant of the second kind	109

11.3.	An expansion for differential polynomials like $\mathcal{P}_k$ in (11.11)	109
11.4.	Formula for S that involves $F_{p,q,r,s,t}$ of (11.23)	111
11.5.	Reformulation in (11.29) for $F_{p,q,r,s,t}$ of (11.23)	112
11.6.	Initial reformulation for $L_{h,i,j}$	113
11.7.	Final reformulation for $L_{h,i,j}$	116
11.8.	Simplification for $F_{p,q,r,s,t}$ that yields $S \equiv R$	121

Part 4. Relative Invariants of a Given Weight   125

Chapter 12. Representations Involving $C_{p,q,r}(P, Q)$ 127
12.1.	The relative invariants in $\mathcal{R}_{4,1}$ for $\mathcal{C}_{4,1}$ of weight $s \leq 12$	127
12.2.	The relative invariants in $\mathcal{R}_{5,1}$ for $\mathcal{C}_{5,1}$ of weight $s \leq 12$	128
12.3.	The relative invariants in $\mathcal{R}_{2,2}$ for $\mathcal{C}_{2,2}$ of weight $s \leq 12$	130

Chapter 13. Computer Algebra for $\mathcal{V}_{3,1;s}$, $\mathcal{V}_{4,1;s}$, $\mathcal{V}_{5,1;s}$, ...   133
13.1.	The relative invariants of weight s in $\mathcal{R}_{m,1}$ for $\mathcal{C}_{m,1}$	133
13.2.	Simple verifications for Section 4.8 about $\mathcal{V}_{3,1;s}$	139
13.3.	Representation of $C_{p,q,r}(P, Q)$ in $\mathcal{R}_{m,1}$ with respect to (4.9)	139
13.4.	Alternative verifications for Section 4.8 about $\mathcal{V}_{3,1;s}$	140
13.5.	Verifications for Section 12.1 about $\mathcal{V}_{4,1;s}$	141
13.6.	Verifications for Section 12.2 about $\mathcal{V}_{5,1;s}$	142
13.7.	Observations about versions of *Mathematica*	143

Chapter 14. Computer Algebra for $\mathcal{V}_{2,2;s}$, $\mathcal{V}_{3,2;s}$, $\mathcal{V}_{4,2;s}$, ...   145
14.1.	The relative invariants of weight s in $\mathcal{R}_{m,2}$ for $\mathcal{C}_{m,2}$	145
14.2.	Representation of $C_{p,q,r}(P, Q)$ in $\mathcal{R}_{m,2}$ with respect to (4.9)	151
14.3.	Verifications for Section 12.3 about $\mathcal{V}_{2,2;s}$	152

Part 5. Modifications Required for Developments before 1989   155

Chapter 15. Suitable Formulas for Transformations
of Homogeneous Linear Differential Equations   157
15.1.	Introduction.	157
15.2.	Consequences due to an improved notation	158
15.3.	Previously missing essential formula for older research	159

Chapter 16. Computer Algebra with Formulas (15.9)–(15.18)   161
16.1.	Computer representations for $c_i^*(z)$ and $c_i^{**}(\zeta)$	161
16.2.	Applications based on the representations for $c_i^*(z)$ and $c_i^{**}(\zeta)$	162

Chapter 17. Computer Algebra with Formulas (15.1)–(15.8)   167
17.1.	Computer-algebra representations of $C_i^*(z)$ and $C_i^{**}(\zeta)$.	167
17.2.	Computer-algebra verifications.	168

Chapter 18. Suitable Context for Older Notation   171
18.1.	Symbolism and terminology	171
18.2.	Our viewpoint abut the older Cockle-semi-invariants	172
18.3.	Original introduction of $\widehat{G}_i(z)$, $\widehat{G}_i^*(z)$, and $\widehat{G}_i^*(z) \equiv \widehat{G}_i(z)$	175
18.4.	Results of Forsyth in the context for Sections 18.1 and 18.2	176
18.5.	Computer-algebra verification of (18.23)	178
18.6.	Several observations	180

18.7. Computer-algebra verification of (18.24)   180
18.8. Brief summary   182

Bibliography   183

Index   187

Preface

The subject of relative invariants for ordinary differential equations has been completely redeveloped in a series of publications begun in 1989. Now, there are satisfactory solutions to the principal unsolved problems that provided interest for researchers after Edmund Laguerre found a relative invariant in [**37, 38**] of 1879 for third-order homogeneous linear differential equations and the French Academy of Sciences encouraged extensions of his research. In particular, Georges-Henri Halphen won the 1880 Grand Prize of the French Academy of Sciences for research about invariants published in [**32**] and Henri Poincaré received honorable mention for his competitive submission to them in 1880.

Explicit formulas for all of the basic relative invariants of homogeneous linear differential equations of each fixed order were found and presented in [**19**] of 2002. For a type of nonlinear differential equation studied by Paul Appell in [**4**] of 1889, he discovered one of its three basic relative invariants. The other two were obtained for [**20**] of 2007 and all three appear in [**20**, page 13, Theorem 1.8] of 2007.

As a remarkable generalization not anticipated by earlier researchers, all of the basic relative invariants were discovered and presented by explicit formulas in [**20**, pages 257, 264, 275–276] for a type of ordinary differential equation involving two integral parameters m and n, where m is the order of the equation and n is its degree when its left member is regarded as a homogeneous polynomial in the various derivatives of the dependent variable. In particular, when $n = 1$, the formulas specialize to yield the ones in [**19**] for the basic relative invariants of homogeneous linear differential equations of each order $m \geq 3$; and, when $m = n = 2$, they specialize to yield the ones in [**20**] for the three basic relative invariants of the nonlinear equations Paul Apple studied in [**4**].

To complete the research involving the preceding results, a construction was developed in [**21**] of 2014 where, under general conditions, it combines two relative invariants of respective weights p and q for the same type of equation to produce a relative invariant of weight $p + q + r$, for any integer $r \geq 0$. Examples were also given in [**21**] to illustrate how, starting with the basic relative invariants for a given type of equation, that construction can be repeatedly applied to obtain linearly independent relative invariants of a given weight whose linear combinations yield all of the relative invariants having that weight.

This revision of [**21**] includes Chapters 15 and 18 as new ones to show why, after a flurry of intense interest during the years 1879–1891, the subject remained in limbo until 1989. In particular, these chapters make precise the principal difficulties earlier researchers failed to overcome.

Roger Chalkley

CHAPTER 1

Historical Introduction

Anyone knowledgeable about the differential calculus can effortlessly develop an interest in the subject of invariants by interacting with the computer-algebra aspects of Chapters 16 and 17 before becoming involved with details.

The initial discovery of a *relative invariant* was made by Edmund Laguerre in [**37, 38**] of 1879 when he found one for monic third-order homogeneous linear differential equations. The principal research about relative invariants before 1989 was performed by Edmund Laguerre, Georges-Henri Halphen, Andrew Forsyth, and Paul Appell during the years 1879–1889. For Laguerre's use of *relative*, see page 39.

1.1. Notation to avoid

Chapters 15 and 18 provide a detailed explanation why research about relative invariants languished during the years from 1890 through 1988. The main cause was that authors who wrote papers about monic homogeneous linear differential equations before 1989 used notation like

$$(1.0) \quad y^{(m)}(z) + \sum_{j=1}^{m} \binom{m}{j} d_j(z) y^{(m-j)}(z) = 0, \quad \text{with binomial coefficients } \binom{m}{j}.$$

Chapter 15 shows how that notation hindered the discovery of adequate formulas for the coefficients of equations resulting from a change of the independent variable. The abandonment of that notation in [**14, 16, 17**] during 1989–1993 was responsible for the advances in [**19, 20**]. Throughout, we shall avoid notation like that of (1.0) except for Chapters 15, 17, and 18 where early details are examined.

1.2. Relative invariant of Edmund Laguerre

To recall a result of Edmund Laguerre, we note that: when $c_1(z)$, $c_2(z)$, $c_3(z)$ are any three meromorphic functions on a region Ω of the complex plane and $\rho(z)$ is a not-identically-zero meromorphic function on Ω, there are unique meromorphic functions $c_1^*(z)$, $c_2^*(z)$, $c_3^*(z)$ on Ω such that the monic third-order homogeneous linear differential equation

$$(1.1) \quad y'''(z) + c_1(z) y''(z) + c_2(z) y'(z) + c_3(z) y(z) = 0, \quad \text{on } \Omega,$$

is transformed by the substitution

$$(1.2) \quad y(z) = \rho(z) v(z)$$

into the monic third-order homogeneous linear differential equation

$$(1.3) \quad v'''(z) + c_1^*(z) v''(z) + c_2^*(z) v'(z) + c_3^*(z) v(z) = 0, \quad \text{on } \Omega.$$

Each of $c_1^*(z)$, $c_2^*(z)$, $c_3^*(z)$ can be expressed in terms of $c_1(z)$, $c_2(z)$, $c_3(z)$ and the derivatives of $\rho(z)$ by simple hand-written computations that yield

$$(*)\quad \begin{bmatrix} c_3^*(z) - \frac{1}{3} c_1^*(z)\, c_2^*(z) \\ + \frac{2}{27}\left(c_1^*(z)\right)^3 - \frac{1}{2} c_2^{*(1)}(z) \\ + \frac{1}{3} c_1^*(z)\, c_1^{*(1)}(z) \\ + \frac{1}{6} c_1^{*(2)}(z) \end{bmatrix} \equiv \begin{bmatrix} c_3(z) - \frac{1}{3} c_1(z)\, c_2(z) \\ + \frac{2}{27}\left(c_1(z)\right)^3 - \frac{1}{2} c_2^{(1)}(z) \\ + \frac{1}{3} c_1(z)\, c_1^{(1)}(z) \\ + \frac{1}{6} c_1^{(2)}(z) \end{bmatrix},$$

for each z in Ω.

When $c_1(z)$, $c_2(z)$, $c_3(z)$ are any three meromorphic functions defined on a region Ω and $z = f(\zeta)$ is a univalent analytic function on a region Ω^{**} with $f(\Omega^{**}) = \Omega$, there are unique meromorphic functions $c_1^{**}(\zeta)$, $c_2^{**}(\zeta)$, $c_3^{**}(\zeta)$ on Ω^{**} such that the substitution

(1.4) $\qquad z = f(\zeta), \quad \text{with} \quad u(\zeta) = y(f(\zeta)),$

transforms (1.1) into

(1.5) $\quad u'''(\zeta) + c_1^{**}(\zeta)\, u''(\zeta) + c_2^{**}(\zeta)\, u'(\zeta) + c_3^{**}(\zeta)\, u(\zeta) = 0, \quad \text{on } \Omega^{**}.$

Each of $c_1^{**}(\zeta)$, $c_2^{**}(\zeta)$, $c_3^{**}(\zeta)$ can be expressed in terms of $c_1(z)$, $c_2(z)$, $c_3(z)$, as well as $z = f(\zeta)$ and derivatives of $f(\zeta)$ by simple computations that yield

$$(**)\quad \begin{bmatrix} c_3^{**}(\zeta) - \frac{1}{3} c_1^{**}(\zeta)\, c_2^{**}(\zeta) \\ + \frac{2}{27}\left(c_1^{**}(\zeta)\right)^3 - \frac{1}{2} c_2^{**(1)}(\zeta) \\ + \frac{1}{3} c_1^{**}(\zeta)\, c_1^{**(1)}(\zeta) \\ + \frac{1}{6} c_1^{**(2)}(\zeta) \end{bmatrix} \equiv (f'(\zeta))^3 \begin{bmatrix} c_3(f(\zeta)) - \frac{1}{3} c_1(f(\zeta))\, c_2(f(\zeta)) \\ + \frac{2}{27}\left(c_1(f(\zeta))\right)^3 - \frac{1}{2} c_2^{(1)}(f(\zeta)) \\ + \frac{1}{3} c_1(f(\zeta))\, c_1^{(1)}(f(\zeta)) \\ + \frac{1}{6} c_1^{(2)}(f(\zeta)) \end{bmatrix},$$

for each ζ in Ω^{**}.

We represent the relative invariant associated with $(*)$ and $(**)$ by

(1.6) $\quad \mathcal{I}_{3,1;\,3} \equiv \boldsymbol{w}_3 - \frac{1}{3}\boldsymbol{w}_1\boldsymbol{w}_2 + \frac{2}{27}(\boldsymbol{w}_1)^3 - \frac{1}{2}\boldsymbol{w}_2^{(1)} + \frac{1}{3}\boldsymbol{w}_1\boldsymbol{w}_1^{(1)} + \frac{1}{6}\boldsymbol{w}_1^{(2)}$

and regard it as a differential polynomial into which substitutions can be made. Thus, with $\boldsymbol{w}_i = \boldsymbol{w}_i^{(0)}$, if $\mathcal{I}_{3,1;\,3}(z)$, $\mathcal{I}_{3,1;\,3}^*(z)$, and $\mathcal{I}_{3,1;\,3}^{**}(\zeta)$ are respectively obtained by replacing each $\boldsymbol{w}_i^{(j)}$ in $\mathcal{I}_{3,1;\,3}$ with the corresponding $c_i^{(j)}(z)$ from (1.1), $c_i^{*(j)}(z)$ from (1.3), and $c_i^{**(j)}(\zeta)$ from (1.5), then $(*)$ and $(**)$ may be written as

$\mathcal{I}_{3,1;\,3}^*(z) \equiv \mathcal{I}_{3,1;\,3}(z), \quad \text{on } \Omega, \quad \text{and} \quad \mathcal{I}_{3,1;\,3}^{**}(\zeta) \equiv (f'(\zeta))^3\, \mathcal{I}_{3,1;\,3}(f(\zeta)), \quad \text{on } \Omega^{**}.$

We note that the relative invariant of Edmund Laguerre presented in [37, page 117] of 1879, or in [33, page 421], corresponds to the notation $2\mathcal{I}_{3,1;\,3}$.

Composites of some (1.2) and some (1.4) yield the transformations of (1.1).

1.3. Terminology for homogeneous linear differential equations

When $c_1(z)$, $c_2(z)$, ..., $c_m(z)$ are meromorphic functions on a region Ω of the complex plane and $\rho(z)$ is a not-identically-zero meromorphic function on Ω, there are unique meromorphic functions $c_1^*(z)$, $c_2^*(z)$, ..., $c_m^*(z)$ on Ω such that the monic mth-order homogeneous linear differential equation

(1.7) $\qquad y^{(m)}(z) + \sum_{j=1}^{m} c_j(z)\, y^{(m-j)}(z) = 0, \quad \text{on } \Omega,$

is transformed by the substitution
$$y(z) = \rho(z)\, v(z) \tag{1.8}$$
into the monic mth-order homogeneous linear differential equation
$$v^{(m)}(z) + \sum_{i=1}^{m} c_i^*(z)\, v^{(m-i)}(z) = 0, \quad \text{on } \Omega. \tag{1.9}$$
We set $n = 1$ in (3.4) of page 19 to obtain $c_i^*(z)$ for (1.9) as in (15.12) of page 158.

In terms of meromorphic functions $c_1(z), c_2(z), \ldots, c_m(z)$ on a region Ω and a univalent analytic function $z = f(\zeta)$ on a region Ω^{**} with $f(\Omega^{**}) = \Omega$, there are unique meromorphic functions $c_1^{**}(\zeta), c_2^{**}(\zeta), \ldots, c_m^{**}(\zeta)$ on Ω^{**} such that the substitution
$$z = f(\zeta), \quad \text{with} \quad u(\zeta) = y(f(\zeta)), \tag{1.10}$$
transforms (1.7) into
$$u^{(m)}(\zeta) + \sum_{i=1}^{m} c_i^{**}(\zeta)\, u^{(m-i)}(\zeta) = 0, \quad \text{on } \Omega^{**}. \tag{1.11}$$

Satisfactory formulas for the $c_i^{**}(\zeta)$ were available only after the investigations in [14, 16, 17] of 1989–1993 that lead to [19, page 136, Theorem A.3] of 2002. To verify that, see the beginning of Section 15.1 on page 157. Here, for $c_i^{**}(\zeta)$ in (1.11), set $n = 1$ in (3.21)–(3.24) on page 24 and obtain (15.16)–(15.18) on page 159.

For any polynomial $\boldsymbol{I}$ over the field $\mathbb{Q}$ of rational numbers in variables $\boldsymbol{w}_i^{(j)}$ having $1 \leq i \leq m$ and $j \geq 0$, let $I(z)$ on Ω, $I^*(z)$ on Ω, and $I^{**}(\zeta)$ on Ω^{**} denote the functions respectively obtained by replacing each $\boldsymbol{w}_i^{(j)}$ in $\boldsymbol{I}$ with the corresponding $c_i^{(j)}(z)$ from (1.7), $c_i^{*(j)}(z)$ from (1.9), and $c_i^{**(j)}(\zeta)$ from (1.11). Then, $\boldsymbol{I}$ is a *relative invariant* for mth-order homogeneous linear differential equations when $\boldsymbol{I}$ effectively involves at least one $\boldsymbol{w}_i^{(j)}$ and there is a fixed positive integer s such that
$$I^*(z) \equiv I(z), \quad \text{on } \Omega, \quad \text{and} \quad I^{**}(\zeta) \equiv \bigl(f'(\zeta)\bigr)^s I(f(\zeta)), \quad \text{on } \Omega^{**}, \tag{1.12}$$
for each equation (1.7) as well as each (1.8) and (1.10).

1.4. Relative invariants of Georges-Henri Halphen

For each fixed integer $m \geq 3$, the polynomial
$$\boldsymbol{\mathcal{I}}_{m,1;3} \equiv \boldsymbol{w}_3 - \frac{m-2}{m}\boldsymbol{w}_1\boldsymbol{w}_2 + \frac{(m-1)(m-2)}{3m^2}(\boldsymbol{w}_1)^3 - \frac{m-2}{2}\boldsymbol{w}_2^{(1)} \tag{1.13}$$
$$+ \frac{(m-1)(m-2)}{2m}\boldsymbol{w}_1\boldsymbol{w}_1^{(1)} + \frac{(m-1)(m-2)}{12}\boldsymbol{w}_1^{(2)}$$
is a relative invariant with $s = 3$ for the equations (1.7) of order m. This is a consequence of Theorem 4.6 on page 32. It was published in a different form by G.-H. Halphen in [32, page 127] of 1884. Namely, by rewriting his expression for V in [32, page 127, Equation (10)] or [35, page 112, Equation (10)] with respect to the coefficients of (1.7), we find that
$$V \equiv \left(\frac{-12}{m(m-1)(m-2)}\right) \mathcal{I}_{m,1;3}(z), \tag{1.14}$$
where, with $\boldsymbol{w}_i = \boldsymbol{w}_i^{(0)}$, $\mathcal{I}_{m,1;3}(z)$ is the function on Ω obtained by replacing each $\boldsymbol{w}_i^{(j)}$ in (1.13) with the corresponding $c_i^{(j)}(z)$ from (1.7).

In particular, for $m = 3$, (1.13) yields (1.6).

By setting $m = 4$ in (1.13), we see that the differential polynomial

(1.15) $\quad \mathcal{I}_{4,1;3} \equiv w_3 - \frac{1}{2} w_1 w_2 + \frac{1}{8}(w_1)^3 - w_2^{(1)} + \frac{3}{4} w_1 w_1^{(1)} + \frac{1}{2} w_1^{(2)}$

is a relative invariant with $s = 3$ for the differential equations

(1.16) $\quad y^{(4)}(z) + c_1(z) y^{(3)}(z) + c_2(z) y^{(2)}(z) + c_3(z) y^{(1)}(z) + c_4(z) y(z) = 0.$

In [**31**, page 331, Equation (9)] or [**35**, page 469, Equation (9)], G.-H. Halphen had already indicated a relative invariant with $s = 3$ for the equations (1.16). His expression is equal to $(-1/2)\mathcal{I}_{4,1;3}(z)$. For the relative invariants presented here thus far, the computations required to directly verify their properties can be done as hand-written ones without great effort.

G.-H. Halphen used [**31**, page 339, line 3] or [**35**, page 474, line 23] of 1883 to make plausible the existence of a relative invariant with $s = 4$ for (1.16). We used Theorem 4.6 with computer algebra to conclude that the differential polynomial

(1.17) $\quad \mathcal{I}_{4,1;4} \equiv w_4 - \frac{1}{4} w_1 w_3 - \frac{1}{2} w_3^{(1)} - \frac{9}{100}(w_2)^2 + \frac{1}{5} w_2^{(2)} + \frac{13}{100}(w_1)^2 w_2$
$\quad\quad\quad + \frac{27}{100} w_1^{(1)} w_2 + \frac{1}{4} w_1 w_2^{(1)} - \frac{39}{1600}(w_1)^4 - \frac{39}{200}(w_1)^2 w_1^{(1)}$
$\quad\quad\quad - \frac{33}{200}\left(w_1^{(1)}\right)^2 - \frac{3}{20} w_1 w_1^{(2)} - \frac{1}{20} w_1^{(3)}$

is a relative invariant with $s = 4$ for the equations (1.16). For merely a verification, see Example 16.3 on page 164. Section 12.1 shows that each relative invariant having $s = 4$ for the equations (1.16) is expressible in the form $\gamma \mathcal{I}_{4,1;4}$, for some nonzero rational number γ.

1.5. Infinitesimal Transformations of Andrew Forsyth

To find explicit expressions for the coefficients of various relative invariants, Forsyth recognized in [**28**, pages 394–401] that computations would be considerably simplified if transformations of the type (1.10) were replaced by corresponding infinitesimal transformations where higher order infinitesimals could be neglected. His viewpoint was expressed in a footnote to [**28**, page 394] as follows.

> "The functions are shown by this process to be invariants only for an infinitesimal, but otherwise perfectly general, transformation; but the immediate purpose is to obtain the numerical coefficients and not to prove the property of general invariance, which, otherwise known, could be derived by the principle of cumulative variations." (Andrew Forsyth)

Indeed, for each integer s satisfying $3 \leq s \leq 7$ and for monic homogeneous linear differential equations of order $m \geq s$, his process yields a relative invariant that satisfies (1.12) with that s. Namely, after correcting the tiny misprint that we noticed for [**28**, page 401, Equation (18)] in [**19**, page 79] and describe for $s = 7$ on page 176 of Chapter 18, we used computer algebra in [**19**] to verify that Forsyth's expressions could be identified with $\gamma_{m,s} \mathcal{I}_{m,1;s}$, for $3 \leq s \leq 7$, where $\mathcal{I}_{m,1;s}$ is given by (4.17) of page 32 with $m = m$ (as any integer $\geq s$), $n = 1$, and $e_1 = s$. Since $\mathcal{I}_{m,1;s}$ was shown in [**19**] to be a relative invariant corresponding to (1.12) with $s \geq 3$ and $m \geq s$, the properties of Forsyth's expressions follow from that. Details about this are presented in Sections 18.6 and 18.7 of Chapter 18.

A direct verification for each of the formulas [**28**, pages 399–401, (14)–(18)] in the proof of Theorem 18.7 required a previously unavailable transformation formula.

1.6. Laguerre-Forsyth canonical forms

Andrew Forsyth established and applied in [**28**, pages 403–407] of 1888 the result that: for any homogeneous linear differential equation of order $m \geq 2$ having meromorphic coefficients on a region, there is a subregion on which the restriction of the equation can be transformed into a homogeneous linear differential equation of order m in which the coefficients of the derivatives of order $m-1$ and $m-2$ are zero. This process is described as a local transformation for a homogeneous linear differential equation to a Laguerre-Forsyth canonical form.

By using infinitesimal transformations with reductions to Laguerre-Forsyth canonical forms, Forsyth obtained expressions in [**28**, pages 404–407] that yield identities various relative invariants would give when restricted to transformations of one Laguerre-Forsyth canonical form into another. The corresponding invariants for this restrictive context were descriptively termed *linear invariants* by Forsyth to distinguish them from the true relative invariants of Laguerre and Halphen.

Without need for infinitesimal transformations, the preceding ideas were thoroughly redeveloped in [**19**, pages 39–49] and used in [**19**, Chapters 7–9] to prove the result [**19**, page 6, Main Theorem] that presented explicit formulas for all of the basic relative invariants of homogeneous linear differential equations. The concept of a Laguerre-Forsyth canonical form was extended in [**20**, pages 47–65, 265–274] to more general types of ordinary differential equations and their properties were essential for the verification in [**20**] of the results presented here in Theorem 4.6 on page 32 and Theorem 4.8 on page 34.

1.7. Differential equations of Paul Appell

In [**4**] of 1889, Paul Appell studied the differential equations expressible as

$$(1.18) \quad \left(y''(z)\right)^2 + 2\,c_{0,1}(z)\,y''(z)\,y'(z) + 2\,c_{0,2}(z)\,y''(z)\,y(z) + c_{1,1}(z)\left(y'(z)\right)^2$$
$$+ 2\,c_{1,2}(z)\,y'(z)\,y(z) + c_{2,2}(z)\left(y(z)\right)^2 = 0,$$

where the $c_{i,j}(z)$ are meromorphic functions on a region Ω of the complex plane. For any not-identically-zero meromorphic function $\rho(z)$ on Ω, there are unique meromorphic functions $c_{i,j}^*(z)$ on Ω such that the substitution

$$(1.19) \quad y(z) \equiv \rho(z)\,v(z)$$

transforms (1.18) into

$$(1.20) \quad \left(v''(z)\right)^2 + 2\,c_{0,1}^*(z)\,v''(z)\,v'(z) + 2\,c_{0,2}^*(z)\,v''(z)\,v(z) + c_{1,1}^*(z)\left(v'(z)\right)^2$$
$$+ 2\,c_{1,2}^*(z)\,v'(z)\,v(z) + c_{2,2}^*(z)\left(v(z)\right)^2 = 0, \quad \text{on } \Omega.$$

Also, for any univalent analytic function $z = f(\zeta)$ on a region Ω^{**} with $f(\Omega^{**}) = \Omega$, there are unique meromorphic functions $c_{i,j}^{**}(\zeta)$ on Ω^{**} such that the substitution

$$(1.21) \quad z = f(\zeta), \quad \text{with} \quad u(\zeta) = y\big(f(\zeta)\big),$$

transforms (1.18) into

$$(1.22) \quad \left(u''(\zeta)\right)^2 + 2\,c_{0,1}^{**}(\zeta)\,u''(\zeta)\,u'(\zeta) + 2\,c_{0,2}^{**}(\zeta)\,u''(\zeta)\,u(\zeta) + c_{1,1}^{**}(z)\left(u'(\zeta)\right)^2$$
$$+ 2\,c_{1,2}^{**}(\zeta)\,u'(\zeta)\,u(\zeta) + c_{2,2}^{**}(\zeta)\left(u(\zeta)\right)^2 = 0, \quad \text{on } \Omega^{**}.$$

Simple handwritten computations show that (1.18) and (1.20) yield

$$(1.23) \quad c_{1,1}^*(z) - \left(c_{0,1}^*(z)\right)^2 \equiv c_{1,1}(z) - \left(c_{0,1}(z)\right)^2, \quad \text{on } \Omega,$$

while (1.18) and (1.22) give

$$(1.24) \quad c_{1,1}^{**}(\zeta) - \left(c_{0,1}^{**}(\zeta)\right)^2 \equiv \left(f'(\zeta)\right)^2 \left[c_{1,1}(f(\zeta)) - \left(c_{0,1}(f(\zeta))\right)^2\right], \quad \text{on } \Omega^{**}.$$

In this manner, Paul Appell found the relative invariant representable by

$$(1.25) \quad \mathcal{I}_{2,2;\,1,1} \equiv \boldsymbol{w}_{1,1} - (\boldsymbol{w}_{0,1})^2.$$

For details about other relative invariants closely related to Appell's research in [4], see Chapter 7. In particular, for the basic relative invariants, see page 67.

As motivation for the notation (1.29) and (1.30), we note that the differential equations (1.18) is the special case $m = 2$ of

$$(1.26) \quad \left(y^{(m)}(z)\right)^2 + \sum_{\substack{0 \le j_1, j_2 \le m \\ (j_1,j_2) \ne (0,0)}} c_{j_1,j_2}(z) \prod_{\nu=1}^{2} y^{(m-j_\nu)}(z) = 0,$$

where $c_{0,0}(z) \equiv 1$ and the coefficients $c_{j_1,j_2}(z)$ are meromorphic functions on some region Ω of the complex plane such that

$$(1.27) \quad c_{j_{\pi(1)},\,j_{\pi(2)}}(z) \equiv c_{j_1,j_2}(z), \quad \text{on } \Omega,$$

for $0 \le j_1, j_2 \le m$ and any permutation π of $\{1, 2\}$.

1.8. Recent developments

Earlier researchers lacked several tools essential for our work. There are now adequate transformation formulas from [19] and [20]; e.g., see Chapter 3 as well as Chapter 15. Also, computer algebra enabled us to discover several key identities for [19, 20]. Moreover, modern algebra provides a precise context.

From 1889 until 2002, the principal unsolved problems were the following ones.

PROBLEM 1. For general systems of differential equations, characterize and find explicit formulas for all relative invariants that have the structure of (1.6) for the equations (1.1), the structures of (1.15) and (1.17) for the equations (1.16), the structure of (1.13) for the equations (1.7), as well as the structure of (1.25) for the equations (1.18).

PROBLEM 2. When a satisfactory solution to PROBLEM 1 is found and those special relative invariants are termed basic ones, discover how any relative invariant can be expressed in terms of the basic ones.

Problem 1 was solved in [19] and [20]. Namely, we characterized and found explicit formulas in [19] for all of the basic relative invariants for homogeneous linear differential equations. Then, after doing the same in [20] for nonlinear differential equations like (1.18) and (1.26), we were able to unify in [20, Part 4] those diverse results by characterizing and finding explicit formulas for all of the basic relative invariants of differential equations obtained as special rewritten versions of ones like

$$(1.28) \quad \left(y^{(m)}(z)\right)^n + \sum_{\substack{0 \le i_1 \le i_2 \le \cdots \le i_n \le m \\ (i_1,i_2,\ldots,i_n) \ne (0,0,\ldots,0)}} a_{i_1,i_2,\ldots,i_n}(z) \prod_{\nu=1}^{n} y^{(m-i_\nu)}(z) = 0, \quad \text{on } \Omega,$$

where m, n are positive integers and each $a_{i_1,i_2,\ldots,i_n}(z)$ is a meromorphic function.

Chapter 2 explains the desirability and technique for rewriting each (1.28) as

$$(1.29) \quad \left(y^{(m)}(z)\right)^n + \sum_{\substack{0 \leq j_1, j_2, \ldots, j_n \leq m \\ (j_1, j_2, \ldots, j_n) \neq (0, 0, \ldots, 0)}} c_{j_1, j_2, \ldots, j_n}(z) \prod_{\nu=1}^{n} y^{(m-j_\nu)}(z) = 0, \quad \text{on } \Omega,$$

where m, n are positive integers and the $c_{j_1, j_2, \ldots, j_n}(z)$ are meromorphic functions on some region Ω of the complex plane such that

$$(1.30) \quad c_{0, 0, \ldots, 0}(z) \equiv 1 \quad \text{and} \quad c_{j_{\pi(1)}, j_{\pi(2)}, \ldots, j_{\pi(n)}}(z) \equiv c_{j_1, j_2, \ldots, j_n}(z),$$

for $0 \leq j_1, j_2, \ldots, j_n \leq m$ and any permutation π of $\{1, 2, \ldots, n\}$.

Each pair of positive integers m, n specifies a collection of equation having the form (1.29)–(1.30). For $n = 1$ and $m = m$, they are the homogeneous linear ones of (1.7) and possess relative invariants only when $m \geq 3$; while, for $m = n = 2$, they specialize to the form (1.18). Chapter 4 begins with (1.29)–(1.30) as its subject.

In terms of (1.29)–(1.30) and a not-identically-zero meromorphic function $\rho(z)$ on Ω, Theorem 3.1 specifies meromorphic functions $c^*_{m, n; j_1, j_2, \ldots, j_n}(z)$ on Ω such that the substitution $y(z) = \rho(z) v(z)$ transforms (1.29)–(1.30) into

$$(1.31) \quad \left(y^{(m)}(z)\right)^n + \sum_{\substack{0 \leq j_1, j_2, \ldots, j_n \leq m \\ (j_1, j_2, \ldots, j_n) \neq (0, 0, \ldots, 0)}} c^*_{m, n; j_1, j_2, \ldots, j_n}(z) \prod_{\nu=1}^{n} y^{(m-j_\nu)}(z) = 0, \quad \text{on } \Omega,$$

where

$$(1.32) \quad c^*_{m, n; 0, 0, \ldots, 0}(z) \equiv 1 \quad \text{and} \quad c^*_{m, n; j_{\pi(1)}, j_{\pi(2)}, \ldots, j_{\pi(n)}}(z) \equiv c^*_{m, n; j_1, j_2, \ldots, j_n}(z),$$

for $0 \leq j_1, j_2, \ldots, j_n \leq m$ and any permutation π of $\{1, 2, \ldots, n\}$.

However, the dependence of $c^*_{m, n; j_1, j_2, \ldots, j_n}(z)$ on m and n can be implied by the context. Thus, we shall henceforth write $c^*_{j_1, j_2, \ldots, j_n}(z)$ for $c^*_{m, n; j_1, j_2, \ldots, j_n}(z)$. Then, (1.3), (1.9), and (1.20) are included as special cases. In particular, see (4.4).

For (1.29)–(1.30) and a univalent analytic function $z = f(\zeta)$ on a region Ω^{**} with $f(\Omega^{**}) = \Omega$, Theorem 3.3 specifies meromorphic functions $c^{**}_{m, n; j_1, j_2, \ldots, j_n}(\zeta)$ on Ω^{**} such that $z = f(\zeta)$, with $u(\zeta) = y(f(\zeta))$, transforms (1.29)–(1.30) into

$$(1.33) \quad \left(u^{(m)}(\zeta)\right)^n + \sum_{\substack{0 \leq j_1, j_2, \ldots, j_n \leq m \\ (j_1, j_2, \ldots, j_n) \neq (0, 0, \ldots, 0)}} c^{**}_{m, n; j_1, j_2, \ldots, j_n}(\zeta) \prod_{\nu=1}^{n} u^{(m-j_\nu)}(\zeta) = 0, \quad \text{on } \Omega^{**},$$

where

$$(1.34) \quad c^{**}_{m, n; 0, 0, \ldots, 0}(\zeta) \equiv 1 \quad \text{and} \quad c^{**}_{m, n; j_{\pi(1)}, j_{\pi(2)}, \ldots, j_{\pi(n)}}(\zeta) \equiv c^{**}_{m, n; j_1, j_2, \ldots, j_n}(\zeta),$$

for $0 \leq j_1, j_2, \ldots, j_n \leq m$ and any permutation π of $\{1, 2, \ldots, n\}$.

The dependence of $c^{**}_{m, n; j_1, j_2, \ldots, j_n}(\zeta)$ on m and n can be inferred from the context. Thus, we shall henceforth write $c^{**}_{j_1, j_2, \ldots, j_n}(\zeta)$ for $c^{**}_{m, n; j_1, j_2, \ldots, j_n}(\zeta)$. This enables (1.5), (1.11), and (1.22) to be included as special cases. Also, see (4.7).

The equations (1.29)–(1.30) possess relative invariants if and only if the fixed positive integers m and n satisfy $(m, n) \neq (1, 1), (2, 1)$. In regard to the situation of historical interest where $m \geq 2$, we characterize the basic relative invariants and present formulas for all of them in Sections 4.3–4.4. For the remarkably simple situation where $m = 1$ and $n \geq 2$, see [20, pages 257–260]. Thus, with PROBLEM 1 solved, the emphasis of this monograph is focused on PROBLEM 2.

1.9. Principal results not in Memoirs [19] and [20]

The subject of relative invariants for differential equations was thoroughly redeveloped in [19] of 2002 and [20] of 2007. There, the concept of a basic relative invariant was formalized and we presented a single set of explicit formulas for the construction of the basic relative invariants for a wide variety of equations. The principal problem left unsolved in [19, 20] was that of explicitly constructing the other relative invariants from the basic ones. At that time, methods of combining two relative invariants to construct others had not been investigated deeply.

In this revision of [21], we continue without alteration the examination of the construction presented on page 36 that uses relative invariants $\boldsymbol{P}$ and $\boldsymbol{Q}$ of respective weights p and q for the same type of equation to construct, for each integer $r \geq 0$, a differential-polynomial combination $\boldsymbol{C}_{p,q,r}(\boldsymbol{P}, \boldsymbol{Q})$ of $\boldsymbol{P}$ and $\boldsymbol{Q}$ over the field $\mathbb{Q}$ of rational numbers such that: $\boldsymbol{C}_{p,q,r}(\boldsymbol{P}, \boldsymbol{Q})$ is a relative invariant of weight $p+q+r$ if and only if $\boldsymbol{C}_{p,q,r}(\boldsymbol{P}, \boldsymbol{Q}) \not\equiv 0$.

For $r \geq 2$, Theorem 4.10 of page 36 establishes that $\boldsymbol{C}_{p,q,r}(\boldsymbol{P}, \boldsymbol{Q})$ is a relative invariant of weight $p+q+r$ if and only if r is an even integer or $\boldsymbol{P}$ and $\boldsymbol{Q}$ are linearly independent over $\mathbb{Q}$.

For $r = 1$, we have $\boldsymbol{C}_{p,q,1}(\boldsymbol{P}, \boldsymbol{Q}) \equiv \boldsymbol{P}\boldsymbol{Q}^{(1)} - (q/p)\boldsymbol{P}^{(1)}\boldsymbol{Q}$. Proposition 8.1 on page 71 shows that $\boldsymbol{C}_{p,q,1}(\boldsymbol{P}, \boldsymbol{Q})$ is a relative invariant of weight $p+q+1$ if and only if $\boldsymbol{P}^q$ and $\boldsymbol{Q}^p$ are linearly independent over $\mathbb{Q}$. To interpret this, see page 73.

For $r = 0$, $\boldsymbol{C}_{p,q,0}(\boldsymbol{P}, \boldsymbol{Q})$ is the relative invariant $\boldsymbol{P}\boldsymbol{Q}$ of weight $p+q$.

Part 1 of this monograph provides the general perspective of [20]. Part 2 gives a proof for Theorem 4.10. An alternate proof is given in Part 3. The results about $\boldsymbol{C}_{p,q,r}(\boldsymbol{P}, \boldsymbol{Q})$ are used in Part 4 with the basic relative invariants for several types of equations to construct all the relative invariants of a given weight for those equations. Part 5 clarifies the problems faced by researchers before 1989 when difficulties were created through the use of notation like that in (1.0).

1.10. Subsidiary details

To define semi-invariants of the first and second kinds for the equations (1.7), let $\boldsymbol{I}$ denote a polynomial over $\mathbb{Q}$ in variables $\boldsymbol{w}_i^{(j)}$, with $1 \leq i \leq m$ and $j \geq 0$, such that $\boldsymbol{I}$ effectively involves some $\boldsymbol{w}_i^{(j)}$. When the first condition $I^*(z) \equiv I(z)$ of (1.12) is satisfied, $\boldsymbol{I}$ is said to be a *semi-invariant of the first kind* for the equations (1.7). When the second condition $I^{**}(\zeta) \equiv \bigl(f'(\zeta)\bigr)^s I\bigl(f(\zeta)\bigr)$ of (1.12) is satisfied for some positive integer s, $\boldsymbol{I}$ is said to be a *semi-invariant of the second kind* for the equations (1.7). Thus, $\boldsymbol{I}$ is a *relative invariant* for the equations (1.7) if and only if $\boldsymbol{I}$ is both a semi-invariant of the first kind and a semi-invariant of the second kind for them. This terminology was introduced by Edmund Laguerre in [37, 38, 33]. For more detail about it, see Section 4.9.

The research of James Cockle in numerous papers typified by [22] of 1862 yielded semi-invariants of the first kind for homogeneous linear differential equations of various orders. Example 17.1 of page 168 is a reformulation of his result in [22]. His research in [23] of 1875 can be verified to give a semi-invariant of the second kind with $s = 2$ for each homogeneous linear differential equation of order $m \geq 2$. Example 17.2 of page 168 provides details. For additional information about the semi-invariantss of James Cockle, see [7].

As noted on page 4, Andrew Forsyth introduced infinitesimal transformations in [28] of 1888 to derive the coefficients for formulas thought likely to specify relative invariants of respective weights $s = 3, 4, 5, 6, 7$ for homogeneous linear differential equations of order $m \geq s$. Because his expressions illustrate well how our explicit formulas for the coefficients of transformed equations can be used to check the validity of older results, we include them as (18.17)–(18.21) on page 176. After a tiny misprint in [28, page 401, (18)] was corrected for (18.21), the results of Section 18.5 verify directly that (18.17)–(18.21) yield relative invariants.

Forsyth's use of infinitesimal transformations had a strong influence on later research even though his formulas corresponding to (18.17)–(18.21) were insufficient for the purpose of discovering a general pattern that would yield a relative invariant of any weight $s \geq 3$ for homogeneous linear differential equations of order $m \geq s$. In particular, Francisco Brioschi used infinitesimal transformations in [8, page 235] of 1891 where he presented (18.17)–(18.21) in a different form. But, his expression that corresponds to (18.21) is thoroughly incorrect. Ludwig Schlesinger employed infinitesimal transformations in [47, page 196] of 1897 when he included without alteration the formulas of [8, page 235]. Infinitesimal transformations were the focus for developments in [7] of 1899 and [53] of 1906. Moreover, they also appeared in [26] of 1903 for the nonlinear equations (1.18).

The most important part of the research done by Andrew Forsyth in [28] of 1888 for invariants of homogeneous linear differential equations was his discovery of the explicit simplified form that various relative invariants would assume when restricted to transformations of one Laguerre-Forsyth canonical form into another such form. For these expressions descriptively termed *linear invariants*, various researchers implied that they may be a key item for future progress. We have already mentioned in Section 1.6 our own indebtedness to Forsyth for those ideas.

For particular applications, various authors have proposed that a given (1.7) be locally transformed into a Laguerre-Forsyth canonical form to which Forsyth's linear invariants could then be applied. This nonconstructive procedure was suggested in [8, 40, 47, 7, 53, 41, 42, 52, 48, 27, 43, 44]. Because the term *relative invariant* has occasionally been incorrectly applied to situations of the preceding type that depend on a local transformation, we employed *global* as a modifier of *relative invariants* in the titles of [19] and [20]. The importance of distinguishing global properties from local ones was made clear by František Neuman in [45] of 1991.

While the equations (1.18) of Paul Appell were studied with respect to relative invariants in [4, 26] and [20, pages 13–17], other results about them appear in [1, 2, 3, 4, 51, 9, 10, 12, 24, 25, 49, 13, 15, 18, 50] and [20, Chapter 19]. Of unusual interest for the equations (1.18) are three relative invariants $\boldsymbol{D}_2$ in (7.2), $\boldsymbol{E}_6$ in (7.10), and $\boldsymbol{E}_7$ in (7.12) that enable us to check, as indicated on page 66, whether any given (1.18) satisfies the solvability condition (7.4) on page 65.

All three of the basic relative invariants for the equations (1.18) were initially discovered in [20, page 13] of 2007 by first finding in [20, Chapters 7–13] all of the basic relative invariants of the equations (1.26)–(1.27) for any $m \geq 2$ and then setting $m = 2$. Here, instead of regarding (1.18) as the special case of (1.26)–(1.27) having $m = 2$, we can view (1.18) as the special case of (4.1)–(4.2) having $m = 2$ and $n = 2$. Thus, we can simply set $m = 2$ and $n = 2$ in Theorem 4.6 on page 32 or in Theorem 4.8 on page 34 to obtain (7.14), (7.15), and (7.16) of page 67.

1.11. Instructive observations

MathSciNet is currently incapable of directing anyone to a publication having satisfactory formulas for all of the coefficients of the monic equation that results from a change of the independent variable in any given monic homogeneous linear differential equation of arbitrary order m. However, the satisfactory formulas of that kind developed and presented in [**19**, pages 135–137] of 2002 were essential for each of the principal advances made in [**19**], [**20**], and [**21**]. For that reason, Chapter 15 has been included as an aid for readers who are unable to use a mathematics library in the manner described in
   http://homepages.uc.edu/~chalklr/Library.pdf
and who may therefore find it difficult to believe that there was extremely little progress about our subject from 1890 through 1988. Chapter 15 shows how the use without exception prior to 1989 of binomial coefficients as in (15.1), (15.3), and (15.6) on pages 157–158 is sufficient to explain why earlier researchers failed to discover adequate formulas for the coefficients of equations resulting from changes of the independent variable.

Chapter 16 is written as if it were a separate expository paper designed to awaken interest in a truly fascinating subject. It shows how easy it is to make interesting discoveries based on satisfactory transformation formulas and the use of computer algebra without need for additional details.

Chapter 17 demonstrates how the current existence of adequate formulas for the coefficients of transformed equations enables one to check the accuracy of results in older publications that involve the notation of (1.0) with binomial coefficients. In particular, this technique is employed in Section 18.5 to verify Theorem 18.7.

Chapter 18 introduces a suitable symbolism and precise definitions for research done before 1989 in order to show how its absence was undoubtedly a serious handicap not only to researchers but also to mathematical historians who were unable to popularize the remarkable results of 1879–1889 with precise statements.

In view of the long history of our subject, we are pleased that the single set of formulas given in Theorem 4.6 enables all of the basic relative invariants to be explicitly obtained for such a wide variety of situations. Various systems of computer algebra can easily incorporate them. Each *Mathematica* notebook in this monograph has an evaluation done with Version 7.0.1 that can be downloaded by using the Google internet browser *Chrome* to visit the web page
   http://homepages.uc.edu/~chalklr/Notebooks.htm
and make a selection. Various other internet browsers may be unable to download these notebooks. The same evaluations for them are also produced by other versions of *Mathematica* such as Versions 8.0.1, 9.0.1, 10.1, and 11.2.

Part 1

General Perspective

CHAPTER 2

The Importance of Suitable Notation

2.1. The representation (2.1) for the equations of interest

Each differential equation considered for transformations in this monograph can be uniquely written, apart from the selections z and $y(z)$, in the form

$$(2.1) \quad \left(y^{(m)}(z)\right)^n + \sum_{\substack{0 \leq i_1 \leq i_2 \leq \cdots \leq i_n \leq m \\ (i_1, i_2, \ldots, i_n) \neq (0, 0, \ldots, 0)}} a_{i_1, i_2, \ldots, i_n}(z) \prod_{\nu=1}^{n} y^{(m-i_\nu)}(z) = 0,$$

with $a_{0, 0, \ldots, 0}(z) \equiv 1$,

where m, n are positive integers and the $a_{i_1, i_2, \ldots, i_n}(z)$ are meromorphic functions on a region Ω of the complex plane. When $n = 1$, this notation reduces to that of (1.7) on page 2 for homogeneous linear differential equations. However, for $n \geq 2$, the computations of pages 21 and 26 show that the formulas for the coefficients of transformed equations take a simple form only when (2.1) is expressible as

$$(2.2) \quad \sum_{i_1=0}^{m} \sum_{i_2=0}^{m} \cdots \sum_{i_n=0}^{m} c_{i_1, i_2, \ldots, i_n}(z) \prod_{\nu=1}^{n} y^{(m-i_\nu)}(z) = 0, \quad \text{with } c_{0, 0, \ldots, 0}(z) \equiv 1,$$

where the coefficients are meromorphic functions on Ω. To make the coefficients for (2.2) unique, we impose the additional condition that they satisfy

$$(2.3) \quad c_{i_{\pi(1)}, i_{\pi(2)}, \ldots, i_{\pi(n)}}(z) \equiv c_{i_1, i_2, \ldots, i_n}(z),$$

for $0 \leq i_1, i_2, \ldots, i_n \leq m$ and any permutation π of $\{1, 2, \ldots, n\}$.

Thus, our responsibility in this chapter is to show that each equation of the form (2.1) is uniquely expressible as (2.2)–(2.3).

2.2. A transitional form (2.4) for (2.1)

Let A denote the set of n-tuples of integers defined by

$$A = \left\{(i_1, i_2, \ldots, i_n) \mid \ 0 \leq i_1 \leq i_2 \leq \cdots \leq i_n \leq m\right\}$$

and let B denote the set of $(m+1)$-tuples of integers defined by

$$B = \left\{(k_0, k_1, \ldots, k_m) \mid \ 0 \leq k_0, k_1, \ldots, k_m \text{ and } k_0 + k_1 + \cdots + k_m = n\right\}.$$

PROPOSITION 2.1. *There are meromorphic functions $b_{k_0, k_1, \ldots, k_m}(z)$ on Ω such that equation (2.1) can be rewritten as*

$$(2.4) \quad \sum_{(k_0, k_1, \ldots, k_m) \in B} b_{k_0, k_1, \ldots, k_m}(z) \prod_{\mu=0}^{m} \left(y^{(m-\mu)}(z)\right)^{k_\mu} = 0, \quad \text{with } b_{n, 0, \ldots, 0}(z) \equiv 1.$$

PROOF. Each $(i_1, i_2, \ldots, i_n)$ in A is uniquely representable in the form
$$(2.5) \qquad (i_1, i_2, \ldots, i_n) = (\underbrace{0, 0, \ldots, 0}_{k_0}, \underbrace{1, 1, \ldots, 1}_{k_1}, \ldots, \underbrace{m, m, \ldots, m}_{k_m}),$$
where $k_0, k_1, \ldots, k_m$ are nonnegative integers such that $k_0 + k_1 + \cdots + k_m = n$. Let ϕ be the function from A to B defined by
$$\phi((i_1, i_2, \ldots, i_n)) = (k_0, k_1, \ldots, k_m), \quad \text{for } (i_1, i_2, \ldots, i_n) \text{ in } A,$$
where, for $0 \leq \mu \leq m$, k_μ is the number of components in $(i_1, i_2, \ldots, i_n)$ that are equal to μ. Clearly, ϕ is one-to-one and onto. Let ψ be the inverse function for ϕ. Then, ψ is a one-to-one function from B onto A. For each $(k_0, k_1, \ldots, k_m)$ in B, we use (2.1) to define $b_{k_0, k_1, \ldots, k_m}(z)$ as the meromorphic function on Ω having
$$(2.6) \qquad b_{k_0, k_1, \ldots, k_m}(z) \equiv a_{i_1, i_2, \ldots, i_n}(z),$$
where $(i_1, i_2, \ldots, i_n) = \psi((k_0, k_1, \ldots, k_m))$.

In particular, for $b_{n, 0, \ldots, 0}(z)$ with $k_0 = n$, we use (2.6) and (2.1) to obtain
$$(2.7) \qquad b_{n, 0, \ldots, 0}(z) \equiv a_{0, 0, \ldots, 0}(z) \equiv 1, \quad \text{on } \Omega.$$

We employ A to write (2.1) in the equivalent form
$$(2.8) \qquad \sum_{(i_1, i_2, \ldots, i_n) \in A} a_{i_1, i_2, \ldots, i_n}(z) \prod_{\nu=1}^{n} y^{(m-i_\nu)}(z) = 0, \quad \text{with } a_{0, 0, \ldots, 0}(z) \equiv 1.$$

For any $(k_0, k_1, \ldots, k_m)$ in B with $\psi((k_0, k_1, \ldots, k_m)) = (i_1, i_2, \ldots, i_n)$, we use (2.5) to see that
$$b_{k_0, k_1, \ldots, k_m}(z) \prod_{\mu=0}^{m} \left(y^{(m-\mu)}(z) \right)^{k_\mu} \equiv a_{i_1, i_2, \ldots, i_n}(z) \prod_{\nu=1}^{n} y^{(m-i_\nu)}(z).$$

Morever, as the m-tuple $(k_0, k_1, \ldots, k_m)$ ranges through B, the corresponding n-tuple $(i_1, i_2, \ldots, i_n) = \psi((k_0, k_1, \ldots, k_m))$ ranges through A. Thus, we obtain
$$(2.9) \qquad \sum_{(k_0, k_1, \ldots, k_m) \in B} b_{k_0, k_1, \ldots, k_m}(z) \prod_{\mu=0}^{m} \left(y^{(m-\mu)}(z) \right)^{k_\mu}$$
$$\equiv \sum_{(i_1, i_2, \ldots, i_n) \in A} a_{i_1, i_2, \ldots, i_n}(z) \prod_{\nu=1}^{n} y^{(m-i_\nu)}(z).$$

In view of (2.9), (2.8) for (2.1), and (2.7), we see that (2.4) is a rewriting of (2.1) when the coefficients of (2.4) are defined by (2.6). This completes the proof. $\square$

2.3. A rewritten form (2.10)–(2.11) for (2.1) that is ideal

THEOREM 2.2. *There are meromorphic functions $c_{i_1, i_2, \ldots, i_n}(z)$ on Ω such that (2.1) can be rewritten as*
$$(2.10) \qquad \sum_{0 \leq i_1, i_2, \ldots, i_n \leq m} c_{i_1, i_2, \ldots, i_n}(z) \prod_{\nu=1}^{n} y^{(m-i_\nu)}(z) = 0, \quad \text{with } c_{0, 0, \ldots, 0}(z) \equiv 1,$$
where
$$(2.11) \qquad c_{i_{\pi(1)}, i_{\pi(2)}, \ldots, i_{\pi(n)}}(z) \equiv c_{i_1, i_2, \ldots, i_n}(z),$$
for $0 \leq i_1, i_2, \ldots, i_n \leq m$ and any permutation π of $\{1, 2, \ldots, n\}$.

2.3. A REWRITTEN FORM (2.10)–(2.11) FOR (2.1) THAT IS IDEAL

PROOF. For each $(k_0, k_1, \ldots, k_m)$ in B, let $\mathcal{S}_{k_0, k_1, \ldots, k_m}$ be the set of n-tuples that are obtained from all possible permutations of the components of the n-tuple

$$(2.12) \qquad (j_1, j_2, \ldots, j_n) = (\underbrace{0, 0, \ldots, 0}_{k_0}, \underbrace{1, 1, \ldots, 1}_{k_1}, \ldots, \underbrace{m, m, \ldots, m}_{k_m}).$$

Moreover, if $(k_0, k_1, \ldots, k_m)$ and $(l_0, l_1, \ldots, l_m)$ are unequal m-tuples of B, then $\mathcal{S}_{k_0, k_1, \ldots, k_m}$ and $\mathcal{S}_{l_0, l_1, \ldots, l_m}$ are disjoint sets. For each $(i_1, i_2, \ldots, i_n)$ in the set

$$(2.13) \qquad \mathcal{S} = \{(i_1, i_2, \ldots, i_n) \mid \text{ for integers satisfying } 0 \leq i_1, i_2, \ldots, i_n \leq m\},$$

there is a unique m-tuple $(k_0, k_1, \ldots, k_m)$ in B such that $(i_1, i_2, \ldots, i_n)$ is an element of $\mathcal{S}_{k_0, k_1, \ldots, k_m}$. This yields the formula

$$(2.14) \qquad \mathcal{S} = \bigcup_{(k_0, k_1, \ldots, k_m) \in B} \mathcal{S}_{k_0, k_1, \ldots, k_m}$$

Since we have $k_0 + k_1 + \cdots + k_m = n$, the number of n-tuples in $\mathcal{S}_{k_0, k_1, \ldots, k_m}$ equals

$$(2.15) \qquad \binom{n}{k_0, k_1, \ldots, k_m} = \frac{n!}{(k_0!)(k_1!) \cdots (k_m!)}.$$

Merely to fix ideas in mind, we count elements in the members of (2.14) and find

$$(m+1)^n = \sum_{(k_0, k_1, \ldots, k_m) \in B} \binom{n}{k_0, k_1, \ldots, k_m} = \sum_{\substack{0 \leq k_0, k_1, \ldots, k_m \\ k_0 + k_1 + \cdots + k_m = n}} \frac{n!}{(k_0!)(k_1!) \cdots (k_m!)}.$$

For any $(i_1, i_2, \ldots, i_n)$ in $\mathcal{S}$ of (2.13), a meromorphic function $c_{i_1, i_2, \ldots, i_n}(z)$ is defined on Ω in terms of (2.4) by

$$(2.16) \qquad c_{i_1, i_2, \ldots, i_n}(z) \equiv \frac{b_{k_0, k_1, \ldots, k_m}(z)}{\binom{n}{k_0, k_1, \ldots, k_m}},$$

where $(k_0, k_1, \ldots, k_m)$ in B is such that $(i_1, i_2, \ldots, i_n) \in \mathcal{S}_{k_0, k_1, \ldots, k_m}$

In particular, for $c_{0, 0, \ldots, 0}(z)$, we have $k_0 = n$ so that (2.16) and (2.4) yield

$$(2.17) \qquad c_{0, 0, \ldots, 0}(z) \equiv \frac{b_{n, 0, \ldots, 0}(z)}{\binom{n}{n, 0, \ldots, 0}} \equiv 1, \quad \text{on } \Omega.$$

Moreover, when $(i_1, i_2, \ldots, i_n)$ in $\mathcal{S}$ belongs to $\mathcal{S}_{k_0, k_1, \ldots, k_m}$ and π is a permutation of $\{1, 2, \ldots, n\}$, we see that $(i_{\pi(1)}, i_{\pi(2)}, \ldots, i_{\pi(n)})$ belongs to $\mathcal{S}_{k_0, k_1, \ldots, k_m}$ and

$$(2.18) \qquad c_{i_{\pi(1)}, i_{\pi(2)}, \ldots, i_{\pi(n)}}(z) \equiv \frac{b_{k_0, k_1, \ldots, k_m}(z)}{\binom{n}{k_0, k_1, \ldots, k_m}} \equiv c_{i_1, i_2, \ldots, i_n}(z).$$

The number of n-tuples in $\mathcal{S}_{k_0, k_1, \ldots, k_m}$ is given by (2.15), Thus, (2.18) yields

$$(2.19) \qquad \sum_{(i_1, i_2, \ldots, i_n) \in \mathcal{S}_{k_0, k_1, \ldots, k_m}} c_{i_1, i_2, \ldots, i_n}(z) \equiv b_{k_0, k_1, \ldots, k_m}(z), \quad \text{for } (k_0, k_1, \ldots, k_m) \text{ in } B.$$

Each $(i_1, i_2, \ldots, i_n)$ in $\mathcal{S}_{k_0, k_1, \ldots, k_m}$ is obtained from (2.12) by a permutation of the components for (2.12). We use this with (2.19) to deduce

$$(2.20) \quad \sum_{(i_1, i_2, \ldots, i_n) \in \mathcal{S}_{k_0, k_1, \ldots, k_m}} c_{i_1, i_2, \ldots, i_n}(z) \prod_{\nu=1}^{n} y^{(m-i_\nu)}(z)$$

$$\equiv \sum_{(i_1, i_2, \ldots, i_n) \in \mathcal{S}_{k_0, k_1, \ldots, k_m}} c_{i_1, i_2, \ldots, i_n}(z) \prod_{\mu=0}^{m} \left(y^{(m-\mu)}(z)\right)^{k_\mu}$$

$$\equiv b_{k_0, k_1, \ldots, k_m}(z) \prod_{\mu=0}^{m} \left(y^{(m-\mu)}(z)\right)^{k_\mu}.$$

In view of (2.14) and (2.20), we find that

$$(2.21) \quad \sum_{0 \leq i_1, i_2, \ldots, i_n \leq m} c_{i_1, i_2, \ldots, i_n}(z) \prod_{\nu=1}^{n} y^{(m-i_\nu)}(z)$$

$$\equiv \sum_{(i_1, i_2, \ldots, i_n) \in \mathcal{S}} c_{i_1, i_2, \ldots, i_n}(z) \prod_{\nu=1}^{n} y^{(m-i_\nu)}(z)$$

$$\equiv \sum_{(k_0, k_1, \ldots, k_m) \in B} \sum_{(i_1, i_2, \ldots, i_n) \in \mathcal{S}_{k_0, k_1, \ldots, k_m}} c_{i_1, i_2, \ldots, i_n}(z) \prod_{\nu=1}^{n} y^{(m-i_\nu)}(z)$$

$$\equiv \sum_{(k_0, k_1, \ldots, k_m) \in B} b_{k_0, k_1, \ldots, k_m}(z) \prod_{\mu=0}^{m} \left(y^{(m-\mu)}(z)\right)^{k_\mu}.$$

By comparing (2.21) and (2.17) with (2.4), we see that (2.10) is a rewriting of (2.4) and (2.1) when the coefficient of (2.10) are defined by (2.16). Moreover, we use (2.18) to deduce that these coefficients satisfy (2.11). This completes the proof. □

In particular, we see that (2.10)–(2.11) can be rewritten as (2.2)–(2.3).

2.4. Uniqueness for (2.10) subject to (2.11)

PROPOSITION 2.3. *There is one and only one equation of the form* (2.10) *that satisfies* (2.11) *and is a rewriting of* (2.1).

PROOF. Suppose there are meromorphic functions $d_{i_1, i_2, \ldots, i_n}(z)$ on Ω such that the equation

$$(2.22) \quad \sum_{0 \leq i_1, i_2, \ldots, i_n \leq m} d_{i_1, i_2, \ldots, i_n}(z) \prod_{\nu=1}^{n} y^{(m-i_\nu)}(z) = 0, \quad \text{with } d_{0, 0, \ldots, 0}(z) \equiv 1,$$

is a rewriting of (2.1) and its coefficients satisfy

$$(2.23) \quad d_{i_{\pi(1)}, i_{\pi(2)}, \ldots, i_{\pi(n)}}(z) \equiv d_{i_1, i_2, \ldots, i_n}(z),$$

for $0 \leq i_1, i_2, \ldots, i_n \leq m$ and any permutation π of $\{1, 2, \ldots, n\}$.

In terms of $c_{i_1, i_2, \ldots, i_n}(z)$ for (2.10)–(2.11) given by (2.16), we introduce

$$(2.24) \quad e_{i_1, i_2, \ldots, i_n}(z) \equiv d_{i_1, i_2, \ldots, i_n}(z) - c_{i_1, i_2, \ldots, i_n}(z), \quad \text{for } 0 \leq i_1, i_2, \ldots, i_n \leq m.$$

Let $L = 0$ be an abbreviation for the equation in (2.1), let $L_1 = 0$ be an abbreviation for the equation in (2.10), let $L_2 = 0$ be an abbreviation for the equation in (2.22), and let $y_0(z)$ be any meromorphic function on a subregion of Ω. Since $L_1 = 0$ and $L_2 = 0$ are rewritings of $L = 0$, the three functions obtained by substituting $y_0^{(k)}(z)$

for $y^{(k)}(z)$ in $L = 0$, in $L_1 = 0$, and in $L_2 = 0$ are equal. Therefore, $y_0(z)$ is a solution of the equation $(L_2 - L_1) = 0$ that (2.24) enables us to write as

$$(2.25) \qquad \sum_{0 \le i_1, i_2, \ldots, i_n \le m} e_{i_1, i_2, \ldots, i_n}(z) \prod_{\nu=1}^{n} y^{(m-i_\nu)}(z) = 0.$$

We use (2.24), (2.23), and (2.11) to see that the coefficients of (2.25) satisfy

$$(2.26) \qquad e_{i_{\pi(1)}, i_{\pi(2)}, \ldots, i_{\pi(n)}}(z) \equiv e_{i_1, i_2, \ldots, i_n}(z),$$

for $0 \le i_1, i_2, \ldots, i_n \le m$ and any permutation π of $\{1, 2, \ldots, n\}$.

When the various terms of (2.25) are collected to rewrite (2.25) in the form

$$(2.27) \qquad \sum_{0 \le j_1 \le j_2 \le \cdots \le j_n \le m} f_{j_1, j_2, \ldots, j_n}(z) \prod_{\nu=1}^{n} y^{(m-i_\nu)}(z) = 0,$$

the condition (2.26) shows that each coefficient $e_{i_1, i_2, \ldots, i_n}(z)$ of (2.25) is expressible as the product of some nonzero rational number and some coefficient $f_{j_1, j_2, \ldots, j_n}(z)$ of (2.27). Since each meromorphic function on each subregion of Ω is a solution of (2.25) and is therefore also a solution of (2.27), we can conclude that each coefficient $f_{j_1, j_2, \ldots, j_n}(z)$ for (2.27) is identically zero. Consequently, each coefficient of (2.25) is identically zero and (2.24) yields

$$(2.28) \qquad d_{i_1, i_2, \ldots, i_n}(z) \equiv c_{i_1, i_2, \ldots, i_n}(z), \quad \text{for } 0 \le i_1, i_2, \ldots, i_n \le m.$$

Thus, there is only one rewritten version of (2.1) having the form of (2.10) subject to (2.11) and we have found that the coefficients of (2.10)–(2.11) are given by (2.16). This completes the proof. $\square$

2.5. Representations in Chapter 4 for the context about invariants

For the subject of invariants, it is desirable to begin with (2.10) in the form

$$(2.29) \qquad \left(y^{(m)}(z)\right)^n + \sum_{\substack{0 \le i_1, i_2, \ldots, i_n \le m \\ (i_1, i_2, \ldots, i_n) \ne (0, 0, \ldots, 0)}} c_{i_1, i_2, \ldots, i_n}(z) \prod_{\nu=1}^{n} y^{(m-i_\nu)}(z) = 0, \quad \text{on } \Omega,$$

where $c_{0, 0, \ldots, 0}(z) \equiv 1$ and the coefficients are meromorphic functions that satisfy

$$(2.30) \qquad c_{i_{\pi(1)}, i_{\pi(2)}, \ldots, i_{\pi(n)}}(z) \equiv c_{i_1, i_2, \ldots, i_n}(z),$$

for $0 \le i_1, i_2, \ldots, i_n \le m$ and any permutation π of $\{1, 2, \ldots, n\}$.

This display of the term $\left(y^{(m)}(z)\right)^n$ indicates the main features of the equation.

In terms of the coefficients for (2.29) and any not-identically-zero meromorphic function $\rho(z)$ on Ω, Theorem 3.1 on page 19 shows that (3.4) and (3.5) specify meromorphic functions $c^*_{i_1, i_2, \ldots, i_n}(z)$ on Ω such that the substitution

$$(2.31) \qquad y(z) = \rho(z)\,v(z)$$

transforms (2.29)-(2.30) into the equation

$$(2.32) \qquad \left(v^{(m)}(z)\right)^n + \sum_{\substack{0 \le i_1, i_2, \ldots, i_n \le m \\ (i_1, i_2, \ldots, i_n) \ne (0, 0, \ldots, 0)}} c^*_{i_1, i_2, \ldots, i_n}(z) \prod_{\nu=1}^{n} v^{(m-i_\nu)}(z) = 0, \quad \text{on } \Omega,$$

where $c^*_{0, 0, \ldots, 0}(z) \equiv 1$ and the coefficients satisfy

$$(2.33) \qquad c^*_{i_{\pi(1)}, i_{\pi(2)}, \ldots, i_{\pi(n)}}(z) \equiv c^*_{i_1, i_2, \ldots, i_n}(z),$$

for $0 \le i_1, i_2, \ldots, i_n \le m$ and any permutation π of $\{1, 2, \ldots, n\}$.

The substitution (2.31) has been termed a change of the dependent variable.

A change of the independent variable for (2.29) is accomplished by the inverse function of a univalent analytic function on Ω.

Let $\zeta = g(z)$ be a univalent analytic function on Ω and let its inverse function be $z = f(\zeta)$. Then, $z = f(\zeta)$ is a univalent analytic function on $\Omega^{**} = g(\Omega)$ and must therefore satisfy $f'(\zeta) \neq 0$, for each ζ in Ω^{**}. It also yields $f(\Omega^{**}) = \Omega$.

In terms of the coefficients for (2.29) and $z = f(\zeta)$, Theorem 3.3 on page 24 shows that (3.21)–(3.24) specify meromorphic functions $c^{**}_{i_1, i_2, \ldots, i_n}(\zeta)$ on Ω^{**} such that the substitution

$$(2.34) \qquad z = f(\zeta), \quad \text{with} \quad u(\zeta) = y\big(f(\zeta)\big),$$

transforms (2.29)–(2.30) into the equation

$$(2.35) \qquad \big(u^{(m)}(\zeta)\big)^n + \sum_{\substack{0 \leq i_1, i_2, \ldots, i_n \leq m \\ (i_1, i_2, \ldots, i_n) \neq (0, 0, \ldots, 0)}} c^{**}_{i_1, i_2, \ldots, i_n}(\zeta) \prod_{\nu=1}^{n} u^{(m-i_\nu)}(\zeta) = 0, \quad \text{on } \Omega^{**},$$

where $c^{**}_{0, 0, \ldots, 0}(\zeta) \equiv 1$ and the coefficients satisfy

$$(2.36) \qquad c^{**}_{i_{\pi(1)}, i_{\pi(2)}, \ldots, i_{\pi(n)}}(\zeta) \equiv c^{**}_{i_1, i_2, \ldots, i_n}(\zeta),$$

for $0 \leq i_1, i_2, \ldots, i_n \leq m$ and any permutation π of $\{1, 2, \ldots, n\}$.

The special case $n = 1$ of Theorems 3.1 and 3.3 yields the transformation formulas developed in [**19**, pages 133–137] and used in [**19**] to find all of the basic relative invariants for homogeneous linear differential equations.

The special case $n = 2$ of Theorems 3.1 and 3.3 yields the transformation formulas that were developed in [**20**, pages 23–33] and used in [**20**] to find all of the basic relative invariants for the differential equations (2.29)–(2.30) having $n = 2$. In particular, when $n = 2$ and $m = 2$, they specialize to give transformation formulas for the equations (1.18) on page 5.

For $m, n \geq 1$, Theorems 3.1 and 3.3 yield the transformation formulas for (2.29)–(2.30) that were presented in [**20**, pages 249–251].

CHAPTER 3

Coefficients of Transformed Equations

3.1. The coefficients $c^*_{i_1, i_2, \ldots, i_n}(z)$ for (2.32)–(2.33) and (4.4)–(4.5)

In conformity with Theorem 2.2 on page 14, let the coefficients of

$$(3.1) \qquad \sum_{0 \leq i_1, i_2, \ldots, i_n \leq m} c_{i_1, i_2, \ldots, i_n}(z) \prod_{\nu=1}^{n} y^{(m-i_\nu)}(z) = 0, \quad \text{on } \Omega,$$

be meromorphic functions on a region Ω that satisfy

$$(3.2) \qquad c_{0, 0, \ldots, 0}(z) \equiv 1 \quad \text{and} \quad c_{i_{\pi(1)}, i_{\pi(2)}, \ldots, i_{\pi(n)}}(z) \equiv c_{i_1, i_2, \ldots, i_n}(z),$$

for $0 \leq i_1, i_2, \ldots, i_n \leq m$ and any permutation π of $\{1, 2, \ldots, n\}$.

Let $\rho(z)$ be a not-identically-zero meromorphic function on Ω. We use (3.1)–(3.2) and $\rho(z)$ to specify the differential equation

$$(3.3) \qquad \sum_{0 \leq i_1, i_2, \ldots, i_n \leq m} c^*_{i_1, i_2, \ldots, i_n}(z) \prod_{\nu=1}^{n} v^{(m-i_\nu)}(z) = 0, \quad \text{on } \Omega,$$

as the unique one of that form whose coefficients are given by

$$(3.4) \qquad c^*_{i_1, i_2, \ldots, i_n}(z) \equiv \sum_{j_1=0}^{i_1} \sum_{j_2=0}^{i_2} \cdots \sum_{j_n=0}^{i_n} \mathcal{A}^{i_1, i_2, \ldots, i_n}_{j_1, j_2, \ldots, j_n}(z) \, c_{j_1, j_2, \ldots, j_n}(z),$$

on Ω for $0 \leq i_1, i_2, \ldots, i_n \leq m$,

in terms of

$$(3.5) \qquad \mathcal{A}^{i_1, i_2, \ldots, i_n}_{j_1, j_2, \ldots, j_n}(z) \equiv \prod_{\nu=1}^{n} \left[\binom{m - j_\nu}{i_\nu - j_\nu} \frac{\rho^{(i_\nu - j_\nu)}(z)}{\rho(z)} \right],$$

where $0 \leq j_\nu \leq i_\nu \leq m$, for $1 \leq \nu \leq n$.

THEOREM 3.1. *The coefficients of (3.3) satisfy*

$$(3.6) \qquad c^*_{0, 0, \ldots, 0}(z) \equiv 1 \quad \text{and} \quad c^*_{i_{\pi(1)}, i_{\pi(2)}, \ldots, i_{\pi(n)}}(z) \equiv c^*_{i_1, i_2, \ldots, i_n}(z), \quad \text{on } \Omega,$$

for $0 \leq i_1, i_2, \ldots, i_n \leq m$ and any permutation π of $\{1, 2, \ldots, n\}$.

Moreover, the substitution

$$(3.7) \qquad y(z) = \rho(z) \, v(z)$$

transforms (3.1)–(3.2) into (3.3) in the sense that: if meromorphic functions $y_0(z)$, $v_0(z)$ on a subregion $\mathcal{U}$ of Ω satisfy $y_0(z) \equiv \rho(z) v_0(z)$, on $\mathcal{U}$, then $y_0(z)$ is a solution of (3.1) if and only if $v_0(z)$ is a solution of (3.3).

PROOF. We apply (3.4), (3.5), and $c_{0,0,\ldots,0}(z) \equiv 1$ to deduce

$$c^*_{0,0,\ldots,0}(z) \equiv \sum_{j_1=0}^{0}\sum_{j_2=0}^{0}\cdots\sum_{j_n=0}^{0} \mathcal{A}^{0,0,\ldots,0}_{j_1,j_2,\ldots,j_n}(z)\, c_{j_1,j_2,\ldots,j_n}(z) \equiv 1.$$

Let π denote any permutation of $\{1, 2, \ldots, n\}$. For $s = 1, 2, \ldots, n$, set $k_s = j_{\pi(s)}$. We use this with (3.4), (3.5), and (3.2) to verify that

$$c^*_{i_{\pi(1)}, i_{\pi(2)}, \ldots, i_{\pi(n)}}(z)$$
$$\equiv \sum_{k_1=0}^{i_{\pi(1)}}\sum_{k_2=0}^{i_{\pi(2)}}\cdots\sum_{k_n=0}^{i_{\pi(n)}} \mathcal{A}^{i_{\pi(1)},i_{\pi(2)},\ldots,i_{\pi(n)}}_{k_1,k_2,\ldots,k_n}(z)\, c_{k_1,k_2,\ldots,k_n}(z)$$
$$\equiv \sum_{j_{\pi(1)}=0}^{i_{\pi(1)}}\sum_{j_{\pi(2)}=0}^{i_{\pi(2)}}\cdots\sum_{j_{\pi(n)}=0}^{i_{\pi(n)}} \mathcal{A}^{i_{\pi(1)},i_{\pi(2)},\ldots,i_{\pi(n)}}_{j_{\pi(1)},j_{\pi(2)},\ldots,j_{\pi(n)}}(z)\, c_{j_{\pi(1)},j_{\pi(2)},\ldots,j_{\pi(n)}}(z)$$
$$\equiv \sum_{j_1=0}^{i_1}\sum_{j_2=0}^{i_2}\cdots\sum_{j_n=0}^{i_n} \mathcal{A}^{i_1,i_2,\ldots,i_n}_{j_1,j_2,\ldots,j_n}(z)\, c_{j_1,j_2,\ldots,j_n}(z) \equiv c^*_{i_i,i_2,\ldots,i_n}(z).$$

Consequently, we see that (3.6) is valid.

Let $y_0(z)$ and $v_0(z)$ be meromorphic functions on a subregion $\mathcal{U}$ of Ω that satisfy $y_0(z) \equiv \rho(z) v_0(z)$. We employ this with (3.5) to deduce

$$(3.8) \quad \prod_{\nu=1}^{n} y_0^{(m-j_\nu)}(z) \equiv \prod_{\nu=1}^{n}\left[\sum_{i_\nu=0}^{m-j_\nu} \binom{m-j_\nu}{i_\nu} \rho^{(i_\nu)}(z)\, v_0^{(m-j_\nu-i_\nu)}(z)\right]$$
$$\equiv \prod_{\nu=1}^{n}\left[\sum_{i_\nu=j_\nu}^{m} \binom{m-j_\nu}{i_\nu-j_\nu} \rho^{(i_\nu-j_\nu)}(z)\, v_0^{(m-i_\nu)}(z)\right]$$
$$\equiv \sum_{i_1=j_1}^{m}\sum_{i_2=j_2}^{m}\cdots\sum_{i_n=j_n}^{m}\prod_{\nu=1}^{n}\left[\binom{m-j_\nu}{i_\nu-j_\nu}\rho^{(i_\nu-j_\nu)}(z)\, v_0^{(m-i_\nu)}(z)\right]$$
$$\equiv \sum_{i_1=j_1}^{m}\sum_{i_2=j_2}^{m}\cdots\sum_{i_n=j_n}^{m} (\rho(z))^n \mathcal{A}^{i_1,i_2,\ldots,i_n}_{j_1,j_2,\ldots,j_n}(z) \prod_{\nu=1}^{n} v_0^{(m-i_\nu)}(z)$$
$$\equiv \sum_{i_1=j_1}^{m}\sum_{i_2=j_2}^{m}\cdots\sum_{i_n=j_n}^{m} \mathcal{F}^{i_1,i_2,\ldots,i_n}_{j_1,j_2,\ldots,j_n}(z),$$

where

$$(3.9) \quad \mathcal{F}^{i_1,i_2,\ldots,i_n}_{j_1,j_2,\ldots,j_n}(z) \equiv (\rho(z))^n\, \mathcal{A}^{i_1,i_2,\ldots,i_n}_{j_1,j_2,\ldots,j_n}(z) \prod_{\nu=1}^{n} v_0^{(m-i_\nu)}(z).$$

Let $F_1(z)$ be the function on $\mathcal{U}$ obtained by replacing each $y^{(k)}(z)$ in the left member of (3.1) with the corresponding $y_0^{(k)}(z)$ and let $F_2(z)$ be the function on $\mathcal{U}$ obtained by replacing each $v^{(k)}(z)$ in the left member of (3.3) with the corresponding $v_0^{(k)}(z)$. We use the definition of $F_1(z)$ and (3.8), (3.9), (3.4), with the definition of $F_2(z)$

to obtain

$$F_1(z) \equiv \sum_{0 \le j_1, j_2, \ldots, j_n \le m} c_{j_1, j_2, \ldots, j_n}(z) \prod_{\nu=1}^{n} y_0^{(m-j_\nu)}(z)$$

$$\equiv \sum_{j_1=0}^{m} \sum_{j_2=0}^{m} \cdots \sum_{j_n=0}^{m} c_{j_1, j_2, \ldots, j_n}(z) \sum_{i_1=j_1}^{m} \sum_{i_2=j_2}^{m} \cdots \sum_{i_n=j_n}^{m} \mathcal{F}_{j_1, j_2, \ldots, j_n}^{i_1, i_2, \ldots, i_n}(z)$$

$$\equiv \sum_{j_1=0}^{m} \sum_{i_1=j_1}^{m} \sum_{j_2=0}^{m} \sum_{i_2=j_2}^{m} \cdots \sum_{j_n=0}^{m} \sum_{i_n=j_n}^{m} c_{j_1, j_2, \ldots, j_n}(z) \, \mathcal{F}_{j_1, j_2, \ldots, j_n}^{i_1, i_2, \ldots, i_n}(z).$$

$$\equiv \sum_{i_1=0}^{m} \sum_{j_1=0}^{i_1} \sum_{i_2=0}^{m} \sum_{j_2=0}^{i_2} \cdots \sum_{i_n=0}^{m} \sum_{j_n=0}^{i_n} c_{j_1, j_2, \ldots, j_n}(z) \, \mathcal{F}_{j_1, j_2, \ldots, j_n}^{i_1, i_2, \ldots, i_n}(z)$$

$$\equiv \sum_{i_1=0}^{m} \sum_{i_2=0}^{m} \cdots \sum_{i_n=0}^{m} \left[\sum_{j_1=0}^{i_1} \sum_{j_2=0}^{i_2} \cdots \sum_{j_n=0}^{i_n} c_{j_1, j_2, \ldots, j_n}(z) \, \mathcal{F}_{j_1, j_2, \ldots, j_n}^{i_1, i_2, \ldots, i_n}(z) \right]$$

$$\equiv \left(\rho(z) \right)^n \sum_{\substack{0 \le i_1 \le m \\ 0 \le i_2 \le m \\ \vdots \\ 0 \le i_n \le m}} \sum_{\substack{0 \le j_1 \le i_1 \\ 0 \le j_2 \le i_2 \\ \vdots \\ 0 \le j_n \le i_n}} \mathcal{A}_{j_1, j_2, \ldots, j_n}^{i_1, i_2, \ldots, i_n}(z) \, c_{j_1, j_2, \ldots, j_n}(z) \prod_{\nu=1}^{n} v_0^{(m-i_\nu)}(z)$$

$$\equiv \left(\rho(z) \right)^n \left(\sum_{0 \le i_1, i_2, \ldots, i_n \le m} c_{i_1, i_2, \ldots, i_n}^{*}(z) \prod_{\nu=1}^{n} v_0^{(m-i_\nu)}(z) \right)$$

$$\equiv \left(\rho(z) \right)^n F_2(z).$$

Hence, $y_0(z)$ is a solution of (3.1) with (3.2) if and only if $v_0(z)$ is a solution of (3.3) with (3.6). This completes the proof. $\square$

It is in the computation above and in the computation of page 26 where the form (2.2) for (2.29) is remarkably effective while the expressions (2.1) and (2.4) lead to insurmountable complications.

3.2. Perspective

For a monic mth-order homogeneous linear differential equation

$$(3.10) \qquad y^{(m)}(z) + \sum_{i=1}^{m} c_i(z) \, y^{(m-i)}(z) = 0, \quad \text{on } \Omega \text{ with } c_0(z) \equiv 1.$$

having meromorphic coefficients, early researchers developed adequate formulas for the equations resulting from (3.10) by a change of the function from $y(z)$ to $v(z)$ by means of $y(z) = \rho(z) \, v(z)$ as in (3.7). For their notation, see page 157. However, earlier researchers failed to develop adequate formulas for a change of the independent variable. In that regard, see Chapter 15.

It is convenient to view a change of the independent variable from z to ζ for (3.10) as specified by the inverse function $z = f(\zeta)$ of a univalent analytic function

$\zeta = g(z)$ on Ω. Then, $z = f(\zeta)$ is a univalent analytic function on $\Omega^{**} = g(\Omega)$ and it yields $f(\Omega^{**}) = \Omega$. Formally replacing each z in (3.10) with $f(\zeta)$, we obtain

$$(3.11) \quad y^{(m)}(f(\zeta)) + \sum_{i=1}^{m} c_i(f(\zeta)) y^{(m-i)}(f(\zeta)) = 0, \quad \text{on } \Omega^{**} \text{ with } c_0(f(\zeta)) \equiv 1,$$

for which we can introduce $u(\zeta) = y(f(\zeta))$. But, then what do we do about $y^{(k)}(f(\zeta))$, for $1 \le k \le m$?

What we did to answer that question in [**19**, Proposition A.2, pages 135–136] of 2002 after earlier progress in [**14, 16, 17**] is develop the result that we repeat here in Theorem 3.2 for essential applications in Section 3.3.

3.2.1. Key result. Henceforth, let $\zeta = g(z)$ denote a given univalent analytic function defined on a region Ω of the complex plane and let $z = f(\zeta)$ denote the inverse function of $g(z)$. Then, $z = f(\zeta)$ is defined on the region $\Omega^{**} = g(\Omega)$ and it yields $f(\Omega^{**}) = \Omega$. Since $z = f(\zeta)$ is a univalent analytic function on Ω^{**}, it satisfies the condition $f'(\zeta) \ne 0$, for each ζ in Ω^{**}. Consequently, analytic functions $\alpha_{i,j}(\zeta)$ on Ω^{**} are recursively defined, for $i \ge 0$ and any integer j, by

$$(3.12) \quad \alpha_{0,j}(\zeta) \equiv 1, \quad \text{on } \Omega^{**} \text{ for any } j,$$

and

$$(3.13) \quad \alpha_{i+1,j}(\zeta) \equiv \sum_{k=1}^{j} \left[\alpha_{i,k}^{(1)}(\zeta) - (i+k) \frac{f''(\zeta)}{f'(\zeta)} \alpha_{i,k}(\zeta) \right],$$
$$\text{on } \Omega^{**} \text{ for } i \ge 0 \text{ and any } j.$$

THEOREM 3.2. *Let $y_0(z)$ be a meromorphic function on a subregion $\mathcal{U}$ of Ω. Then, the meromorphic function $y_0^{(i)}(f(\zeta))$ on $\mathcal{V} = g(\mathcal{U})$ obtained by substituting $z = f(\zeta)$ in the ith derivative $y_0^{(i)}(z)$ of $y_0(z)$ with respect to z is given by*

$$(3.14) \quad y_0^{(i)}(f(\zeta)) \equiv (f'(\zeta))^{-i} \sum_{j=0}^{i} \alpha_{i-j,j}(\zeta) (y_0 \circ f)^{(j)}(\zeta), \quad \text{for } i \ge 0.$$

PROOF. For $i \ge 0$ and $j \ge 1$, we use (3.13) to deduce

$$(3.15) \quad \alpha_{i+1,j}(\zeta) - \alpha_{i+1,j-1}(\zeta) \equiv \alpha_{i,j}^{(1)}(\zeta) - (i+j) \frac{f''(\zeta)}{f'(\zeta)} \alpha_{i,j}(\zeta).$$

For $i \ge 1$ and $j = 0$, we obtain $\alpha_{i+1,0}(\zeta) \equiv \alpha_{i+1,-1}(\zeta) \equiv \alpha_{i,0}(\zeta) \equiv 0$ via (3.13) and see that (3.15) is valid. For $i = j = 0$, we employ both (3.13) and (3.12) to verify that each term of (3.15) is identically zero. Since (3.15) is valid for any integers i and j that satisfy $i \ge 0$ and $j \ge 0$, we therefore have

$$(3.16) \quad \alpha_{i-j,j}^{(1)}(\zeta) \equiv \alpha_{i+1-j,j}(\zeta) - \alpha_{i+1-j,j-1}(\zeta) + i \frac{f''(\zeta)}{f'(\zeta)} \alpha_{i-j,j}(\zeta),$$
$$\text{for } i \ge j \ge 0,$$

We shall establish (3.14) by proving the equivalent formula

$$(3.17) \quad (f'(\zeta))^i y_0^{(i)}(f(\zeta)) \equiv \sum_{j=0}^{i} \alpha_{i-j,j}(\zeta) (y_0 \circ f)^{(j)}(\zeta), \quad \text{for } i \ge 0.$$

For $i = 0$, (3.12) shows that (3.17) is valid. Let i be a nonnegative integer for which (3.17) is true. We differentiate (3.17) with respect to ζ and use (3.16) as well as (3.17) to obtain

$$\left(f'(\zeta)\right)^{i+1} y_0^{(i+1)}(f(\zeta)) + i\frac{f''(\zeta)}{f'(\zeta)} \left(f'(\zeta)\right)^i y_0^{(i)}(f(\zeta))$$

$$\equiv \sum_{j=0}^{i} \alpha_{i-j,j}^{(1)}(\zeta) \, (y_0 \circ f)^{(j)}(\zeta) + \sum_{j=0}^{i} \alpha_{i-j,j}(\zeta) \, (y_0 \circ f)^{(j+1)}(\zeta)$$

$$\equiv \sum_{j=0}^{i} \alpha_{i+1-j,j}(\zeta) \, (y_0 \circ f)^{(j)}(\zeta) - \sum_{j=0}^{i} \alpha_{i+1-j,j-1}(\zeta) \, (y_0 \circ f)^{(j)}(\zeta)$$

$$+ i\frac{f''(\zeta)}{f'(\zeta)} \sum_{j=0}^{i} \alpha_{i-j,j}(\zeta) \, (y_0 \circ f)^{(j)}(\zeta) + \sum_{j=1}^{i+1} \alpha_{i+1-j,j-1}(\zeta) \, (y_0 \circ f)^{(j)}(\zeta)$$

$$\equiv \sum_{j=0}^{i} \alpha_{i+1-j,j}(\zeta) \, (y_0 \circ f)^{(j)}(\zeta) - \alpha_{i+1,-1}(\zeta) \, (y_0 \circ f)(\zeta)$$

$$+ i\frac{f''(\zeta)}{f'(\zeta)} \left(f'(\zeta)\right)^i y_0^{(i)}(f(\zeta)) + \alpha_{0,i}(\zeta) \, (y_0 \circ f)^{(i+1)}(\zeta).$$

With $\alpha_{i+1,-1}(\zeta) \equiv 0$ from (3.13) and $\alpha_{0,i}(\zeta) \equiv 1 \equiv \alpha_{0,i+1}(\zeta)$ from (3.12), we rewrite the preceding expression to obtain

$$\left(f'(\zeta)\right)^{i+1} y_0^{(i+1)}(f(\zeta)) \equiv \sum_{j=0}^{i+1} \alpha_{i+1-j,j}(\zeta) \, (y_0 \circ f)^{(j)}(\zeta).$$

Thus, (3.17) is valid for each $i \geq 0$. This completes the proof. $\square$

3.3. The coefficients $c^{**}_{i_1, i_2, \ldots, i_n}(\zeta)$ for (2.35)–(2.36) and (4.7)–(4.8)

In conformity with Theorem 2.2 on page 14, let the coefficients of

$$(3.18) \qquad \sum_{0 \leq i_1, i_2, \ldots, i_n \leq m} c_{i_1, i_2, \ldots, i_n}(z) \prod_{\nu=1}^{n} y^{(m-i_\nu)}(z) = 0, \quad \text{on } \Omega,$$

be meromorphic functions on Ω that satisfy

$$(3.19) \qquad c_{0,0,\ldots,0}(z) \equiv 1 \quad \text{and} \quad c_{i_{\pi(1)}, i_{\pi(2)}, \ldots, i_{\pi(n)}}(z) \equiv c_{i_1, i_2, \ldots, i_n}(z),$$

for $0 \leq i_1, i_2, \ldots, i_n \leq m$ and any permutation π of $\{1, 2, \ldots, n\}$.

Let $\zeta = g(z)$ be a univalent analytic function on Ω and let its inverse function be denoted by $z = f(\zeta)$ on $\Omega^{**} = g(\Omega)$. Since $z = f(\zeta)$ is a univalent analytic function on Ω^{**}, it satisfies $f'(\zeta) \neq 0$ for each ζ in Ω^{**}. We use (3.18)–(3.19) and $z = f(\zeta)$ to specify the differential equation

$$(3.20) \qquad \sum_{0 \leq i_1, i_2, \ldots, i_n \leq m} c^{**}_{i_1, i_2, \ldots, i_n}(\zeta) \prod_{\nu=1}^{n} u^{(m-i_\nu)}(\zeta) = 0, \quad \text{on } \Omega^{**},$$

as the unique one of that form whose coefficients are given by

$$
(3.21) \quad c^{**}_{i_1, i_2, \ldots, i_n}(\zeta) \equiv \sum_{j_1=0}^{i_1} \sum_{j_2=0}^{i_2} \cdots \sum_{j_n=0}^{i_n} \mathcal{B}^{i_1, i_2, \ldots, i_n}_{j_1, j_2, \ldots, j_n}(\zeta) \, c_{j_1, j_2, \ldots, j_n}\bigl(f(\zeta)\bigr),
$$

for $0 \leq i_1, i_2, \ldots, i_n \leq m$,

$$
(3.22) \quad \mathcal{B}^{i_1, i_2, \ldots, i_n}_{j_1, j_2, \ldots, j_n}(\zeta) \equiv \bigl(f'(\zeta)\bigr)^{j_1+j_2+\cdots+j_n} \prod_{\nu=1}^{n} \alpha_{i_\nu - j_\nu, m - i_\nu}(\zeta), \quad \text{on } \Omega^{**},
$$

for $0 \leq j_1 \leq i_1,\ 0 \leq j_2 \leq i_2,\ \ldots,\ 0 \leq j_n \leq i_n$,

$$
(3.23) \quad \alpha_{0,j}(\zeta) \equiv 1, \quad \text{on } \Omega^{**}, \text{ for any } j,
$$

and

$$
(3.24) \quad \alpha_{i,j}(\zeta) \equiv \sum_{k=1}^{j} \left[\alpha^{(1)}_{i-1, k}(\zeta) - (i + k - 1) \frac{f''(\zeta)}{f'(\zeta)} \alpha_{i-1, k}(\zeta) \right],
$$

on Ω^{**}, for $i \geq 1$ and any j,

THEOREM 3.3. *The coefficients of* (3.20) *satisfy*

$$
(3.25) \quad c^{**}_{0, 0, \ldots, 0}(\zeta) \equiv 1 \quad \text{and} \quad c^{**}_{i_{\pi(1)}, i_{\pi(2)}, \ldots, i_{\pi(n)}}(\zeta) \equiv c^{**}_{i_1, i_2, \ldots, i_n}(\zeta),
$$

for $0 \leq i_1, i_2, \ldots, i_n \leq m$ *and any permutation* π *of* $\{1, 2, \ldots, n\}$.

Moreover, the substitution

$$
(3.26) \quad z = f(\zeta), \quad \text{with} \quad u(\zeta) = y\bigl(f(\zeta)\bigr),
$$

transforms (3.18)–(3.19) *into* (3.20) *in the sense that: if meromorphic functions* $y_0(z)$ *on a subregion* $\mathcal{U}$ *of* Ω *and* $u_0(\zeta)$ *on a subregion* $\mathcal{V}$ *of* Ω^{**} *are related by*

$$
(3.27) \quad u_0(\zeta) = y_0\bigl(f(\zeta)\bigr), \quad \text{on } \mathcal{V} = g(\mathcal{U}),
$$

then $y_0(z)$ *is a solution of* (3.18) *if and only if* $u_0(\zeta)$ *is a solution of* (3.20).

OBSERVATION. We see that $\zeta = g(z)$ on Ω and $z = f(\zeta)$ on $\Omega^{**} = g(\Omega)$ establish a bijection between the meromorphic functions on subregions of Ω and the meromorphic functions on subregions of Ω^{**}. Namely, for $y_0(z)$ on a subregion $\mathcal{U}$ of Ω, there corresponds $u_0(\zeta) = (y_0 \circ f)(\zeta)$ on the subregion $\mathcal{V} = g(\mathcal{U})$ of Ω^{**}. While, to $u_0(\zeta)$ on a subregion $\mathcal{V}$ of Ω^{**}, there is $y_0(z) = (u_0 \circ g)(z)$ on $\mathcal{U} = f(\mathcal{V})$.

PROOF. We use (3.21), (3.22), (3.23), and $c_{0, 0, \ldots, 0}(z) \equiv 1$ on Ω to obtain

$$
c^{**}_{0, 0, \ldots, 0}(\zeta) \equiv \sum_{j_1=0}^{0} \sum_{j_2=0}^{0} \cdots \sum_{j_n=0}^{0} \mathcal{B}^{0, 0, \ldots, 0}_{j_1, j_2, \ldots, j_n}(\zeta) \, c_{j_1, j_2, \ldots, j_n}\bigl(f(\zeta)\bigr) \equiv 1.
$$

Let π denote any permutation of $\{1, 2, \ldots, n\}$. For $s = 1, 2, \ldots, n$, we introduce $k_s = j_{\pi(s)}$. We use this with (3.21), (3.22), and (3.19) to deduce

3.3. THE COEFFICIENTS $c^{**}_{i_1, i_2, \ldots, i_n}(\zeta)$ FOR (2.35)–(2.36) AND (4.7)–(4.8)

$$c^{**}_{i_{\pi(1)}, i_{\pi(2)}, \ldots, i_{\pi(n)}}(\zeta)$$

$$\equiv \sum_{k_1=0}^{i_{\pi(1)}} \sum_{k_2=0}^{i_{\pi(2)}} \cdots \sum_{k_n=0}^{i_{\pi(n)}} \mathcal{B}^{i_{\pi(1)}, i_{\pi(2)}, \ldots, i_{\pi(n)}}_{k_1, k_2, \ldots, k_n}(\zeta) \, c_{k_1, k_2, \ldots, k_n}(f(\zeta))$$

$$\equiv \sum_{j_{\pi(1)}=0}^{i_{\pi(1)}} \sum_{j_{\pi(2)}=0}^{i_{\pi(2)}} \cdots \sum_{j_{\pi(n)}=0}^{i_{\pi(n)}} \mathcal{B}^{i_{\pi(1)}, i_{\pi(2)}, \ldots, i_{\pi(n)}}_{j_{\pi(1)}, j_{\pi(2)}, \ldots, j_{\pi(n)}}(\zeta) \, c_{j_{\pi(1)}, j_{\pi(2)}, \ldots, j_{\pi(n)}}(f(\zeta))$$

$$\equiv \sum_{j_1=0}^{i_1} \sum_{j_2=0}^{i_2} \cdots \sum_{j_n=0}^{i_n} \mathcal{B}^{i_1, i_2, \ldots, i_n}_{j_1, j_2, \ldots, j_n}(\zeta) \, c_{j_1, j_2, \ldots, j_n}(f(\zeta)) \equiv c^{**}_{i_i, i_2, \ldots, i_n}(\zeta).$$

This yields (3.25).

Let $y_0(z)$ and $u_0(\zeta)$ be meromorphic functions related as indicated in (3.27). For $1 \leq \nu \leq n$ and $0 \leq j_\nu \leq m$, we use Theorem 3.2 to obtain

$$(3.28) \quad y_0^{(m-j_\nu)}(f(\zeta)) \equiv (f'(\zeta))^{-(m-j_\nu)} \sum_{k_\nu=0}^{m-j_\nu} \alpha_{m-j_\nu-k_\nu, k_\nu}(\zeta) \, (y_0 \circ f)^{(k_\nu)}(\zeta).$$

We find that (3.28), (3.27), and (3.22) yield

$$(3.29) \quad (f'(\zeta))^{mn} \prod_{\nu=1}^n y_0^{(m-j_\nu)}(f(\zeta))$$

$$\equiv (f'(\zeta))^{mn} \prod_{\nu=1}^n \left[(f'(\zeta))^{-(m-j_\nu)} \sum_{k_\nu=0}^{m-j_\nu} \alpha_{m-j_\nu-k_\nu, k_\nu}(\zeta) \, (y_0 \circ f)^{(k_\nu)}(\zeta) \right]$$

$$\equiv (f'(\zeta))^{j_1+j_2+\cdots+j_n} \prod_{\nu=1}^n \left[\sum_{i_\nu=j_\nu}^{m} \alpha_{i_\nu-j_\nu, m-i_\nu}(\zeta) \, (y_0 \circ f)^{(m-i_\nu)}(\zeta) \right]$$

$$\equiv \sum_{i_1=j_1}^{m} \sum_{i_2=j_2}^{m} \cdots \sum_{i_n=j_n}^{m} (f'(\zeta))^{j_1+j_2+\cdots+j_n} \prod_{\nu=1}^n \left[\alpha_{i_\nu-j_\nu, m-i_\nu}(\zeta) \, u_0^{(m-i_\nu)}(\zeta) \right]$$

$$\equiv \sum_{i_1=j_1}^{m} \sum_{i_2=j_2}^{m} \cdots \sum_{i_n=j_n}^{m} \mathcal{B}^{i_1, i_2, \ldots, i_n}_{j_1, j_2, \ldots, j_n}(\zeta) \prod_{\nu=1}^n u_0^{(m-i_\nu)}(\zeta)$$

$$\equiv \sum_{i_1=j_1}^{m} \sum_{i_2=j_2}^{m} \cdots \sum_{i_n=j_n}^{m} \mathcal{G}^{i_1, i_2, \ldots, i_n}_{j_1, j_2, \ldots, j_n}(\zeta),$$

where

$$(3.30) \quad \mathcal{G}^{i_1, i_2, \ldots, i_n}_{j_1, j_2, \ldots, j_n}(\zeta) \equiv \mathcal{B}^{i_1, i_2, \ldots, i_n}_{j_1, j_2, \ldots, j_n}(\zeta) \prod_{\nu=1}^n u_0^{(m-i_\nu)}(\zeta).$$

Let $G_1(z)$ be the function on $\mathcal{U}$ obtained by replacing each $y^{(k)}(z)$ in the left member of (3.18) with the corresponding $y_0^{(k)}(z)$ and let $G_2(\zeta)$ be the function on $\mathcal{V}$ obtained by replacing each $u(\zeta)$ in the left member of (3.20) with the corresponding $u_0(\zeta)$. By employing the definition of $G_1(z)$ with (3.29), (3.30), (3.21), and the definition of $G_2(\zeta)$, we verify that

$$\left(f'(\zeta)\right)^{mn} G_1\bigl(f(\zeta)\bigr)$$

$$\equiv \left(f'(\zeta)\right)^{mn} \left[\sum_{0 \leq j_1, j_2, \ldots, j_n \leq m} c_{j_1, j_2, \ldots, j_n}(z) \prod_{\nu=1}^{n} y_0^{(m-j_\nu)}(z)\right]_{z=f(\zeta)}$$

$$\equiv \sum_{j_1=0}^{m} \sum_{j_2=0}^{m} \cdots \sum_{j_n=0}^{m} c_{j_1, j_2, \ldots, j_n}\bigl(f(\zeta)\bigr) \left(f'(\zeta)\right)^{mn} \prod_{\nu=1}^{n} y_0^{(m-j_\nu)}\bigl(f(\zeta)\bigr)$$

$$\equiv \sum_{j_1=0}^{m} \sum_{j_2=0}^{m} \cdots \sum_{j_n=0}^{m} c_{j_1, j_2, \ldots, j_n}\bigl(f(\zeta)\bigr) \sum_{i_1=j_1}^{m} \sum_{i_2=j_2}^{m} \cdots \sum_{i_n=j_n}^{m} \mathcal{G}_{j_1, j_2, \ldots, j_n}^{i_1, i_2, \ldots, i_n}(\zeta)$$

$$\equiv \sum_{j_1=0}^{m} \sum_{i_1=j_1}^{m} \sum_{j_2=0}^{m} \sum_{i_2=j_2}^{m} \cdots \sum_{j_n=0}^{m} \sum_{i_n=j_n}^{m} c_{j_1, j_2, \ldots, j_n}\bigl(f(\zeta)\bigr) \mathcal{G}_{j_1, j_2, \ldots, j_n}^{i_1, i_2, \ldots, i_n}(\zeta)$$

$$\equiv \sum_{i_1=0}^{m} \sum_{j_1=0}^{i_1} \sum_{i_2=0}^{m} \sum_{j_2=0}^{i_2} \cdots \sum_{i_n=0}^{m} \sum_{j_n=0}^{i_n} c_{j_1, j_2, \ldots, j_n}\bigl(f(\zeta)\bigr) \mathcal{G}_{j_1, j_2, \ldots, j_n}^{i_1, i_2, \ldots, i_n}(\zeta)$$

$$\equiv \sum_{i_1=0}^{m} \sum_{i_2=0}^{m} \cdots \sum_{i_n=0}^{m} \left[\sum_{j_1=0}^{i_1} \sum_{j_2=0}^{i_2} \cdots \sum_{j_n=0}^{i_n} c_{j_1, j_2, \ldots, j_n}\bigl(f(\zeta)\bigr) \mathcal{G}_{j_1, j_2, \ldots, j_n}^{i_1, i_2, \ldots, i_n}(\zeta)\right]$$

$$\equiv \sum_{\substack{0 \leq i_1 \leq m \\ 0 \leq i_2 \leq m \\ \cdots \\ 0 \leq i_n \leq m}} \sum_{\substack{0 \leq j_1 \leq i_1 \\ 0 \leq j_2 \leq i_2 \\ \cdots \\ 0 \leq j_n \leq i_n}} \mathcal{B}_{j_1, j_2, \ldots, j_n}^{i_1, i_2, \ldots, i_n}(\zeta) \, c_{j_1, j_2, \ldots, j_n}\bigl(f(\zeta)\bigr) \prod_{\nu=1}^{n} u_0^{(m-i_\nu)}(\zeta)$$

$$\equiv \sum_{i_1=0}^{m} \sum_{i_2=0}^{m} \cdots \sum_{i_n=0}^{m} c_{i_1, i_2, \ldots, i_n}^{**}(\zeta) \prod_{\nu=1}^{n} u_0^{(m-i_\nu)}(\zeta)$$

$$\equiv \sum_{0 \leq i_1, i_2, \ldots, i_n \leq m} c_{i_1, i_2, \ldots, i_n}^{**}(\zeta) \prod_{\nu=1}^{n} u_0^{(m-i_\nu)}(\zeta)$$

$$\equiv G_2(\zeta).$$

Consequently, $G_2(\zeta)$ is identically zero on $\mathcal{V}$ if and only if $G_1\bigl(f(\zeta)\bigr)$ is identically zero on $\mathcal{V}$. But, $G_1\bigl(f(\zeta)\bigr)$ is identically zero on $\mathcal{V}$ if and only if $G_1(z)$ is identically zero on $\mathcal{U}$. Thus, $y_0(z)$ is a solution of the equation (3.18) if and only if $u_0(\zeta)$ is a solution of the equation (3.20). This completes the proof. $\square$

3.4. Observations about modifications versus abstractions

With addition and multiplication for functions, the set $\mathfrak{F}_\Omega$ of meromorphic functions on a region Ω forms a field. Also, the usual differentiation $f(z) \mapsto f'(z)$ for functions in $\mathfrak{F}_\Omega$ is a derivation for $\mathfrak{F}_\Omega$ and with it $\mathfrak{F}_\Omega$ is a particular example of an *ordinary differential field of characteristic zero* as defined in [**36**]

We could refer to coefficients like those of (3.1) and (3.3) as elements of $\mathfrak{F}_\Omega$ and refer to coefficients like those of (3.20) as elements of $\mathfrak{F}_{\Omega^{**}}$. However, we have deliberately avoided doing so because the present context enables one to clearly see how modifications as in [**19**, pages 94–103] and [**20**, pages 135–140] can be made for other situations such as where the coefficients may be real-valued functions of a real variable under various restrictions.

In particular, one should not attempt as in [**29**] and [**30**] to create a generalized theory for an arbitrary ordinary differential field $\mathfrak{F}$ of characteristic zero that would include the transformation formulas of Theorems 3.1 and 3.3 when $\mathfrak{F}$ is specialized to be $\mathfrak{F}_\Omega$. That is because there is nothing that can be defined for an arbitrary ordinary differential field $\mathfrak{F}$ of characteristic zero that specializes when $\mathfrak{F}$ is $\mathfrak{F}_\Omega$ to yield the various univalent analytic functions $\zeta = g(z)$ in $\mathfrak{F}_\Omega$ whose inverse functions specify the transformations (3.26).

3.5. $\mathcal{C}_{m,n}$ as a set of elements

In Section 4.1 on page 29, we describe $\mathcal{C}_{m,n}$ as the set of differential equations that can be represented in the form (4.1) subject to (4.2) on some region Ω. Here, we note that a definition for $\mathcal{C}_{m,n}$ can be given that does not require selecting notation for the independent variable and a typical function of that variable.

The ordinary differential equations of our study can be identified with the elements of the set $\mathcal{C}_{m,n}$ defined for $m, n \geq 1$ in terms of the index set

$$\mathfrak{I} = \{(i_1, i_2, \ldots, i_n) \mid i_1, i_2, \ldots, i_n \text{ are integers and } 0 \leq i_1, i_2, \ldots, i_n \leq m\}$$

by the requirement that: $\mathcal{H}$ is an element of $\mathcal{C}_{m,n}$ if and only if

$$(3.31) \qquad \mathcal{H} = \{c_{i_1, i_2, \ldots, i_n}\}_{(i_1, i_2, \ldots, i_n) \in \mathfrak{I}},$$

where each $c_{i_1, i_2, \ldots, i_n}$ is a meromorphic function on a region Ω such that

$$(3.32) \qquad c_{0, 0, \ldots, 0} \equiv 1 \quad \text{as well as} \quad c_{i_{\pi(1)}, i_{\pi(2)}, \ldots, i_{\pi(n)}} \equiv c_{i_1, i_2, \ldots, i_n},$$

for $0 \leq i_1, i_2, \ldots, i_n \leq m$ and any permutation π of $\{1, 2, \ldots, n\}$.

Each element of $\mathcal{C}_{m,n}$ specifies a unique ordinary differential equation of the general type (4.1)–(4.2). But it can be represented in numerous ways depending on notation chosen for the independent variable and a typical function of that variable. With the selection of z and y, we obtain the representation (4.1)–(4.2) for $\mathcal{H}$.

For an element $\mathcal{H}$ of $\mathcal{C}_{m,n}$ defined on Ω by (3.31)–(3.32), let ρ denote a not-identically-zero meromorphic function on Ω. Since the functions $c^*_{i_1, i_2, \ldots, i_n}$ defined by (3.4)–(3.5) satisfy (3.6), the corresponding

$$(3.33) \qquad \mathcal{H}^* = \{c^*_{i_1, i_2, \ldots, i_n}\}_{(i_1, i_2, \ldots, i_n) \in \mathfrak{I}}$$

is therefore an element of $\mathcal{C}_{m,n}$. Thus, ρ transforms $\mathcal{H}$ into $\mathcal{H}^*$ by the technique of Theorem 3.1. The selection of z and v gives the representation (4.4)–(4.5).

For an element $\mathcal{H}$ of $\mathcal{C}_{m,n}$ defined on Ω by (3.31)–(3.32), let g be a univalent analytic function on Ω and let f on $\Omega^{**} = g(\Omega)$ denote the inverse function for g. Since the functions $c^{**}_{i_1, i_2, \ldots, i_n}$ defined on Ω^{**} by (3.21)–(3.24) satisfy (3.25), the corresponding

$$(3.34) \qquad \mathcal{H}^{**} = \{c^{**}_{i_1, i_2, \ldots, i_n}\}_{(i_1, i_2, \ldots, i_n) \in \mathfrak{I}}$$

is therefore an element of $\mathcal{C}_{m,n}$. Hence, f transforms $\mathcal{H}$ into $\mathcal{H}^{**}$ by the technique of Theorem 3.3. The selection of ζ and u yields the representation (4.7)–(4.8).

CHAPTER 4

Consistent Reformulation from 1989 Onward

4.1. Transformations for the equations of $\mathcal{C}_{m,n}$

In terms of fixed positive integers m and n, we let $\mathcal{C}_{m,n}$ denote the collection of ordinary differential equations that are expressible in the form

$$(4.1) \qquad \left(y^{(m)}(z)\right)^n + \sum_{\substack{0 \leq i_1, i_2, \ldots, i_n \leq m \\ (i_1, i_2, \ldots, i_n) \neq (0, 0, \ldots, 0)}} c_{i_1, i_2, \ldots, i_n}(z) \prod_{\nu=1}^{n} y^{(m-i_\nu)}(z) = 0, \quad \text{on } \Omega,$$

where the coefficients $c_{i_1, i_2, \ldots, i_n}(z)$ are meromorphic functions on some region Ω of the complex plane such that

$$(4.2) \qquad c_{0, 0, \ldots, 0}(z) \equiv 1 \quad \text{and} \quad c_{i_{\pi(1)}, i_{\pi(2)}, \ldots, i_{\pi(n)}}(z) \equiv c_{i_1, i_2, \ldots, i_n}(z),$$

for $0 \leq i_1, i_2, \ldots, i_n \leq m$ and any permutation π of $\{1, 2, \ldots, n\}$.

For observations about this notation, see Chapter 2 and Section 3.5.

For any meromorphic $\rho(z) \not\equiv 0$ on Ω, Theorem 3.1 on page 19 shows that (3.4) specifies meromorphic functions $c^*_{i_1, i_2, \ldots, i_n}(z)$ on Ω such that the substitution

$$(4.3) \qquad y(z) = \rho(z)\, v(z)$$

transforms (4.1)–(4.2) into the differential equation

$$(4.4) \qquad \left(v^{(m)}(z)\right)^n + \sum_{\substack{0 \leq i_1, i_2, \ldots, i_n \leq m \\ (i_1, i_2, \ldots, i_n) \neq (0, 0, \ldots, 0)}} c^*_{i_1, i_2, \ldots, i_n}(z) \prod_{\nu=1}^{n} v^{(m-i_\nu)}(z) = 0, \quad \text{on } \Omega,$$

where

$$(4.5) \qquad c^*_{0, 0, \ldots, 0}(z) \equiv 1 \quad \text{and} \quad c^*_{i_{\pi(1)}, i_{\pi(2)}, \ldots, i_{\pi(n)}}(z) \equiv c^*_{i_1, i_2, \ldots, i_n}(z),$$

for $0 \leq i_1, i_2, \ldots, i_n \leq m$ and any permutation π of $\{1, 2, \ldots, n\}$.

For any univalent analytic function $z = f(\zeta)$ defined on a region Ω^{**} with $f(\Omega^{**}) = \Omega$, Theorem 3.3 on page 24 shows that (3.21) specifies meromorphic functions $c^{**}_{i_1, i_2, \ldots, i_n}(\zeta)$ on Ω^{**} such that the substitution

$$(4.6) \qquad z = f(\zeta), \quad \text{with} \quad u(\zeta) = y\bigl(f(\zeta)\bigr),$$

transforms (4.1)–(4.2) into the differential equation

$$(4.7) \qquad \left(u^{(m)}(\zeta)\right)^n + \sum_{\substack{0 \leq i_1, i_2, \ldots, i_n \leq m \\ (i_1, i_2, \ldots, i_n) \neq (0, 0, \ldots, 0)}} c^{**}_{i_1, i_2, \ldots, i_n}(\zeta) \prod_{\nu=1}^{n} u^{(m-i_\nu)}(\zeta) = 0, \quad \text{on } \Omega^{**},$$

where

$$(4.8) \qquad c^{**}_{0, 0, \ldots, 0}(\zeta) \equiv 1 \quad \text{and} \quad c^{**}_{i_{\pi(1)}, i_{\pi(2)}, \ldots, i_{\pi(n)}}(\zeta) \equiv c^{**}_{i_1, i_2, \ldots, i_n}(\zeta),$$

for $0 \leq i_1, i_2, \ldots, i_n \leq m$ and any permutation π of $\{1, 2, \ldots, n\}$.

4.2. The ring $\mathcal{R}_{m,n}$ of differential polynomials

Given $m, n \geq 1$, we assume that the symbols

(4.9) $\quad \boldsymbol{w}^{(k)}_{i_1, i_2, \ldots, i_n}, \quad$ for $\ 1 \leq i_n \leq m, \ 0 \leq i_1 \leq i_2 \leq \cdots \leq i_n, \ $ and $\ k \geq 0,$

are algebraically independent variables over the field $\mathbb{Q}$ of rational numbers and we introduce $\mathcal{R}_{m,n}$ as the ring of polynomials in these variables over $\mathbb{Q}$. We agree that $\boldsymbol{w}_{i_1, i_2, \ldots, i_n} \equiv \boldsymbol{w}^{(0)}_{i_1, i_2, \ldots, i_n}$. A *derivation* $'$ for $\mathcal{R}_{m,n}$ is a rule that assigns a unique element $\boldsymbol{F}'$ in $\mathcal{R}_{m,n}$ to each element $\boldsymbol{F}$ of $\mathcal{R}_{m,n}$ in such a manner that

$$(\boldsymbol{F} + \boldsymbol{G})' \equiv \boldsymbol{F}' + \boldsymbol{G}' \ \text{ and } \ (\boldsymbol{F}\boldsymbol{G})' \equiv \boldsymbol{F}'\boldsymbol{G} + \boldsymbol{F}\boldsymbol{G}', \quad \text{for any } \boldsymbol{F}, \boldsymbol{G} \text{ in } \mathcal{R}_{m,n}.$$

As a consequence, any derivation $'$ for $\mathcal{R}_{m,n}$ yields $\gamma' = 0$, for each γ in $\mathbb{Q}$.

Henceforth, we let $'$ denote the unique derivation for $\mathcal{R}_{m,n}$ such that

(4.10) $\quad \left(\boldsymbol{w}^{(k)}_{i_1, i_2, \ldots, i_n}\right)' = \boldsymbol{w}^{(k+1)}_{i_1, i_2, \ldots, i_n}, \quad$ for each $\boldsymbol{w}^{(k)}_{i_1, i_2, \ldots, i_n}$ of (4.9).

To verify the existence and uniqueness of $'$, we combine (4.10) with a specialization of either [**5**, page 139, Proposition 4] or [**6**, page A.V.130, Theorem 1].

With $'$, $\mathcal{R}_{m,n}$ is an *ordinary differential ring* in the terminology of [**36, 46**] and its *constants*, the $\boldsymbol{F}$ in $\mathcal{R}_{m,n}$ satisfying $\boldsymbol{F}' \equiv 0$, are the elements of $\mathbb{Q}$. Thus, the *nonconstant polynomials of* $\mathcal{R}_{m,n}$ are the elements of $\mathcal{R}_{m,n}$ not in $\mathbb{Q}$.

For any $\boldsymbol{F}$ in $\mathcal{R}_{m,n}$ and $r \geq 0$, we write $\boldsymbol{F}^{(0)} \equiv \boldsymbol{F}$ and $\boldsymbol{F}^{(r+1)} \equiv \left(\boldsymbol{F}^{(r)}\right)'$.

An element of $\mathcal{R}_{m,n}$ is said to be a *differential-polynomial combination over* $\mathbb{Q}$ *of* $\boldsymbol{F}, \boldsymbol{G}, \ldots$ in $\mathcal{R}_{m,n}$ when it is expressible as a polynomial combination over $\mathbb{Q}$ of $\boldsymbol{F}, \boldsymbol{F}^{(1)}, \boldsymbol{F}^{(2)}, \ldots, \boldsymbol{G}, \boldsymbol{G}^{(1)}, \boldsymbol{G}^{(2)}, \ldots$. In this sense, the elements of $\mathcal{R}_{m,n}$ are *differential polynomials* in the variables $\boldsymbol{w}_{i_1, i_2, \ldots, i_n}$ of (4.9) having $k = 0$.

To simplify the writing of (4.23) and (4.37), we introduce

(4.11) $$\boldsymbol{w}_{\underbrace{0, \ldots, 0}_{n}} \equiv 1.$$

To further simplify the writing of (4.23) and (4.37), we introduce $\boldsymbol{X}_{j_1, j_2, \ldots, j_n}$ for $0 \leq j_1, j_2, \ldots, j_n \leq m$ in terms of (4.9) and (4.11) through

(4.12) $\quad \boldsymbol{X}_{j_1, j_2, \ldots, j_n} \equiv \boldsymbol{w}_{j_{\sigma(1)}, j_{\sigma(2)}, \ldots, j_{\sigma(n)}}, \quad$ where σ is a permutation

of $\{1, 2, \ldots, n\}$ such that $j_{\sigma(1)} \leq j_{\sigma(2)} \leq \cdots \leq j_{\sigma(n)}$.

NOTATION 4.0. For any (4.1), (4.3), (4.6), and $\boldsymbol{F}$ in $\mathcal{R}_{m,n}$, let $F(z)$ denote the function on Ω obtained by replacing each $\boldsymbol{w}^{(k)}_{i_1, i_2, \ldots, i_n}$ in $\boldsymbol{F}$ with the corresponding $c^{(k)}_{i_1, i_2, \ldots, i_n}(z)$ from (4.1); let $F^*(z)$ be the function on Ω obtained by replacing $\boldsymbol{w}^{(k)}_{i_1, i_2, \ldots, i_n}$ in $\boldsymbol{F}$ with $c^{*(k)}_{i_1, i_2, \ldots, i_n}(z)$ from (4.4); and let $F^{**}(\zeta)$ be the function on Ω^{**} obtained by replacing $\boldsymbol{w}^{(k)}_{i_1, i_2, \ldots, i_n}$ in $\boldsymbol{F}$ with $c^{**(k)}_{i_1, i_2, \ldots, i_n}(\zeta)$ from (4.7).

Suppose that $\boldsymbol{F}$ is initially given as a differential-polynomial combination over $\mathbb{Q}$ of symbols $\boldsymbol{X}_{j_1, j_2, \ldots, j_n}$ and variables $\boldsymbol{w}^{(k)}_{i_1, i_2, \ldots, i_n}$ from (4.9). Then, instead of first using (4.12) and (4.11) to express $\boldsymbol{F}$ as a polynomial over $\mathbb{Q}$ in the variables (4.9), we can obtain $F(z)$ by directly replacing each $\boldsymbol{X}^{(k)}_{j_1, j_2, \ldots, j_n}$ with $c^{(k)}_{j_1, j_2, \ldots, j_n}(z)$ from (4.1) and each $\boldsymbol{w}^{(k)}_{i_1, i_2, \ldots, i_n}$ with $c^{(k)}_{i_1, i_2, \ldots, i_n}(z)$ from (4.1). This is a consequence of (4.2), (4.12), $c_{0, 0, \ldots, 0}(z) \equiv 1$, and (4.11). Similar observations apply to $F^*(z)$ with respect to (4.5) as well as apply to $F^{**}(\zeta)$ with respect to (4.8).

DEFINITION 4.1. A polynomial $\boldsymbol{F}$ in $\mathcal{R}_{m,n}$ is a *semi-invariant of the first kind for* $\mathcal{C}_{m,n}$ when it is not a constant and yields
$$F^*(z) \equiv F(z), \tag{4.13}$$
for each (4.1)–(4.2) of $\mathcal{C}_{m,n}$ and each (4.3) as a *transformation of the first kind*.

DEFINITION 4.2. A polynomial $\boldsymbol{F}$ in $\mathcal{R}_{m,n}$ is a *semi-invariant of the second kind for* $\mathcal{C}_{m,n}$ when it is not a constant and there is an integer s such that
$$F^{**}(\zeta) \equiv \bigl(f'(\zeta)\bigr)^s F\bigl(f(\zeta)\bigr), \tag{4.14}$$
for each (4.1)–(4.2) of $\mathcal{C}_{m,n}$ and each (4.6) as a *transformation of the second kind*.

DEFINITION 4.3. A polynomial $\boldsymbol{F}$ in $\mathcal{R}_{m,n}$ is a *relative invariant for* $\mathcal{C}_{m,n}$ when it is both a semi-invariant of the first kind for $\mathcal{C}_{m,n}$ and a semi-invariant of the second kind for $\mathcal{C}_{m,n}$.

DEFINITION 4.4. The *weight* of the variable $\boldsymbol{w}^{(k)}_{i_1,i_2,\ldots,i_n}$ in (4.9) is the positive integer $i_1 + i_2 + \cdots + i_n + k$; the *weight* of a nonzero rational number is 0; and the *weight* of a nonzero monomial in $\mathcal{R}_{m,n}$ is the sum of the weights of its factors. A polynomial in $\mathcal{R}_{m,n}$ is *isobaric of weight* s when it is nonzero and each of its nonzero monomials has weight s.

If $\boldsymbol{F}$ in $\mathcal{R}_{m,n}$ is a semi-invariant of the second kind for $\mathcal{C}_{m,n}$, then $\boldsymbol{F}$ is an isobaric polynomial, there is a unique positive integer s for $\boldsymbol{F}$ that satisfies (4.14), and s for $\boldsymbol{F}$ in (4.14) *is equal to the weight of* $\boldsymbol{F}$. This result from [20, page 252] is established on pages 43–44 for its restatement in Theorem 5.4.

DEFINITION 4.5. A polynomial $\boldsymbol{F}$ in $\mathcal{R}_{m,n}$ is *basic* when:
(1) there are integers $e_1, e_2, \ldots, e_n$ subject to $0 \leq e_1 \leq e_2 \leq \cdots \leq e_n \leq m$ as well as either $e_n \geq 3$ or $n \geq 2$ and $e_{n-1} \geq 1$ such that the coefficient of the variable $\boldsymbol{w}_{e_1,e_2,\ldots,e_n}$ in $\boldsymbol{F}$ is 1;
(2) $\boldsymbol{F}$ is isobaric of weight $e_1 + e_2 + \cdots + e_n$;
(3) if $\boldsymbol{w}^{(k)}_{i_1,i_2,\ldots,i_n}$ is any variable from (4.9) that $\boldsymbol{F}$ effectively involves, then $i_\nu \leq e_\nu$, for $1 \leq \nu \leq n$; and
(4) if $\boldsymbol{w}^{(k)}_{i_1,i_2,\ldots,i_n}$ and $\boldsymbol{w}^{(l)}_{j_1,j_2,\ldots,j_n}$ are any two variables from (4.9) such that the coefficient of the product $\boldsymbol{w}^{(k)}_{i_1,i_2,\ldots,i_n}\boldsymbol{w}^{(l)}_{j_1,j_2,\ldots,j_n}$ in $\boldsymbol{F}$ is nonzero, then either (i) $\boldsymbol{w}^{(k)}_{i_1,i_2,\ldots,i_n}$ satisfies $1 \leq i_n \leq 2$ and $i_\nu = 0$, for $1 \leq \nu \leq n-1$, or (ii) $\boldsymbol{w}^{(l)}_{j_1,j_2,\ldots,j_n}$ satisfies $1 \leq j_n \leq 2$ and $j_\nu = 0$, for $1 \leq \nu \leq n-1$.

A *basic polynomial* $\boldsymbol{F}$ in $\mathcal{R}_{m,n}$ has *index* $(e_1, e_2, \ldots, e_n)$ when $e_1, e_2, \ldots, e_n$ are the unique integers that satisfy the first three of the previous four conditions. Naturally, a *basic relative invariant* is a relative invariant that is a basic polynomial.

Examples of basic relative invariants are given by $\mathcal{I}_{3,1;\,3}$ in (1.6) for $\mathcal{C}_{3,1}$, $\mathcal{I}_{4,1;\,3}$ and $\mathcal{I}_{4,1;\,4}$ in (1.15) and (1.17) for $\mathcal{C}_{4,1}$, as well as $\mathcal{I}_{2,2;\,1,1}$ in (1.25) as one of three for $\mathcal{C}_{2,2}$. Their corresponding indexes are indicated by the integers following the semicolon in the subscript notation.

All of the basic relative invariants for the various $\mathcal{C}_{m,n}$ having $m \geq 2$ are given by explicit formulas in Sections 4.3 and 4.4. For $m = 1$, see [20, Chapter 22].

4.3. The basic relative invariants in $\mathcal{R}_{m,n}$ for $\mathcal{C}_{m,n}$ when $m \geq 2$

There are no relative invariants for $\mathcal{C}_{2,1}$; see [**19**, page 145, Proposition A.15]. Thus, we focus here on a fixed $\mathcal{C}_{m,n}$ having $m \geq 2$, $n \geq 1$, and $(m, n) \neq (2, 1)$. The corresponding possible indices $(e_1, e_2, \ldots, e_n)$ for Definition 4.5 satisfy

(4.15) $\quad 0 \leq e_1 \leq \cdots \leq e_n \leq m$ and either $e_n \geq 3$ or $n \geq 2$ and $e_{n-1} \geq 1$.

The number $\mathcal{N}_{m,n}$ of such indices is given by Proposition 5.10 on page 47 as

$$\mathcal{N}_{m,n} = \binom{m+n}{n} - 3.$$

Thus, in view of Theorem 4.6, $\mathcal{N}_{m,n}$ is also the number of basic relative invariants in $\mathcal{R}_{m,n}$ for $\mathcal{C}_{m,n}$.

Throughout, we set $e_0 = 0$ and rewrite (4.15) in the equivalent form

(4.16) $\quad e_0 = 0 \leq e_1 \leq e_2 \leq \cdots \leq e_n \leq m$ and $e_{n-1} = 0$ implies $e_n \geq 3$.

The introduction of $e_0 = 0$ is made so that (4.17) does not have a vacuous sum when $n = 1$ and $m \geq 3$.

THEOREM 4.6. *For any integers $e_1, e_2, \ldots, e_n$ that satisfy (4.16), there is a unique basic relative invariant in $\mathcal{R}_{m,n}$ for $\mathcal{C}_{m,n}$ having index $(e_1, e_2, \ldots, e_n)$ and it is given with respect to (4.9), (4.11), and (4.12) by*

$$(4.17) \quad \boldsymbol{\mathcal{I}}_{m,n;\, e_1,\ldots, e_n} \equiv \sum_{h_0=0}^{e_0} \sum_{h_1=0}^{e_1} \cdots \sum_{h_{n-1}=0}^{e_{n-1}} \boldsymbol{I}_{m,n;\, e_1,\ldots, e_n;\, h_1,\ldots, h_{n-1},\, e_1+\cdots+e_n},$$

where $\boldsymbol{I}_{m,n;\, e_1,\ldots, e_n;\, h_1,\ldots, h_{n-1}, h_n}$ is defined recursively in terms of

$$(4.18) \quad \boldsymbol{a}_{m,n} \equiv \frac{1}{\binom{m+1}{3}} \Big[\boldsymbol{w}_{\underbrace{0,\ldots,0}_{n-1},2} - \frac{m-1}{2} \boldsymbol{w}^{(1)}_{\underbrace{0,\ldots,0}_{n-1},1} - \frac{m-1}{2m} (\boldsymbol{w}_{\underbrace{0,\ldots,0}_{n-1},1})^2 \Big],$$

$$(4.19) \quad \boldsymbol{d}_{m,n} \equiv \frac{1}{m(m-1)} \boldsymbol{w}_{\underbrace{0,\ldots,0}_{n-1},1},$$

(4.20) $\quad \boldsymbol{K}_{m,n;\, i,j} \equiv 0, \quad$ *for $i \leq -1$ and any j,*

(4.21) $\quad \boldsymbol{K}_{m,n;\, 0,j} \equiv 1, \quad$ *for any j,*

$$(4.22) \quad \boldsymbol{K}_{m,n;\, i,j} \equiv \sum_{k=j+1}^{m} \Big[\boldsymbol{K}^{(1)}_{m,n;\, i-1,k} - (m-1)\, \boldsymbol{d}_{m,n}\, \boldsymbol{K}_{m,n;\, i-1,k} \\ + (m+2-i-k)(2-i-k)\, \boldsymbol{a}_{m,n}\, \boldsymbol{K}_{m,n;\, i-2,k} \Big],$$
for $i \geq 1$ and any j,

$$(4.23) \quad \boldsymbol{L}_{m,n;\, i_1,\ldots, i_n} \equiv \sum_{j_1=0}^{i_1} \cdots \sum_{j_n=0}^{i_n} \boldsymbol{K}_{m,n;\, i_1-j_1, j_1} \cdots \boldsymbol{K}_{m,n;\, i_n-j_n, j_n}\, \boldsymbol{X}_{j_1,\ldots, j_n},$$
for $i_1, \ldots, i_n \leq m$,

$$(4.24) \quad H_{m,n;\, e_1,\ldots, e_n;\, h_1,\ldots, h_n} \equiv \prod_{\nu=1}^{n} \Big[\binom{m-h_\nu}{e_\nu - h_\nu} \prod_{r=1}^{e_\nu - h_\nu} (e_\nu - r) \Big],$$
for any $h_1, \ldots, h_n$,

$$(4.25) \quad \boldsymbol{M}_{m,n;\,e_1,\ldots,e_n;\,h_1,\ldots,h_{n-1},\,i}$$
$$\equiv \Big[H_{m,n;\,e_1,\ldots,e_n;\,h_1,\ldots,h_n} \, \boldsymbol{L}_{m,n;\,h_1,\ldots,h_n} \Big]_{h_n = i-(h_1+\cdots+h_{n-1})},$$
$$\text{for } 0 \le h_1 \le e_1, \ldots, 0 \le h_{n-1} \le e_{n-1},$$
$$\text{and } i \le h_1 + \cdots + h_{n-1} + m,$$

$$(4.26) \quad A_{e_1,\ldots,e_n,\,i} \equiv \frac{-1}{e_1 + \cdots + e_n + i - 1}, \quad \text{for } i \ge 1,$$

and

$$(4.27) \quad B_{e_1,\ldots,e_n,\,i} \equiv \frac{e_1 + \cdots + e_n - i}{e_1 + \cdots + e_n + i - 2}, \quad \text{for } i \ge 1,$$

by means of

$$(4.28) \quad \boldsymbol{I}_{m,n;\,e_1,\ldots,e_n;\,h_1,\ldots,h_{n-1},0} \equiv \boldsymbol{I}_{m,n;\,e_1,\ldots,e_n;\,h_1\ldots,h_{n-1},1} \equiv 0,$$
$$\text{for } 0 \le h_1 \le e_1, \ldots, 0 \le h_{n-1} \le e_{n-1},$$

and

$$(4.29) \quad \boldsymbol{I}_{m,n;\,e_1,\ldots,e_n;\,h_1,\ldots,h_{n-1},\,i} \equiv \boldsymbol{M}_{m,n;\,e_1,\ldots,e_n;\,h_1,\ldots,h_{n-1},\,i}$$
$$+ A_{e_1,\ldots,e_n,\,i-1} \, \boldsymbol{I}^{(1)}_{m,n;\,e_1,\ldots,e_n;\,h_1,\ldots,h_{n-1},\,i-1}$$
$$+ B_{e_1,\ldots,e_n,\,i-1} \, \boldsymbol{a}_{m,n} \, \boldsymbol{I}_{m,n;\,e_1,\ldots,e_n;\,h_1,\ldots,h_{n-1},\,i-2},$$
$$\text{for } 0 \le h_1 \le e_1, \ldots, 0 \le h_{n-1} \le e_{n-1},$$
$$\text{and } 2 \le i \le h_1 + \cdots + h_{n-1} + e_n.$$

PROOF. Theorem 4.6 is a direct restatement of [**20**, page 316, Theorem 27.13] in combination with formulas used to establish [**20**, page 314, Theorem 27.7]. □

OBSERVATION 4.7. For $n = 1$ and $3 \le e_1 \le m$, the basic relative invariant $\boldsymbol{\mathcal{I}}_{m,1;\,e_1}$ of weight e_1 in $\mathcal{R}_{m,n}$ for $\mathcal{C}_{m,1}$ is given by (4.17) as $\boldsymbol{\mathcal{I}}_{m,1;\,e_1} \equiv \boldsymbol{I}_{m,1;\,e_1;\,e_1}$ with n replaced by 1 in (4.18)–(4.29). In particular, (4.18) and (4.19) become

$$\boldsymbol{a}_{m,1} \equiv \frac{1}{\binom{m+1}{3}} \Big[\boldsymbol{w}_2 - \tfrac{m-1}{2} \boldsymbol{w}^{(1)}_1 - \tfrac{m-1}{2m} (\boldsymbol{w}_1)^2 \Big] \quad \text{and} \quad \boldsymbol{d}_{m,1} \equiv \frac{1}{m(m-1)} \boldsymbol{w}_1;$$

formulas (4.24) and (4.25) yield $\quad \boldsymbol{M}_{m,1;\,e_1;\,i} \equiv \binom{m-i}{e_1-i} \Big[\prod_{r=1}^{e_1-i} (e_1 - r) \Big] \boldsymbol{L}_{m,n;\,i};$

also, (4.28) gives $\boldsymbol{I}_{m,1;\,e_1;\,0} \equiv \boldsymbol{I}_{m,1;\,e_1;\,1} \equiv 0$; and, for (4.29), we have

$$\boldsymbol{I}_{m,1;\,e_1;\,i} \equiv \boldsymbol{M}_{m,1;\,e_1;\,i} + A_{e_1,\,i-1} \, \boldsymbol{I}^{(1)}_{m,1;\,e_1;\,i-1} + B_{e_1,\,i-1} \, \boldsymbol{a}_{m,1} \, \boldsymbol{I}_{m,1;\,e_1;\,i-2},$$
$$\text{when } 2 \le i \le e_1.$$

4.4. Alternative formulas for $\mathcal{I}_{m,n;\,e_1,\,\ldots,\,e_n}$ when $m \geq 2$

For $m \geq 2$, $n \geq 1$, and $(m, n) \neq (2, 1)$, each basic relative invariant in $\mathcal{R}_{m,n}$ for $\mathcal{C}_{m,n}$ is given explicitly in Theorem 4.8 by a technique that is independent of the method used for Theorem 4.6 of the preceding section. Thus, we have a desirable check on either procedure by employing both of them together.

THEOREM 4.8. *For $m \geq 2$, $n \geq 1$, $(m, n) \neq (2, 1)$, and integers $e_1, \ldots, e_n$ that satisfy (4.16), there is a unique basic relative invariant in $\mathcal{R}_{m,n}$ for $\mathcal{C}_{m,n}$ having index $(e_1, \ldots, e_n)$ and it is given by*

$$(4.30) \quad \mathcal{I}_{m,n;\,e_1,\,\ldots,\,e_n} \equiv \mathcal{J}_{m,n;\,e_1,\,\ldots,\,e_n},$$

where

$$(4.31) \quad \mathcal{J}_{m,n;\,e_1,\,\ldots,\,e_n} \equiv \sum_{h_0=0}^{e_0} \sum_{h_1=0}^{e_1} \cdots \sum_{h_{n-1}=0}^{e_{n-1}} \boldsymbol{J}_{m,n;\,e_1,\,\ldots,\,e_n;\,h_1,\,\ldots,\,h_{n-1},\,e_1+\cdots+e_n}$$

and $\boldsymbol{J}_{m,n;\,e_1,\,\ldots,\,e_n;\,h_1,\,\ldots,\,h_{n-1},\,i}$ is defined recursively in terms of

$$(4.32) \quad \boldsymbol{b}_{m,n} \equiv \frac{1}{\binom{m+1}{3}} \left[\boldsymbol{w}_{\underbrace{0,\,\ldots,\,0}_{n-1},\,2} - \frac{m-2}{3} \boldsymbol{w}^{(1)}_{\underbrace{0,\,\ldots,\,0}_{n-1},\,1} - \frac{(3m-1)(m-2)}{6m(m-1)} (\boldsymbol{w}_{\underbrace{0,\,\ldots,\,0}_{n-1},\,1})^2 \right]$$

$$(4.33) \quad \boldsymbol{d}_{m,n} \equiv \frac{1}{m(m-1)} \boldsymbol{w}_{\underbrace{0,\,\ldots,\,0}_{n-1},\,1},$$

$$(4.34) \quad \boldsymbol{U}_{m,n;\,i,j} \equiv 0, \quad \text{for } i \leq -1 \text{ and any } j,$$

$$(4.35) \quad \boldsymbol{U}_{m,n;\,0,j} \equiv 1, \quad \text{for any } j,$$

$$(4.36) \quad \boldsymbol{U}_{m,n;\,i,j} \equiv \sum_{k=j+1}^{m} \left[\boldsymbol{U}^{(1)}_{m,n;\,i-1,k} + 2(i+k-m-1)\,\boldsymbol{d}_{m,n}\,\boldsymbol{U}_{m,n;\,i-1,k} \\ + (m+2-i-k)(2-i-k)\,\boldsymbol{b}_{m,n}\,\boldsymbol{U}_{m,n;\,i-2,k} \right],$$

$$\text{for } i \geq 1 \text{ and any } j,$$

$$(4.37) \quad \boldsymbol{V}_{m,n;\,i_1,\,\ldots,\,i_n} \equiv \sum_{j_1=0}^{i_1} \cdots \sum_{j_n=0}^{i_n} \boldsymbol{U}_{m,n;\,i_1-j_1,j_1} \cdots \boldsymbol{U}_{m,n;\,i_n-j_n,j_n}\,\boldsymbol{X}_{j_1,\,\ldots,\,j_n},$$

$$\text{for } i_1,\,\ldots,\,i_n \leq m,$$

$$(4.38) \quad H_{m,n;\,e_1,\,\ldots,\,e_n;\,h_1,\,\ldots,\,h_n} \equiv \prod_{\nu=1}^{n} \left[\binom{m-h_\nu}{e_\nu-h_\nu}^{e_\nu-h_\nu} \prod_{r=1}^{e_\nu-h_\nu} (e_\nu - r) \right],$$

$$\text{for any } h_1,\,\ldots,\,h_n,$$

(4.39) $$\boldsymbol{W}_{m,n;\,e_1,...,e_n;\,h_1,...,h_{n-1},\,i}$$
$$\equiv \Big[H_{m,n;\,e_1,...,e_n;\,h_1,...,h_n} \boldsymbol{V}_{m,n;\,h_1,...,h_n} \Big]_{h_n = i - (h_1 + \cdots + h_{n-1})},$$
$$\text{for } 0 \leq h_1 \leq e_1, \ldots, 0 \leq h_{n-1} \leq e_{n-1},$$
$$\text{and } i \leq h_1 + \cdots + h_{n-1} + m,$$

(4.40) $$A_{e_1,...,e_n,i} \equiv \frac{-1}{e_1 + \cdots + e_n + i - 1}, \quad \text{for } i \geq 1,$$

(4.41) $$B_{e_1,...,e_n,i} \equiv \frac{e_1 + \cdots + e_n - i}{e_1 + \cdots + e_n + i - 2}, \quad \text{for } i \geq 1,$$

by means of

(4.42) $$\boldsymbol{J}_{m,n;\,e_1,...,e_n;\,h_1,...,h_{n-1},0} \equiv \boldsymbol{J}_{m,n;\,e_1,...,e_n;\,h_1,...,h_{n-1},1} \equiv 0,$$
$$\text{for } 0 \leq h_1 \leq e_1, \ldots, 0 \leq h_{n-1} \leq e_{n-1},$$

and

(4.43) $$\boldsymbol{J}_{m,n;\,e_1,...,e_n;\,h_1,...,h_{n-1},\,i}$$
$$\equiv \begin{bmatrix} \boldsymbol{W}_{m,n;\,e_1,...,e_n;\,h_1,...,h_{n-1},\,i} \\ + A_{e_1,...,e_n,\,i-1} \begin{pmatrix} \boldsymbol{J}^{(1)}_{m,n;\,e_1,...,e_n;\,h_1,...,h_{n-1},\,i-1} \\ + 2(i-1)\,\boldsymbol{d}_{m,n}\,\boldsymbol{J}_{m,n;\,e_1,...,e_n;\,h_1,...,h_{n-1},\,i-1} \end{pmatrix} \\ + B_{e_1,...,e_n,\,i-1}\,\boldsymbol{b}_{m,n}\,\boldsymbol{J}_{m,n;\,e_1,...,e_n;\,h_1,...,h_{n-1},\,i-2} \end{bmatrix},$$
$$\text{for } 0 \leq h_1 \leq e_1, \ldots, 0 \leq h_{n-1} \leq e_{n-1},$$
$$\text{and } 2 \leq i \leq h_1 + \cdots + h_{n-1} + e_n.$$

PROOF. Theorem 4.8 is a direct restatement of [**20**, page 316, Theorem 27.13] in combination with formulas used to establish [**20**, page 314, Theorem 27.7]. □

OBSERVATION 4.9. For $n = 1$ and $3 \leq e_1 \leq m$, the basic relative invariant $\boldsymbol{\mathcal{I}}_{m,1;\,e_1}$ of weight e_1 for $\mathcal{C}_{m,1}$ is given by (4.30)–(4.31) as $\boldsymbol{\mathcal{I}}_{m,1;\,e_1} \equiv \boldsymbol{J}_{m,1;\,e_1;\,e_1}$ with n replaced by 1 in (4.32)–(4.43). Then, (4.32) and (4.33) become

$$\boldsymbol{b}_{m,1} \equiv \frac{1}{\binom{m+1}{3}}\Big[\boldsymbol{w}_2 - \tfrac{m-2}{3}\boldsymbol{w}_1^{(1)} - \tfrac{(3m-1)(m-2)}{6m(m-1)}(\boldsymbol{w}_1)^2\Big] \quad \text{and} \quad \boldsymbol{d}_{m,1} \equiv \frac{1}{m(m-1)}\boldsymbol{w}_1;$$

formulas (4.38) and (4.39) yield $\quad \boldsymbol{W}_{m,1;\,e_1;\,i} \equiv \binom{m-i}{e_1-i}\Big[\prod_{r=1}^{e_1-i}(e_1 - r)\Big]\boldsymbol{V}_{m,1;\,i};$

also, (4.42) gives $\boldsymbol{J}_{m,1;\,e_1;\,0} \equiv \boldsymbol{J}_{m,1;\,e_1;\,1} \equiv 0$; and, for (4.43), we have

$$\boldsymbol{J}_{m,1;\,e_1;\,i} \equiv \boldsymbol{W}_{m,1;\,e_1;\,i} + A_{e_1,\,i-1}\Big[\boldsymbol{J}^{(1)}_{m,1;\,e_1;\,i-1} + 2(i-1)\,\boldsymbol{d}_{m,1}\,\boldsymbol{J}_{m,1;\,e_1;\,i-1}\Big]$$
$$+ B_{e_1,\,i-1}\,\boldsymbol{b}_{m,1}\,\boldsymbol{J}_{m,1;\,e_1;\,i-2}, \qquad \text{when } 2 \leq i \leq e_1.$$

4.5. Principal technique for combining relative invariants

For $m \geq 2$, $n \geq 1$, $r \geq 0$, elements $\boldsymbol{P}$, $\boldsymbol{Q}$ of $\mathcal{R}_{m,n}$, and positive integers p, q, a differential polynomial $\boldsymbol{C}_{p,q,r}(\boldsymbol{P}, \boldsymbol{Q})$ is defined in $\mathcal{R}_{m,n}$ by

$$(4.44) \qquad \boldsymbol{C}_{p,q,r}(\boldsymbol{P}, \boldsymbol{Q}) \equiv \sum_{s=0}^{r} \sum_{t=0}^{r-s} \boldsymbol{A}_{p,q,r,s,t}\, \boldsymbol{P}^{(t)} \boldsymbol{Q}^{(r-s-t)},$$

where $\boldsymbol{A}_{p,q,r,s,t}$ is specified in $\mathcal{R}_{m,n}$ through

$$(4.45) \qquad \boldsymbol{a}_{m,n} \equiv \frac{1}{\binom{m+1}{3}} \Big[\boldsymbol{w}_{\underbrace{0,\ldots,0}_{n-1},2} - \tfrac{m-1}{2} \boldsymbol{w}^{(1)}_{\underbrace{0,\ldots,0}_{n-1},1} - \tfrac{m-1}{2m}(\boldsymbol{w}_{\underbrace{0,\ldots,0}_{n-1},1})^2 \Big],$$

$$(4.46) \qquad \boldsymbol{B}_{h,i,j} \equiv 0, \quad \text{for } i \leq -1 \text{ and any } h,\, j,$$

$$(4.47) \qquad \boldsymbol{B}_{h,0,j} \equiv 1, \quad \text{for any } h,\, j,$$

$$(4.48) \qquad \boldsymbol{B}_{h,i,j} \equiv \sum_{k=i-1}^{j-1} \Big[\boldsymbol{B}^{(1)}_{h,i-1,k} - k(2h+k-1)\boldsymbol{a}_{m,n}\, \boldsymbol{B}_{h,i-2,k-1} \Big],$$
$$\text{for } i \geq 1 \text{ and any } h,\, j,$$

$$(4.49) \qquad \mathfrak{C}_{p,q,r,\mu} \equiv (-1)^{\mu}\, \frac{\binom{r}{\mu}\binom{2q+r-1}{\mu}}{\binom{2p+\mu-1}{\mu}}, \quad \text{for } 0 \leq \mu \leq r,$$

and

$$(4.50) \qquad \boldsymbol{A}_{p,q,r,s,t} \equiv \sum_{k=0}^{s} \mathfrak{C}_{p,q,r,t+k}\, \boldsymbol{B}_{p,k,t+k}\, \boldsymbol{B}_{q,s-k,r-t-k},$$
$$\text{for } 0 \leq s \leq r \text{ and } 0 \leq t \leq r-s.$$

THEOREM 4.10. *For $r \geq 2$, suppose that $\boldsymbol{P}$ and $\boldsymbol{Q}$ are relative invariants in $\mathcal{R}_{m,n}$ for $\mathcal{C}_{m,n}$ of respective weights p and q. Then, $\boldsymbol{C}_{p,q,r}(\boldsymbol{P}, \boldsymbol{Q})$ in (4.44) is a relative invariant of weight $p + q + r$ for $\mathcal{C}_{m,n}$ if and only if either r is an even integer or $\boldsymbol{P}$ and $\boldsymbol{Q}$ are linearly independent over $\mathbb{Q}$.*

PROOF. Theorem 9.14 on page 95 and Theorem 11.11 on page 123 independently establish that $\boldsymbol{C}_{p,q,r}(\boldsymbol{P}, \boldsymbol{Q})$ is a relative invariant of weight $p+q+r$ in $\mathcal{R}_{m,n}$ for $\mathcal{C}_{m,n}$ if and only if it is nonzero. In view of Theorem 10.11 on page 103, it is nonzero if and only if either r is an even integer or $\boldsymbol{P}$ and $\boldsymbol{Q}$ are linearly independent over $\mathbb{Q}$. We combine these results to complete the proof. $\square$

For relative invariant $\boldsymbol{P}$ and $\boldsymbol{Q}$ in $\mathcal{R}_{m,n}$ of respective weights p and q, we have

$$(4.51) \qquad \boldsymbol{C}_{p,q,1}(\boldsymbol{P}, \boldsymbol{Q}) \equiv \boldsymbol{P}\boldsymbol{Q}^{(1)} - \frac{q}{p}\boldsymbol{P}^{(1)}\boldsymbol{Q}$$

and

$$(4.52) \qquad \boldsymbol{C}_{p,q,0}(\boldsymbol{P}, \boldsymbol{Q}) \equiv \boldsymbol{P}\boldsymbol{Q}.$$

Of course, the product in (4.52) is a relative invariant in $\mathcal{R}_{m,n}$ of weight $p + q$. Propositions 8.1 on page 71 shows that the right member of (4.51) is a relative invariant of weight $p + q + 1$ if and only if $\boldsymbol{P}^q$ and $\boldsymbol{Q}^p$ are linearly independent over $\mathbb{Q}$. This condition is examined in Proposition 8.2 on page 73.

4.6. Several immediate applications of Theorem 4.10

For $m \geq 2$ and $n \geq 1$, we assume that $\boldsymbol{P}$ and $\boldsymbol{Q}$ are relative invariants in $\mathcal{R}_{m,n}$ for $\mathcal{C}_{m,n}$ of respective weights p and q. Then, we must have $(m, n) \neq (2, 1)$.

4.6.1. The construction via (4.44) when $r = 2$. We easily employ (4.44) and (4.46)–(4.50) to verify that

$$\boldsymbol{C}_{p,q,2}(\boldsymbol{P}, \boldsymbol{Q}) \equiv \boldsymbol{P}\boldsymbol{Q}^{(2)} + \alpha \, \boldsymbol{P}^{(1)} \boldsymbol{Q}^{(1)} + \beta \, \boldsymbol{P}^{(2)} \boldsymbol{Q} + \gamma \, \boldsymbol{a}_{m,n} \, \boldsymbol{P}\boldsymbol{Q} \tag{4.53}$$

where

$$\alpha = \frac{-(2q+1)}{p}, \quad \beta = \frac{q(2q+1)}{p(2p+1)}, \quad \text{and} \quad \gamma = \frac{-4q(p+q+1)}{2p+1}. \tag{4.54}$$

Theorem 4.10 shows that (4.53) is a relative invariant in $\mathcal{R}_{m,n}$ for $\mathcal{C}_{m,n}$ of weight $p+q+2$. In particular, when $\boldsymbol{P} \equiv \boldsymbol{Q}$ and therefore $p = q$, we rewrite (4.53)–(4.54) to see that a relative invariant $\boldsymbol{F}_{m,n}$ for $\mathcal{C}_{m,n}$ of weight $2p+2$ is given in $\mathcal{R}_{m,n}$ by

$$\boldsymbol{F}_{m,n} \equiv \boldsymbol{P}\boldsymbol{P}^{(2)} - \left(\frac{2p+1}{2p}\right) \left(\boldsymbol{P}^{(1)}\right)^2 - 2p\,\boldsymbol{a}_{m,n}\,\boldsymbol{P}^2. \tag{4.55}$$

For $m \geq 3$ and $n = 1$, the right members of (4.51) and (4.55) yield constructions for relative invariants of homogeneous linear differential equations that were familiar to Andrew Forsyth in 1888; namely, see [**28**, page 409, Equation (vii)] and the formula of [**28**, page 418]. However, he did not have constructions based on the right member of (4.53) with (4.54).

The right member of (4.53) with (4.54) first appeared in [**20**, page 147] of 2007 for the context $m \geq 2$ and $n = 2$.

4.6.2. The construction via (4.44) when $r \geq 3$. Due to the complexity of the computations for (4.44) based on (4.46)–(4.50) when $r \geq 3$, we present a computer program in Section 6.5 to efficiently expand (4.44). For $r = 3$, it yields

$$\boldsymbol{C}_{p,q,3}(\boldsymbol{P}, \boldsymbol{Q}) \equiv \boldsymbol{P}\boldsymbol{Q}^{(3)} + x_1 \, \boldsymbol{P}^{(1)} \boldsymbol{Q}^{(2)} + x_2 \, \boldsymbol{P}^{(2)} \boldsymbol{Q}^{(1)} + x_3 \, \boldsymbol{P}^{(3)} \boldsymbol{Q} \tag{4.56}$$
$$+ x_4 \, \boldsymbol{a}_{m,n} \, \boldsymbol{P}\boldsymbol{Q}^{(1)} + x_5 \, \boldsymbol{a}_{m,n} \, \boldsymbol{P}^{(1)} \boldsymbol{Q} + x_6 \, \boldsymbol{a}_{m,n}^{(1)} \, \boldsymbol{P}\boldsymbol{Q},$$

in $\mathcal{R}_{m,n}$, where

$$x_1 = \frac{-3(q+1)}{p}, \qquad x_4 = \frac{-4(3pq + 3q^2 + p + 6q + 2)}{2p+1}, \tag{4.57}$$

$$x_2 = \frac{3(q+1)(2q+1)}{p(2p+1)}, \qquad x_5 = \frac{4q(q+1)(3p^2 + 3pq + 6p + q + 2)}{p(p+1)(2p+1)}, \tag{4.58}$$

$$x_3 = \frac{-q(q+1)(2q+1)}{p(p+1)(2p+1)}, \qquad x_6 = \frac{2q(q-p)(2p+2q+3)}{(p+1)(2p+1)}. \tag{4.59}$$

Theorem 4.10 shows that, when $\boldsymbol{P}$ and $\boldsymbol{Q}$ are linearly independent over $\mathbb{Q}$, (4.56) subject to (4.57)–(4.59) is a relative invariant for $\mathcal{C}_{m,n}$ of weight $p+q+3$.

For $r = 4$, a corresponding expansion of $\boldsymbol{C}_{p,q,4}(\boldsymbol{P}, \boldsymbol{Q})$ in $\mathcal{R}_{m,n}$ is given by (6.10)–(6.21). It is a relative invariant for $\mathcal{C}_{m,n}$ of weight $p+q+4$.

Applications in [**20**, pages 147–153] of 2007 motivated the discoveries of (4.53), (4.56), and (6.10) for the special situation where $\mathcal{R}_{m,n}$ has $n = 2$ and $m \geq 2$. For the systematic representations of relative invariants like those presented in Section 4.8 and Chapter 12, the constructions of Theorem 4.10 are essential.

4.7. Representations of relative invariants

For $m \geq 2$, Theorem 5.14 on page 50 shows that each relative invariant in $\mathcal{R}_{m,n}$ for $\mathcal{C}_{m,n}$ is expressible as a differential-polynomial combination over $\mathbb{Q}$ of $\boldsymbol{a}_{m,n}$ and the basic relative invariants in $\mathcal{R}_{m,n}$ for $\mathcal{C}_{m,n}$. The involvement of $\boldsymbol{a}_{m,n}$ can now be accounted for by means of Theorem 4.10. The illustrations of that given in Section 4.8 and in Chapter 12 involve the vector spaces introduced next.

NOTATION 4.11. For each $s \geq 1$, let $\mathcal{V}_{m,n;s}$ denote the set that consists of the zero element of $\mathcal{R}_{m,n}$ together with all the relative invariants in $\mathcal{R}_{m,n}$ for $\mathcal{C}_{m,n}$ that have weight s. Remark 5.8 on page 47 shows that $\mathcal{V}_{m,n;1} = \{0\}$.

With the addition for $\mathcal{R}_{m,n}$ and the scalar multiplication for elements of $\mathcal{V}_{m,n;s}$ by ones in $\mathbb{Q}$, $\mathcal{V}_{m,n;s}$ clearly forms a vector space over $\mathbb{Q}$.

PROPOSITION 4.12. *For $s \geq 1$, $\mathcal{V}_{m,n;s}$ is a finite-dimensional vector space.*

PROOF. For $s \geq 1$, let $\mathcal{W}_{m,n;s}$ denote the subset of $\mathcal{R}_{m,n}$ that consists of 0 together with all of the isobaric polynomials in $\mathcal{R}_{m,n}$ whose weight is s. Then, $\mathcal{W}_{m,n;s}$ is a vector space over $\mathbb{Q}$ with respect to the addition for $\mathcal{R}_{m,n}$ and scalar multiplication for elements in $\mathcal{W}_{m,n;s}$ by ones in $\mathbb{Q}$. Moreover, $\mathcal{V}_{m,n;s}$ is a subspace of $\mathcal{W}_{m,n;s}$. Let $\mathcal{B}_{m,n;s}$ be the subset of $\mathcal{W}_{m,n;s}$ whose elements are the monic monomials having weight s. Then, $\mathcal{B}_{m,n;s}$ is a finite set and a basis for $\mathcal{W}_{m,n;s}$. Hence, $\mathcal{W}_{m,n;s}$ and its subspace $\mathcal{V}_{m,n;s}$ are finite-dimensional vector spaces. This completes the proof. □

For each $s \geq 1$, let $d_{m,n}(s)$ denote the dimension of the vector space $\mathcal{V}_{m,n;s}$. Whenever $d_{m,n}(s) \geq 1$, there are relative invariants $\boldsymbol{E}_{s,1}, \boldsymbol{E}_{s,2}, \ldots, \boldsymbol{E}_{s,d_{m,n}(s)}$ in $\mathcal{R}_{m,n}$ for $\mathcal{C}_{m,n}$ of weight s such that the linear combinations

$$(4.60) \qquad \sum_{i=1}^{d_{m,n}(s)} K_{s,i} \boldsymbol{E}_{s,i}, \quad \text{with } K_{s,1}, K_{s,2}, \ldots, K_{s,d_{m,n}(s)} \text{ in } \mathbb{Q} \text{ not all zero,}$$

uniquely specify the relative invariants of weight s in $\mathcal{R}_{m,n}$. Of course, $d_{m,n}(1) = 0$.

4.8. The relative invariants in $\mathcal{R}_{3,1}$ for $\mathcal{C}_{3,1}$ of weight $s \leq 13$

To illustrate the preceding context when $m = 3$ and $n = 1$, we obtain

$$\boldsymbol{a}_{3,1} \equiv \tfrac{1}{4}\bigl[\boldsymbol{w}_2 - \boldsymbol{w}_1^{(1)} - \tfrac{1}{3}(\boldsymbol{w}_1)^2\bigr]$$

in $\mathcal{R}_{3,1}$ from (4.45) and we use Theorem 4.6 to see that

$$(4.61) \qquad \boldsymbol{\mathcal{I}}_{3,1;3} \equiv \boldsymbol{w}_3 - \tfrac{1}{3}\boldsymbol{w}_1\boldsymbol{w}_2 + \tfrac{2}{27}(\boldsymbol{w}_1)^3 - \tfrac{1}{2}\boldsymbol{w}_2^{(1)} + \tfrac{1}{3}\boldsymbol{w}_1\boldsymbol{w}_1^{(1)} + \tfrac{1}{6}\boldsymbol{w}_1^{(2)}$$

from (1.6) is the only basic relative invariant in $\mathcal{R}_{3,1}$ for $\mathcal{C}_{3,1}$. The technique of Sections 13.2 or 13.4 shows that there are no relative invariants in $\mathcal{R}_{3,1}$ for $\mathcal{C}_{3,1}$ having weight 2, 4, 5, or 7. It also establishes that, for $s = 3, 6, 8, 9, 10, 11, 12,$ and 13, the relative invariants of weight s in $\mathcal{R}_{3,1}$ for $\mathcal{C}_{3,1}$ are uniquely given by

$$(4.62) \qquad \sum_{i=1}^{d_{3,1}(s)} K_{s,i} \boldsymbol{L}_{s,i}, \quad \text{with } K_{s,1}, K_{s,2}, \ldots, K_{s,d_{3,1}(s)} \text{ in } \mathbb{Q} \text{ not all zero,}$$

where:

the weight $s = 3$ has $d_{3,1}(3) = 1$ and $\boldsymbol{L}_{3,1} \equiv \boldsymbol{\mathcal{I}}_{3,1;3}$;

the weight $s = 6$ has $d_{3,1}(6) = 1$ and $\boldsymbol{L}_{6,1} \equiv (\boldsymbol{\mathcal{I}}_{3,1;3})^2$;

the weight $s = 8$ has $d_{3,1}(8) = 1$ and $\boldsymbol{L}_{8,1} \equiv C_{3,3,2}(\boldsymbol{\mathcal{I}}_{3,1;3},\, \boldsymbol{\mathcal{I}}_{3,1;3})$;

the weight $s = 9$ has $d_{3,1}(9) = 1$ and $\boldsymbol{L}_{9,1} \equiv (\boldsymbol{\mathcal{I}}_{3,1;3})^3$;

the weight $s = 10$ has $d_{3,1}(10) = 1$ and $\boldsymbol{L}_{10,1} \equiv C_{3,3,4}(\boldsymbol{\mathcal{I}}_{3,1;3},\, \boldsymbol{\mathcal{I}}_{3,1;3})$;

the weight $s = 11$ has $d_{3,1}(11) = 1$ and $\boldsymbol{L}_{11,1} \equiv C_{3,6,2}(\boldsymbol{\mathcal{I}}_{3,1;3},\, (\boldsymbol{\mathcal{I}}_{3,1;3})^2)$;

the weight $s = 12$ has $d_{3,1}(12) = 3$ as well as

$$\boldsymbol{L}_{12,1} \equiv (\boldsymbol{\mathcal{I}}_{3,1;3})^4,$$
$$\boldsymbol{L}_{12,2} \equiv C_{3,6,3}(\boldsymbol{\mathcal{I}}_{3,1;3},\, (\boldsymbol{\mathcal{I}}_{3,1;3})^2), \quad \text{and}$$
$$\boldsymbol{L}_{12,3} \equiv C_{3,3,6}(\boldsymbol{\mathcal{I}}_{3,1;3},\, \boldsymbol{\mathcal{I}}_{3,1;3});$$

the weight $s = 13$ has $d_{3,1}(13) = 1$ and $\boldsymbol{L}_{13,1} \equiv C_{3,6,4}(\boldsymbol{\mathcal{I}}_{3,1;3},\, (\boldsymbol{\mathcal{I}}_{3,1;3})^2)$.

We note that each of the relative invariants $\boldsymbol{L}_{s,i}$ for (4.62) is obtained from the basic relative invariant $\boldsymbol{\mathcal{I}}_{3,1;3}$ through repetitions of the constructions (4.44), (4.51), and (4.52). They account for the necessary involvement of $\boldsymbol{a}_{3,1}$.

Similar representations in regard to $\mathcal{R}_{4,1}$ for $\mathcal{C}_{4,1}$, $\mathcal{R}_{5,1}$ for $\mathcal{C}_{5,1}$, and $\mathcal{R}_{2,2}$ for $\mathcal{C}_{2,2}$ are presented in Chapter 12.

4.9. The terminology *relative invariant*

When the term *relative invariant* was introduced by Edmund Laguerre in [**37**] of 1879, he used the word *relative* to distinguish his invariant from his conception of an *absolute invariant* whose existence remained to be established.

To give an example of an absolute invariant, we use the relative invariants for (1.16) given in $\mathcal{R}_{4,1}$ by $\boldsymbol{\mathcal{I}}_{4,1;3}$ of (1.15) and $\boldsymbol{\mathcal{I}}_{4,1;4}$ of (1.17) to introduce

$$(4.63) \qquad \boldsymbol{A} \equiv \frac{(\boldsymbol{\mathcal{I}}_{4,1;3})^4}{(\boldsymbol{\mathcal{I}}_{4,1;4})^3}, \quad \text{in the quotient field } \mathcal{Q}_{4,1} \text{ of } \mathcal{R}_{4,1}.$$

After setting $m = 4$ in (1.7), (1.9), and (1.11), we identify (1.7) with (1.16) and employ the notation where: $A(z)$ on Ω, $A^*(z)$ on Ω, and $A^{**}(\zeta)$ on Ω^{**} are the functions respectively obtained by replacing each $w_i^{(j)}$ in $\boldsymbol{A}$ with the corresponding $c_i^{(j)}(z)$ on Ω from (1.7), the corresponding $c_i^{*(j)}(z)$ on Ω from (1.9), as well as the corresponding $c_i^{**(j)}(\zeta)$ on Ω^{**} from (1.11) — when denominators are $\neq 0$. Then, when $\boldsymbol{\mathcal{I}}_{4,1;4}(z) \neq 0$, the properties for $\boldsymbol{\mathcal{I}}_{4,1;3}$ and $\boldsymbol{\mathcal{I}}_{4,1;4}$ analogous to (1.12) yield

$$A^*(z) \equiv \frac{(\boldsymbol{\mathcal{I}}_{4,1;3}^*(z))^4}{(\boldsymbol{\mathcal{I}}_{4,1;4}^*(z))^3} \equiv \frac{(\boldsymbol{\mathcal{I}}_{4,1;3}(z))^4}{(\boldsymbol{\mathcal{I}}_{4,1;4}(z))^3} \equiv A(z), \quad \text{on } \Omega,$$

and

$$A^{**}(\zeta) \equiv \frac{(\boldsymbol{\mathcal{I}}_{4,1;3}^{**}(\zeta))^4}{(\boldsymbol{\mathcal{I}}_{4,1;4}^{**}(\zeta))^3} \equiv \frac{\big((f'(\zeta))^3 \boldsymbol{\mathcal{I}}_{4,1;3}(f(\zeta))\big)^4}{\big((f'(\zeta))^4 \boldsymbol{\mathcal{I}}_{4,1;4}(f(\zeta))\big)^3} \equiv A(f(\zeta)), \quad \text{on } \Omega^{**}.$$

These relations characterize $\boldsymbol{A}$ of (4.63) as an absolute invariant for (1.16).

In [**20**, Chapter 20], the theory of semi-invariants and relative invariants in $\mathcal{R}_{m,n}$ for the equations of $\mathcal{C}_{m,n}$ was extended to an analogous theory about invariants in the quotient field $\mathcal{Q}_{m,n}$ of $\mathcal{R}_{m,n}$ for the equations of $\mathcal{C}_{m,n}$. Natural modifications of Definitions 4.1, 4.2, and 4.3 on page 31 lead to the concepts in $\mathcal{Q}_{m,n}$ of *rational semi-invariants of the first and second kinds* as well as *rational relative invariants* and *absolute invariants*.

The result of [**20**, page 244, Theorem 20.15] shows that: *$\boldsymbol{A}$ in $\mathcal{Q}_{m,n}$ is an absolute invariant for the equations of $\mathcal{C}_{m,n}$ if and only if $\boldsymbol{A}$ is expressible in the form $\boldsymbol{A} = \boldsymbol{F}/\boldsymbol{G}$ where $\boldsymbol{F}$ and $\boldsymbol{G}$ are relative invariants in $\mathcal{R}_{m,n}$ for the equations of $\mathcal{C}_{m,n}$ such that $\boldsymbol{F}$ and $\boldsymbol{G}$ have the same weight and are relatively prime polynomials.*

To employ the single relative invariant $\mathcal{I}_{3,1;3}$ of Edmund Laguerre in (4.61) as well as in (1.6) for the construction of an absolute invariant, we apply the preceding characterization of an absolute invariants to the relative invariants $\boldsymbol{L}_{12,1}$, $\boldsymbol{L}_{12,2}$, and $\boldsymbol{L}_{12,3}$ of Section 4.8 to conclude that: each of the six expressions

$$\boldsymbol{A}_{i,j} \equiv \frac{\boldsymbol{L}_{12,i}}{\boldsymbol{L}_{12,j}}, \quad \text{for } 1 \leq i, j \leq 3 \text{ and } i \neq j,$$

is an absolute invariant in $\mathcal{Q}_{3,1}$ for the differential equations of $\mathcal{C}_{3,1}$.

4.10. Subjects of other chapters

The results in Sections 4.3 and 4.4 about explicit formulas for basic relative invariants were thoroughly verified in the arguments for [**20**, Theorem 27.13] with the supplement of [**19**, Proposition A.15]. Except for those details, we shall make this monograph independent of previous publications by including the necessary background information in Chapters 3 and 5.

Chapter 3 provides a complete derivation for convenient explicit formulas that specify the coefficients of the transformed equations (4.4) and (4.7).

Chapter 5 includes verifications for various properties of semi-invariants and relative invariants used to establish Theorem 4.10.

Chapter 6 enables useful computer representations to be obtained for the basic relative invariants and for the constructions of Theorem 4.10.

Chapter 7 uses research of Paul Appell to illustrate one type of application for Theorem 4.10.

Chapter 8 develops properties of $\boldsymbol{C}_{p,q,1}(\boldsymbol{P}, \boldsymbol{Q})$ in (4.51) where $r = 1$.

Chapters 9 and 11 independently establish that: for $m, r \geq 2$ and the context of Theorem 4.10, $\boldsymbol{C}_{p,q,r}(\boldsymbol{P}, \boldsymbol{Q})$ in (4.44) is a relative invariant in $\mathcal{R}_{m,n}$ for $\mathcal{C}_{m,n}$ if and only if it is not identically zero.

Chapter 10 shows that: for $m, r \geq 2$, $\boldsymbol{C}_{p,q,r}(\boldsymbol{P}, \boldsymbol{Q})$ is not identically zero if and only if either r is an even integer or $\boldsymbol{P}$ and $\boldsymbol{Q}$ are linearly independent over $\mathbb{Q}$.

Chapter 12 illustrates the use of $\boldsymbol{C}_{p,q,r}(\boldsymbol{P}, \boldsymbol{Q})$ to conveniently express relative invariants in terms of basic ones.

Chapters 13 and 14 show how the relative invariants in $\mathcal{R}_{m,n}$ for $\mathcal{C}_{m,n}$ having a given weights s can be deduced and represented with a system of computer algebra. Furthermore, they provide the results summarized in Chapter 12.

Chapters 15 and 18 supply historical perspective about the subject before 1989 and the challenges that hindered progress.

Chapters 16 and 17 provide an effortless way for readers to learn about various identities through interactions with a system of computer algebra.

CHAPTER 5

Supplementary Results

Here, for $m \geq 2$, we show that:

- $\boldsymbol{a}_{m,n}$ in (4.18) and (4.45) is a semi-invariant of the first kind for $\mathcal{C}_{m,n}$;
- $\boldsymbol{b}_{m,n}$ in (4.32) and (11.2) is a semi-invariant of the second kind for $\mathcal{C}_{m,n}$;
- each nonconstant differential-polynomial combination of semi-invariants of the first kind over $\mathbb{Q}$ is a semi-invariant of the first kind;
- if $\boldsymbol{F}$ is a semi-invariant of the second kind for $\mathcal{C}_{m,n}$ as in Definition 4.2, then $\boldsymbol{F}$ is isobaric and its exponent s in (4.14) is equal its weight;
- any relative invariant for $\mathcal{C}_{m,n}$ is expressible as a differential-polynomial combination over $\mathbb{Q}$ of $\boldsymbol{a}_{m,n}$ and the basic relative invariants for $\mathcal{C}_{m,n}$.

5.1. Semi-invariants of the first kind

PROPOSITION 5.1. *Suppose that $\boldsymbol{F}$ is a differential-polynomial combination over $\mathbb{Q}$ of semi-invariants of the first kind in $\mathcal{R}_{m,n}$ for $\mathcal{C}_{m,n}$. Then, with respect to Notation 1.0, $\boldsymbol{F}$ satisfies $F^*(z) \equiv F(z)$.*

PROOF. Let (4.1)–(4.2) be an equation of $\mathcal{C}_{m,n}$ and let (4.3) be a transformation of the first kind for that (4.1)–(4.2) into a corresponding (4.4)–(4.5) in $\mathcal{C}_{m,n}$. Let $\boldsymbol{F}_1, \boldsymbol{F}_2, \ldots, \boldsymbol{F}_\lambda$ in $\mathcal{R}_{m,n}$ be semi-invariants of the first kind for $\mathcal{C}_{m,n}$, let γ in $\mathbb{Q}$ be nonzero, and set

$$(5.1) \qquad \boldsymbol{E} \equiv \gamma \prod_{i=1}^{\lambda} (\boldsymbol{F}_i^{(\mu_i)})^{\nu_i},$$

where μ_i and ν_i are nonnegative integers when $1 \leq i \leq \lambda$. With respect to this context, Definition 4.1 yields $F_i^*(z) \equiv F_i(z)$ and therefore $F_i^{*(\mu_i)}(z) \equiv F_i^{(\mu_i)}(z)$, when $1 \leq i \leq \lambda$. Thus, substitutions in (5.1) give $E^*(z) \equiv E(z)$. Moreover, $\boldsymbol{F}$ is expressible as $\boldsymbol{F} \equiv \boldsymbol{E}_1 + \boldsymbol{E}_2 + \cdots + \boldsymbol{E}_\kappa$, where each $\boldsymbol{E}_j$ is an expression analogous to (5.1) that yields $E_j^*(z) \equiv E_j(z)$, for $1 \leq j \leq \kappa$. Hence, we have $F^*(z) \equiv F(z)$. This completes the proof. □

PROPOSITION 5.2. *For $m \geq 2$ and $n \geq 1$, the polynomial defined in $\mathcal{R}_{m,n}$ by*

$$(5.2) \qquad \boldsymbol{G}_{m,n} \equiv \boldsymbol{w}_{\underbrace{0, \ldots, 0}_{n-1}, 2} - \frac{m-1}{2} \boldsymbol{w}^{(1)}_{\underbrace{0, \ldots, 0}_{n-1}, 1} - \frac{m-1}{2m} (\boldsymbol{w}_{\underbrace{0, \ldots, 0}_{n-1}, 1})^2$$

is an isobaric semi-invariant of the first kind for $\mathcal{C}_{m,n}$ having weight 2.

PROOF. Clearly, $\boldsymbol{G}_{m,n}$ is an isobaric polynomial in $\mathcal{R}_{m,n}$ of weight 2. To show that $\boldsymbol{G}_{m,n}$ is a semi-invariant of the first kind, let (4.1)–(4.2) be an equation of $\mathcal{C}_{m,n}$, let a transformation (4.3) for (4.1)–(4.2) be given, and let (4.4)–(4.5) be the result

of that transformation for which Theorem 3.1 on page 19 provides details. In that context, we set
$$\mathfrak{c}_i(z) \equiv c_{0,\ldots,0,i}(z) \quad \text{and} \quad \mathfrak{c}_i^*(z) \equiv c_{0,\ldots,0,i}^*(z), \quad \text{for } i = 0, 1, 2.$$

We introduce $\mathfrak{w}_1 \equiv w_{0,\ldots,0,1}$ and $\mathfrak{w}_2 \equiv w_{0,\ldots,0,2}$ to rewrite (5.2) as

(5.3) $$\boldsymbol{G}_{m,n} \equiv \mathfrak{w}_2 - \frac{m-1}{2}\mathfrak{w}_1^{(1)} - \frac{m-1}{2m}(\mathfrak{w}_1)^2.$$

In view of (3.4), (3.5) and $\mathfrak{c}_0(z) \equiv 1$, we find that

(5.4) $$\mathfrak{c}_1^*(z) \equiv \mathfrak{c}_1(z) + m\frac{\rho^{(1)}(z)}{\rho(z)}$$

and

(5.5) $$\mathfrak{c}_2^*(z) \equiv \mathfrak{c}_2(z) + (m-1)\frac{\rho^{(1)}(z)}{\rho(z)}\mathfrak{c}_1(z) + \binom{m}{2}\frac{\rho^{(2)}(z)}{\rho(z)}.$$

After substituting $\mathfrak{c}_i^{*(k)}(z)$ for $\mathfrak{w}_i^{(k)}$ in (5.3), we use (5.4) as well as (5.5) and a substitution of $\mathfrak{c}_i^{(k)}(z)$ for $\mathfrak{w}_i^{(k)}$ in (5.3) to obtain

$$G_{m,n}^*(z) \equiv \left[\mathfrak{c}_2^*(z) - \frac{m-1}{2}\mathfrak{c}_1^{*(1)}(z) - \frac{m-1}{2m}(\mathfrak{c}_1^*(z))^2\right]$$

$$\equiv \left[\begin{array}{l} \mathfrak{c}_2(z) + (m-1)\dfrac{\rho^{(1)}(z)}{\rho(z)}\mathfrak{c}_1(z) + \binom{m}{2}\dfrac{\rho^{(2)}(z)}{\rho(z)} \\ -\dfrac{m-1}{2}\left(\mathfrak{c}_1^{(1)}(z) + m\dfrac{\rho^{(2)}(z)}{\rho(z)} - m\left[\dfrac{\rho^{(1)}(z)}{\rho(z)}\right]^2\right) \\ -\dfrac{m-1}{2m}\left((\mathfrak{c}_1(z))^2 + 2m\dfrac{\rho^{(1)}(z)}{\rho(z)}\mathfrak{c}_1(z) + m^2\left[\dfrac{\rho^{(1)}(z)}{\rho(z)}\right]^2\right) \end{array}\right]$$

$$\equiv \left[\mathfrak{c}_2(z) - \frac{m-1}{2}\mathfrak{c}_1^{(1)}(z) - \frac{m-1}{2m}(\mathfrak{c}_1(z))^2\right] \equiv G_{m,n}(z).$$

Thus, $\boldsymbol{G}_{m,n}$ is a semi-invariant of the first kind. This completes the proof. □

COROLLARY 5.3. *The polynomial* $\boldsymbol{a}_{m,n}$ *defined in* $\mathcal{R}_{m,n}$ *by* (4.18) *or* (4.45) *is an isobaric semi-invariant of the first kind for* $\mathcal{C}_{m,n}$ *having weight* 2.

PROOF. This is a consequence of the relation $\boldsymbol{a}_{m,n} \equiv \left(1/\binom{m+1}{3}\right)\boldsymbol{G}_{m,n}$. □

5.2. Semi-invariants of the second kind

The next result generalizes arguments that were advanced in [**32**, page 120] or [**35**, pages 106–107] as well as [**19**, pages 140–142] and [**20**, pages 37–38].

THEOREM 5.4. *Let* $\boldsymbol{F}$ *in* $\mathcal{R}_{m,n}$ *be a semi-invariant of the second kind for* $\mathcal{C}_{m,n}$ *and let s be an integer in terms of which* $\boldsymbol{F}$ *satisfies* (4.14) *of Definition* 4.2. *Then,* $\boldsymbol{F}$ *is isobaric and s is equal to the weight of* $\boldsymbol{F}$.

PROOF. Since F is not a constant, there is a positive integer ω such that

(5.6) $$F \equiv \sum_{\mu=0}^{\omega} F_\mu,$$

where F_ω in $\mathcal{R}_{m,n}$ is isobaric of weight ω and, for $0 \le \mu \le \omega - 1$, either $F_\mu \equiv 0$ or F_μ is an isobaric polynomial in $\mathcal{R}_{m,n}$ of weight μ. For any (4.1)–(4.2) on some Ω, we restrict $\zeta = g(z)$ on Ω to functions of the type $g_t(z) = (1/t)z$, where t is a fixed nonzero rational number. The corresponding inverse function is the univalent analytic function $z = f_t(\zeta) = t\zeta$ on $\Omega_t^{**} = g_t(\Omega)$. For it, (3.23), (3.24), and (3.22) yield $\alpha_{i,j}(\zeta) \equiv 0$, for $i \ge 1$ and any j, as well as

$$\mathcal{B}_{j_1,j_2,\ldots,j_n}^{i_1,i_2,\ldots,i_n}(\zeta) \equiv \begin{cases} t^{i_1+i_2+\cdots+i_n}, & \text{if } (j_1, j_2, \ldots, j_n) = (i_1, i_2, \ldots, i_n), \\ 0, & \text{if } (j_1, j_2, \ldots, j_n) \ne (i_1, i_2, \ldots, i_n). \end{cases}$$

Now, (3.21) gives $c^{**}_{i_1,i_2,\ldots,i_n}(\zeta) \equiv t^{i_1+i_2+\cdots+i_n} c_{i_1,i_2,\ldots,i_n}(t\zeta)$. Thus, we have

(5.7) $$c^{**(k)}_{i_1,i_2,\ldots,i_n}(\zeta) \equiv t^{i_1+i_2+\cdots+i_n+k} c^{(k)}_{i_1,i_2,\ldots,i_n}(t\zeta),$$
$$\text{for } 0 \le i_1, i_2, \ldots, i_n \le m \text{ and } k \ge 0.$$

Of course, substitutions in (5.6) yield

(5.8) (a) $F(z) \equiv \sum_{\mu=0}^{\omega} F_\mu(z)$, on Ω; (b) $F^{**}(\zeta) \equiv \sum_{\mu=0}^{\omega} F_\mu^{**}(\zeta)$, on Ω^{**}.

If T is a nonzero term of F_μ, then T is expressible as

(5.9) $$T \equiv \gamma \prod_{\nu=1}^{\lambda} w^{(k_\nu)}_{i_{1,\nu}, i_{2,\nu}, \ldots, i_{n,\nu}}, \quad \text{with} \quad \sum_{\nu=1}^{\lambda} (i_{1,\nu} + i_{2,\nu} + \cdots + i_{n,\nu} + k_\nu) = \mu,$$

where: γ in $\mathbb{Q}$ is nonzero; $\lambda \ge 0$; $0 \le i_{1,\nu} \le i_{2,\nu}, \le \cdots \le i_{n,\nu} \le m$, $1 \le i_{n,\nu}$, and $k_\nu \ge 0$, for $1 \le \nu \le \lambda$. Thus, (5.9) and (5.7) yield

(5.10) $$T^{**}(\zeta) \equiv \gamma \prod_{\nu=1}^{\lambda} c^{**(k_\nu)}_{i_{1,\nu}, i_{2,\nu}, \ldots, i_{n,\nu}}(\zeta)$$
$$\equiv \left[\prod_{\nu=1}^{\lambda} t^{i_{1,\nu}+i_{2,\nu}+\cdots+i_{n,\nu}+k_\nu}\right] \left[\gamma \prod_{\nu=1}^{\lambda} c^{(k_\nu)}_{i_{1,\nu}, i_{2,\nu}, \ldots, i_{n,\nu}}(t\zeta)\right] \equiv t^\mu T(t\zeta).$$

Since each summand F_μ for (5.6) is either zero or a sum of terms like (5.9) for which (5.10) is valid, we apply (5.10) to deduce

(5.11) $$F_\mu^{**}(\zeta) \equiv t^\mu F_\mu(t\zeta), \quad \text{for } 0 \le \mu \le \omega.$$

Then, we see that (4.14), (5.8)-(b) and (5.11) require

(5.12) $$t^s F(t\zeta) \equiv \left(f_t'(\zeta)\right)^s F\bigl(f_t(\zeta)\bigr) \equiv F^{**}(\zeta) \equiv \sum_{\mu=0}^{\omega} F_\mu^{**}(\zeta) \equiv \sum_{\mu=0}^{\omega} t^\mu F_\mu(t\zeta),$$

for any (4.1)–(4.2) and any nonzero rational number t. We replace $t\zeta$ in (5.12) with z and use (a) of (5.8) to verify that

(5.13) $$t^s \sum_{\mu=0}^{\omega} F_\mu(z) \equiv \sum_{\mu=0}^{\omega} t^\mu F_\mu(z), \quad \text{for any (4.1)–(4.2) and any nonzero } t \text{ in } \mathbb{Q}.$$

To establish

(5.14) $$t^s \sum_{\mu=0}^{\omega} \boldsymbol{F}_\mu \equiv \sum_{\mu=0}^{\omega} t^\mu \boldsymbol{F}_\mu, \quad \text{for any nonzero rational number } t,$$

suppose that t_0 is a nonzero rational number such that (5.14) is not valid for $t = t_0$. Then, the differential polynomial

$$\boldsymbol{D} \equiv (t_0)^s \sum_{\mu=0}^{\omega} \boldsymbol{F}_\mu - \sum_{\mu=0}^{\omega} (t_0)^\mu \boldsymbol{F}_\mu$$

in $\mathcal{R}_{m,n}$ satisfies $\boldsymbol{D} \not\equiv 0$. By applying Corollary 5.13, we see that there is a differential equation (4.1)–(4.2) on some region Ω such that the substitution of $c^{(k)}_{i_1, i_2, \ldots, i_n}(z)$ from it for $w^{(k)}_{i_1, i_2, \ldots, i_n}$ in $\boldsymbol{D}$ yields $\boldsymbol{D}(z) \not\equiv 0$. But, this contradicts (5.13). Hence, (5.14) is valid.

For each $t \neq 0$, the terms of weight ω in the left and right members of (5.14) give $t^s \boldsymbol{F}_\omega \equiv t^\omega \boldsymbol{F}_\omega$. In view of $\boldsymbol{F}_\omega \not\equiv 0$, we obtain $s = \omega$. Regarding (5.14) as an equality of two polynomial functions of t over $\mathcal{R}_{m,n}$, we note that the coefficients of like powers of t must be equal. Thus, we deduce $\boldsymbol{F}_\mu \equiv 0$, for $0 \leq \mu < \omega$, and $\boldsymbol{F} \equiv \boldsymbol{F}_\omega$. Consequently, $\boldsymbol{F}$ is an isobaric polynomial of weight ω and we have $s = \omega$. This completes the proof. $\square$

Let $\boldsymbol{S}_1$ and $\boldsymbol{S}_2$ be semi-invariants of the second kind in $\mathcal{R}_{m,n}$ for $\mathcal{C}_{m,n}$ whose respective weights are s_1 and s_2. Then, $\boldsymbol{S}_1 \boldsymbol{S}_2$ is semi-invariant of the second kind for $\mathcal{C}_{m,n}$ of weight $s_1 + s_2$. Moreover, if $s_2 = s_1$ and α, β are rational numbers such that $\alpha \boldsymbol{S}_1 + \beta \boldsymbol{S}_2 \not\equiv 0$, then $\alpha \boldsymbol{S}_1 + \beta \boldsymbol{S}_2$ is a semi-invariant of the second kind for $\mathcal{C}_{m,n}$. While $\boldsymbol{S}_1^{(1)}$ and $\boldsymbol{S}_2^{(1)}$ are not semi-invariants of the second kind for $\mathcal{C}_{m,n}$, we do have the following result.

PROPOSITION 5.5. *Let $\boldsymbol{S}$ denote a semi-invariant of the second kind in $\mathcal{R}_{m,n}$ for $\mathcal{C}_{m,n}$ whose weight is s and use*

(5.15) $$\boldsymbol{d}_{m,n} \equiv \frac{1}{m(m-1)} \boldsymbol{w}_{\underbrace{0, \ldots, 0}_{n-1}, 1}.$$

from (4.19), or (4.33), to define $\boldsymbol{T}$ in $\mathcal{R}_{m,n}$ by

(5.16) $$\boldsymbol{T} \equiv \boldsymbol{S}^{(1)} + 2s\,\boldsymbol{d}_{m,n}\,\boldsymbol{S}.$$

Then, $\boldsymbol{T}$ is a semi-invariant of the second kind for $\mathcal{C}_{m,n}$ of weight $s + 1$.

PROOF. We employ (3.21), (3.22), (3.23), and (3.24) to obtain

(5.17) $$c^{**}_{\underbrace{0, \ldots, 0}_{n-1}, 1}(\zeta) \equiv \sum_{j_n=0}^{1} \mathcal{B}^{\overbrace{0, \ldots, 0, 1}^{n-1}}_{\underbrace{0, \ldots, 0}_{n-1}, j_n}(\zeta) \; c_{\underbrace{0, \ldots, 0}_{n-1}, j_n}(f(z))$$

$$\equiv f'(\zeta)\,\alpha_{0,m-1}(\zeta)\,c_{\underbrace{0, \ldots, 0}_{n-1}, 1}(f(z)) + \alpha_{1,m-1}(\zeta)$$

$$\equiv f'(\zeta)\,c_{\underbrace{0, \ldots, 0}_{n-1}, 1}(f(z)) - \binom{m}{2}\frac{f''(\zeta)}{f'(\zeta)}.$$

Since (5.15) yields $d^{**}_{m,n}(\zeta) \equiv \dfrac{c^{**}_{0,\ldots,0,1}(\zeta)}{m(m-1)}$ and $d_{m,n}(z) \equiv \dfrac{c_{0,\ldots,0,1}(z)}{m(m-1)}$, we find that (5.17) gives

$$(5.18) \qquad 2d^{**}_{m,n}(\zeta) \equiv 2f'(\zeta)\, d_{m,n}\bigl(f(\zeta)\bigr) - \frac{f''(\zeta)}{f'(\zeta)}.$$

After substituting $c^{**(k)}_{i_1,i_2,\ldots,i_n}(\zeta)$ from (4.7) for $w^{(k)}_{i_1,i_2,\ldots,i_n}$ in $\boldsymbol{T}$ of (5.16), we apply the formulas $S^{**}(\zeta) \equiv \bigl(f'(\zeta)\bigr)^s S\bigl(f(\zeta)\bigr)$ and (5.18) to deduce

$$\begin{aligned}
T^{**}(\zeta) &\equiv S^{**(1)}(\zeta) + 2s\, d^{**}_{m,n}(\zeta)\, S^{**}(\zeta) \\
&= \bigl(f'(\zeta)\bigr)^{s+1} S^{(1)}\bigl(f(\zeta)\bigr) + s\bigl(f'(\zeta)\bigr)^{s-1} f''(\zeta)\, S\bigl(f(\zeta)\bigr) \\
&\quad + 2s\bigl(f'(\zeta)\bigr)^{s+1} d_{m,n}\bigl(f(\zeta)\bigr) S\bigl(f(\zeta)\bigr) - s\bigl(f'(\zeta)\bigr)^{s-1} f''(\zeta)\, S\bigl(f(\zeta)\bigr) \\
&\equiv \bigl(f'(\zeta)\bigr)^{s+1}\left[S^{(1)}\bigl(f(\zeta)\bigr) + 2s\, d_{m,n}\bigl(f(\zeta)\bigr) S\bigl(f(\zeta)\bigr) \right] \\
&\equiv \bigl(f'(\zeta)\bigr)^{s+1} T\bigl(f(\zeta)\bigr), \quad \text{on } \Omega^{**}.
\end{aligned}$$

Since (5.16) shows that $\boldsymbol{T}$ is not a constant, this completes the proof. □

PROPOSITION 5.6. *For $m \geq 2$ and $n \geq 1$, the polynomial defined in $\mathcal{R}_{m,n}$ by*

$$(5.19) \quad \boldsymbol{H}_{m,n} \equiv \underbrace{w_{0,\ldots,0,2}}_{n-1} - \frac{m-2}{3}\underbrace{w^{(1)}_{0,\ldots,0,1}}_{n-1} - \frac{(3m-1)(m-2)}{6m(m-1)}(\underbrace{w_{0,\ldots,0,1}}_{n-1})^2$$

is a semi-invariant of the second kind for $\mathcal{C}_{m,n}$ having weight 2.

PROOF. Clearly, $\boldsymbol{H}_{m,n}$ is an isobaric polynomial in $\mathcal{R}_{m,n}$ of weight 2. To show that $\boldsymbol{H}_{m,n}$ is a semi-invariant of the second kind, let (4.1)–(4.2) be any equation of $\mathcal{C}_{m,n}$ and let (4.6) be any transformation of the second kind for that (4.1)–(4.2) into a corresponding equation (4.7)–(4.8). Section 3.3 provides details. For that context, we set

$$(5.20) \qquad \mathfrak{c}_i(z) \equiv \underbrace{c_{0,\ldots,0,i}(z)}_{n-1} \quad \text{and} \quad \mathfrak{c}^{**}_i(\zeta) \equiv \underbrace{c^{**}_{0,\ldots,0,i}(\zeta)}_{n-1}, \quad \text{for } i = 0, 1, 2.$$

In terms of $\mathfrak{w}_1 \equiv \underbrace{w_{0,\ldots,0,1}}_{n-1}$ and $\mathfrak{w}_2 \equiv \underbrace{w_{0,\ldots,0,2}}_{n-1}$, (5.19) yields

$$(5.21) \qquad \boldsymbol{H}_{m,n} \equiv \mathfrak{w}_2 - \frac{m-2}{3}\mathfrak{w}^{(1)}_1 - \frac{(3m-1)(m-2)}{6m(m-1)}(\mathfrak{w}_1)^2.$$

Formulas (3.23) and (3.24) of page 24 show that: for any j,

$$\alpha_{0,j}(\zeta) \equiv 1,$$

$$\alpha_{1,j}(\zeta) \equiv -\binom{j+1}{2}\frac{f''(\zeta)}{f'(\zeta)},$$

and

$$\alpha_{2,j}(\zeta) \equiv -\binom{j+2}{3}\frac{f'''(\zeta)}{f'(\zeta)} + 3\binom{j+3}{4}\left(\frac{f''(\zeta)}{f'(\zeta)}\right)^2.$$

In view of this, (5.20) with (3.21) and (3.22) from page 24 give

$$\text{(5.22)} \quad \mathfrak{c}_1^{**}(\zeta) \equiv \sum_{j=0}^{1} \mathcal{B}_{0,\ldots,0,j}^{0,\ldots,0,1}(\zeta) \, \mathfrak{c}_j\big(f(\zeta)\big) \equiv f'(\zeta)\,\mathfrak{c}_1\big(f(\zeta)\big) + \alpha_{1,m-1}(\zeta)$$

$$\equiv f'(\zeta)\,\mathfrak{c}_1\big(f(\zeta)\big) - \binom{m}{2}\frac{f''(\zeta)}{f'(\zeta)}$$

and

$$\text{(5.23)} \quad \mathfrak{c}_2^{**}(\zeta) \equiv \sum_{j=0}^{2} \mathcal{B}_{0,\ldots,0,j}^{0,\ldots,0,2}(\zeta) \, \mathfrak{c}_j\big(f(\zeta)\big)$$

$$\equiv \big(f'(\zeta)\big)^2 \mathfrak{c}_2\big(f(\zeta)\big) + f'(\zeta)\,\alpha_{1,m-2}(\zeta)\,\mathfrak{c}_1\big(f(\zeta)\big) + \alpha_{2,m-2}(\zeta)$$

$$\equiv \big(f'(\zeta)\big)^2 \mathfrak{c}_2\big(f(\zeta)\big) - \binom{m-1}{2} f''(\zeta)\,\mathfrak{c}_1\big(f(\zeta)\big)$$

$$- \binom{m}{3}\frac{f'''(\zeta)}{f'(\zeta)} + 3\binom{m+1}{4}\left(\frac{f''(\zeta)}{f'(\zeta)}\right)^2.$$

After substituting $\mathfrak{c}_i^{**(k)}(\zeta)$ for $\mathfrak{w}_i^{(k)}$ in (5.21), we use (5.23) as well as (5.22) and a substitution of $\mathfrak{c}_i^{(k)}(z)$ for $\mathfrak{w}_i^{(k)}$ in (5.21) to obtain

$$H_{m,n}^{**}(\zeta) \equiv \left[\mathfrak{c}_2^{**}(\zeta) - \frac{m-2}{3}\mathfrak{c}_1^{**(1)}(\zeta) - \frac{(3m-1)(m-2)}{6m(m-1)}\big(\mathfrak{c}_1^{**}(\zeta)\big)^2\right]$$

$$\equiv \left[\begin{array}{l} \big(f'(\zeta)\big)^2 \mathfrak{c}_2\big(f(\zeta)\big) - \binom{m-1}{2} f''(\zeta)\,\mathfrak{c}_1\big(f(\zeta)\big) \\[4pt] - \binom{m}{3}\dfrac{f'''(\zeta)}{f'(\zeta)} + 3\binom{m+1}{4}\left(\dfrac{f''(\zeta)}{f'(\zeta)}\right)^2 \\[4pt] - \dfrac{m-2}{3}\big(f'(\zeta)\big)^2 \mathfrak{c}_1^{(1)}\big(f(\zeta)\big) - \dfrac{m-2}{3} f''(\zeta)\,\mathfrak{c}_1\big(f(\zeta)\big) \\[4pt] + \binom{m}{3}\dfrac{f'''(\zeta)}{f'(\zeta)} - \binom{m}{3}\left(\dfrac{f''(\zeta)}{f'(\zeta)}\right)^2 \\[4pt] - \dfrac{(3m-1)(m-2)}{6m(m-1)}\big(f'(\zeta)\big)^2\big(\mathfrak{c}_1(f(\zeta))\big)^2 \\[4pt] + \dfrac{(3m-1)(m-2)}{6} f''(\zeta)\,\mathfrak{c}_1\big(f(\zeta)\big) \\[4pt] - \dfrac{(3m-1)}{4}\binom{m}{3}\left(\dfrac{f''(\zeta)}{f'(\zeta)}\right)^2 \end{array}\right]$$

$$\equiv \big(f'(\zeta)\big)^2 \left[\begin{array}{l} \mathfrak{c}_2\big(f(\zeta)\big) - \dfrac{m-2}{3}\mathfrak{c}_1^{(1)}\big(f(\zeta)\big) \\[4pt] - \dfrac{(3m-1)(m-2)}{6m(m-1)}\big(\mathfrak{c}_1(f(\zeta))\big)^2 \end{array}\right]$$

$$\equiv \big(f'(\zeta)\big)^2 H_{m,n}\big(f(\zeta)\big).$$

Thus, $\boldsymbol{H}_{m,n}$ is a semi-invariant of the second kind. That completes the proof. $\square$

COROLLARY 5.7. *For $m \geq 2$, the polynomial $\boldsymbol{b}_{m,n}$ defined in $\mathcal{R}_{m,n}$ by (4.32) on page 34 is a semi-invariant of the second kind for $\mathcal{C}_{m,n}$ having weight 2.*

PROOF. This is a consequence of the relation $\boldsymbol{b}_{m,n} \equiv \left(1/\binom{m+1}{3}\right)\boldsymbol{H}_{m,n}$. □

REMARK 5.8. An isobaric polynomial $\boldsymbol{F}$ in $\mathcal{R}_{m,n}$ of weight 1 has the form
$$\boldsymbol{F} \equiv \gamma\, \boldsymbol{w}_{\underbrace{0,\,\ldots,\,0}_{n-1},\,1}, \quad \text{for some nonzero } \gamma \text{ in } \mathbb{Q}.$$
Formula (5.4) shows that $\boldsymbol{F}$ is not a semi-invariant of the first kind for $\mathcal{C}_{m,n}$. Thus, *there are no relative invariants of weight 1 in $\mathcal{R}_{m,n}$ for $\mathcal{C}_{m,n}$.*

5.3. The number of basic relative invariants in $\mathcal{R}_{m,n}$ for $\mathcal{C}_{m,n}$

Theorem 4.6 of page 32 shows that: for $m \geq 2$, the number $\mathcal{N}_{m,n}$ of basic relative invariants for $\mathcal{C}_{m,n}$ is equal to the number of sequences $(e_1, e_2, \ldots, e_n)$ of integers $e_1, e_2, \ldots, e_n$ that satisfy $0 \leq e_1 \leq e_2 \leq \cdots \leq e_n \leq m$ and are not equal to $(0, 0, \ldots, 0)$, $(0, 0, \ldots, 1)$, or $(0, 0, \ldots, 2)$.

We shall deduce the formula $\mathcal{N}_{m,n} = \binom{m+n}{n} - 3$ from the following result.

LEMMA 5.9. *For any integers $j \geq 0$ and $n \geq 1$, let $\mathcal{L}_{j,n}$ denote the number of sequences $(i_1, i_2, \ldots, i_n)$ of integers $i_1, i_2, \ldots, i_n$ that satisfy*
$$0 \leq i_1 \leq i_2 \leq \cdots \leq i_n \leq j. \tag{5.24}$$
Then, $\mathcal{L}_{j,n}$ is given by
$$\mathcal{L}_{j,n} \equiv \binom{j+n}{n}, \quad \text{when } j \geq 0 \text{ and } n \geq 1. \tag{5.25}$$

PROOF. For $n = 1$, the number of integers i_1 that satisfy $0 \leq i_1 \leq j$ is $j + 1$. Thus, (5.25) is true for $n = 1$ and any $j \geq 0$. Let n be a positive integer such that (5.25) is valid for that n and any $j \geq 0$. Then, the number of sequences $(i_1, i_2, \ldots, i_n, i_{n+1})$ that satisfy
$$0 \leq i_1 \leq i_2 \leq \cdots \leq i_n \leq i_{n+1} \leq j. \tag{5.26}$$
can be counted in terms of the values for i_{n+1} from 0 through j. This gives
$$\mathcal{L}_{j,n+1} \equiv \mathcal{L}_{0,n} + \mathcal{L}_{1,n} + \cdots + \mathcal{L}_{j,n} \equiv \sum_{i=0}^{j} \binom{i+n}{n}$$
$$\equiv \sum_{i=0}^{j}\left[\binom{i+n+1}{n+1} - \binom{i+n}{n+1}\right] \equiv \binom{j+n+1}{n+1}.$$
Hence, (5.25) is valid for any $n \geq 1$. This completes the proof. □

PROPOSITION 5.10. *For $m \geq 2$ and $n \geq 1$, the number $\mathcal{N}_{m,n}$ of basic relative invariants in $\mathcal{R}_{m,n}$ for $\mathcal{C}_{m,n}$ is given by $\mathcal{N}_{m,n} = \binom{m+n}{n} - 3$.*

PROOF. We set $j = m$ in (5.25) and note that exactly three of the sequences counted by $\binom{m+n}{n}$ must be excluded. □

PROPOSITION 5.11. *For $m \geq 1$ and $n \geq 1$, the number $\mathcal{M}_{m,n}$ of variables $\boldsymbol{w}^{(0)}_{i_1, i_2, \ldots, i_n}$ having $k = 0$ in (4.9) of page 30 is given by $\mathcal{M}_{m,n} = \binom{m+n}{n} - 1$.*

PROOF. Here, with $0 \leq i_1 \leq i_2 \leq \cdots \leq i_n \leq m$ and $i_n \geq 1$, we set $j = m$ in (5.25) and note that only $(0, 0, \ldots, 0)$ counted by $\binom{m+n}{n}$ must be excluded. □

5.4. Nonsolutions of nonzero equations

Given $m, n \geq 1$, we assume that the variables

(5.27) $\quad \boldsymbol{w}^{(k)}_{i_1, i_2, \ldots, i_n},\quad$ for $\quad 1 \leq i_n \leq m,\ 0 \leq i_1 \leq i_2 \leq \cdots \leq i_n,\ $ and $\ k \geq 0,$

from (4.9) are algebraically independent over the field $\mathbb{C}$ of complex numbers and we let $\widehat{\mathcal{R}}_{m,n}$ denote the polynomial ring in these variables over $\mathbb{C}$. Let ′ denote the unique derivation for $\widehat{\mathcal{R}}_{m,n}$ such that the elements of $\mathbb{C}$ are constants of $\widehat{\mathcal{R}}_{m,n}$ and

(5.28) $\quad \bigl(\boldsymbol{w}^{(k)}_{i_1, i_2, \ldots, i_n}\bigr)' = \boldsymbol{w}^{(k+1)}_{i_1, i_2, \ldots, i_n},\quad$ for each $\boldsymbol{w}^{(k)}_{i_1, i_2, \ldots, i_n}$ of (5.27).

To verify the existence and uniqueness of ′, we combine (5.28) with a specialization of either [**5**, page 139, Proposition 4] or [**6**, page A.V.130, Theorem 1]. When the derivation ′ for $\widehat{\mathcal{R}}_{m,n}$ is restricted to the subring $\mathcal{R}_{m,n}$ of $\widehat{\mathcal{R}}_{m,n}$ that consists of the polynomials in the variables (5.27) over the field $\mathbb{Q}$ of rational numbers, it coincides with the derivation ′ introduced for $\mathcal{R}_{m,n}$ in Section 4.2. Thus, with ′, $\widehat{\mathcal{R}}_{m,n}$ is an ordinary differential ring and the differential ring $\mathcal{R}_{m,n}$ introduced in Section 4.2 can be regarded as a differential subring of $\widehat{\mathcal{R}}_{m,n}$. For terminology, see [**36**, **46**].

Proposition 5.11 shows that there are $M = \binom{m+n}{n} - 1$ variables $\boldsymbol{w}^{(0)}_{i_1, i_2, \ldots, i_n}$ from (5.27) having $k = 0$. Let them be designated in some order by

$$\mathfrak{s}_1,\ \mathfrak{s}_2,\ \mathfrak{s}_3,\ \ldots,\ \mathfrak{s}_{M-1},\ \mathfrak{s}_M.$$

Thus, there are functions $\phi_1, \phi_2, \ldots, \phi_n$ and χ such that

(5.29) $\quad \mathfrak{s}_h \equiv \boldsymbol{w}^{(0)}_{\phi_1(h), \phi_2(h), \ldots, \phi_n(h)} \quad$ and $\quad \boldsymbol{w}^{(0)}_{i_1, i_2, \ldots, i_n} \equiv \mathfrak{s}_{\chi(i_1, i_2, \ldots, i_n)}.$

The derivation ′ for $\widehat{\mathcal{R}}_{m,n}$ yields

(5.30) $\quad \mathfrak{s}^{(k)}_h \equiv \boldsymbol{w}^{(k)}_{\phi_1(h), \phi_2(h), \ldots, \phi_n(h)},\quad$ for $1 \leq h \leq M$ and $k \geq 0$.

For $\nu = 1, 2, \ldots, M$, let $\mathfrak{S}_\nu$ denote the ring of polynomials over $\mathbb{C}$ in the variables of (5.30) having $1 \leq h \leq \nu$ and $k \geq 0$. We note that $\mathfrak{S}_M = \widehat{\mathcal{R}}_{m,n}$.

PROPOSITION 5.12. *In terms of an integer ν subject to $1 \leq \nu \leq M$, let $\boldsymbol{Q}$ be a nonzero polynomial in $\mathfrak{S}_\nu$ and let z_0 be a complex number. Then, there are analytic functions $d_1(z), d_2(z), \ldots, d_\nu(z)$ on a neighborhood Ω of z_0 such that the analytic function $Q(z)$ on Ω obtained by replacing each $\mathfrak{s}^{(k)}_h$ in $\boldsymbol{Q}$ with the corresponding $d^{(k)}_h(z)$ satisfies $Q(z) \neq 0$, for each z in Ω.*

PROOF. First, suppose that $\nu = 1$. Then, $\boldsymbol{Q}$ is a nonzero polynomial over $\mathbb{C}$ in the variables $\mathfrak{s}_1, \mathfrak{s}^{(1)}_1, \mathfrak{s}^{(2)}_1, \ldots$. The Euler Γ-function $\Gamma(z)$ is analytic on the complex plane except for poles at $z = 0, -1, -2, \ldots$. That $\Gamma(z)$ is not a solution of any nonzero algebraic differential equation over the field $\mathbb{C}$ of complex numbers is a special case of results known from [**34**] or [**11**]. Thus, we can find a complex number z_1 and a region Ω containing z_0 such that: the function $d_1(z) \equiv \Gamma(z + z_1)$ is analytic on Ω and the analytic function $Q(z)$ on Ω obtained by replacing each $\mathfrak{s}^{(k)}_1$ in $\boldsymbol{Q}$ with the corresponding $d^{(k)}_1(z)$ satisfies $Q(z) \neq 0$, for each z in Ω. Hence, the conclusion is valid for $\nu = 1$.

Induction Hypothesis. *Suppose ν is an integer subject to $1 \leq \nu < M$ such that: for any nonzero polynomial $\boldsymbol{Q}$ in $\mathfrak{S}_\nu$ and any z_0 in $\mathbb{C}$, there are analytic functions $d_1(z), \ldots, d_\nu(z)$ on a region Ω containing z_0 such that the analytic function $Q(z)$ obtained by substituting $d_h^{(k)}(z)$ for $\mathfrak{s}_h^{(k)}$ in $\boldsymbol{Q}$ satisfies $Q(z) \neq 0$, for each z in Ω.*

Let $\widetilde{\boldsymbol{Q}}$ be a nonzero polynomial in $\mathfrak{S}_{\nu+1}$ and let z_0 be a complex number. Then, $\widetilde{\boldsymbol{Q}}$ is expressible as
$$\widetilde{\boldsymbol{Q}} \equiv \sum_{i=1}^{\lambda} \boldsymbol{Q}_i \, \boldsymbol{R}_i \quad \text{with} \quad \boldsymbol{R}_i \equiv \prod_{j=1}^{n_i} \mathfrak{s}_{\nu+1}^{(\omega_{i,j})}, \quad \text{for } 1 \leq i \leq \lambda,$$
where: λ is a positive integer; each of $\boldsymbol{Q}_1, \ldots, \boldsymbol{Q}_\lambda$ is a nonzero polynomial in $\mathfrak{S}_\nu$; $n_1, \ldots, n_\lambda$ are nonnegative integers; for $1 \leq i \leq \lambda$ and $1 \leq j \leq n_i$, $\omega_{i,j}$ is a nonnegative integer; and the polynomials $\boldsymbol{R}_1, \ldots, \boldsymbol{R}_\lambda$ are linearly independent over $\mathbb{C}$.

If $\lambda = 1$ and $n_1 = 0$, then $\boldsymbol{R}_1 \equiv 1$ and $\widetilde{\boldsymbol{Q}} \equiv \boldsymbol{Q}_1$. Since $\widetilde{\boldsymbol{Q}}$ belongs to $\mathfrak{S}_\nu$, the induction hypothesis yields suitable analytic functions $d_1(z), \ldots, d_\nu(z)$ on a region Ω containing z_0. Then, we can arbitrarily select $d_{\nu+1}(z) \equiv 1$ on Ω.

Suppose that $\lambda > 1$ or $\lambda = 1$ and $n_1 \geq 1$. When $\lambda > 1$, n_1 and n_2 are not both zero. Thus, the notation can be adjusted so that $n_1 \geq 1$. The induction applied to $\boldsymbol{Q}_1$ yields analytic functions $d_1(z), \ldots, d_\nu(z)$ on a region $\mathcal{U}_1$ containing z_0 such that the analytic function $Q_1(z)$ on $\mathcal{U}_1$ obtained by replacing $\mathfrak{s}_h^{(k)}$ in $\boldsymbol{Q}_1$ with $d_h^{(k)}(z)$ satisfies $Q_1(z) \neq 0$, for each z in $\mathcal{U}_1$. Also, when $\lambda > 1$, there are corresponding analytic functions $Q_2(z), \ldots, Q_\lambda(z)$ on $\mathcal{U}_1$ obtained by substituting $d_h^{(k)}(z)$ for $\mathfrak{s}_h^{(k)}$ in $\boldsymbol{Q}_2, \ldots, \boldsymbol{Q}_\lambda$. We set $\tau_i = Q_i(z_0)$, for $1 \leq i \leq \lambda$, and
$$\widetilde{\boldsymbol{R}} \equiv \sum_{i=1}^{\lambda} \tau_i \, \boldsymbol{R}_i.$$
Due to the linear independence of $\boldsymbol{R}_1, \ldots, \boldsymbol{R}_\lambda$ over $\mathbb{C}$, we use $\tau_1 = Q_1(z_0) \neq 0$ to see that $\widetilde{\boldsymbol{R}}$ is a nonzero polynomial in the variables $\mathfrak{s}_{\nu+1}^{(0)}, \mathfrak{s}_{\nu+1}^{(1)}, \ldots$ over $\mathbb{C}$. Thus, the argument for $\nu = 1$ gives an analytic function $d_{\nu+1}(z)$ on a subregion $\mathcal{U}_2$ of $\mathcal{U}_1$ containing z_0 such that the function $\widetilde{R}(z)$ on $\mathcal{U}_2$ obtained by substituting $d_{\nu+1}^{(k)}(z)$ for $\mathfrak{s}_{\nu+1}^{(k)}$ in $\widetilde{\boldsymbol{R}}$ satisfies $\widetilde{R}(z) \neq 0$, for each z in $\mathcal{U}_2$. When $1 \leq h \leq \nu+1$ and $k \geq 0$, the result of replacing each $\mathfrak{s}_h^{(k)}$ in $\widetilde{\boldsymbol{Q}}$ with the corresponding $d_h^{(k)}(z)$ yields an analytic function $\widetilde{Q}(z)$ on $\mathcal{U}_2$ having
$$\widetilde{Q}(z_0) = \sum_{i=1}^{\lambda} Q_i(z_0) \, R_i(z_0) = \sum_{i=1}^{\lambda} \tau_i \, R_i(z_0) = \widetilde{R}(z_0) \neq 0.$$
Continuity yields a subregion Ω of $\mathcal{U}_2$ containing z_0 such that $\widetilde{Q}(z) \neq 0$, for each z in Ω. This completes the induction step and the proof. $\square$

COROLLARY 5.13. *Let $\boldsymbol{D}$ be a nonzero element of $\widehat{\mathcal{R}}_{m,n}$ and let z_0 denote a complex number. Then, there is a differential equation (4.1)–(4.2) with analytic coefficients $c_{j_1, j_2, \ldots, j_n}(z)$ on a region Ω containing z_0 such that the analytic function $D(z)$ on Ω obtained by replacing each $\boldsymbol{w}_{i_1, i_2, \ldots, i_n}^{(k)}$ of (5.27) in $\boldsymbol{D}$ with the corresponding $c_{i_1, i_2, \ldots, i_n}^{(k)}(z)$ satisfies $D(z) \neq 0$, for each z in Ω.*

PROOF. Since $\boldsymbol{D}$ is a polynomial in $\mathfrak{S}_M$, Proposition 5.12 shows that there are analytic functions $d_1(z), \ldots, d_M(z)$ on a region Ω containing z_0 such that the replacement of each $\mathfrak{s}_h^{(k)}$ in $\boldsymbol{D}$ with the corresponding $d_h^{(k)}(z)$ yields an analytic function $D(z)$ on Ω that satisfies $D(z) \neq 0$, for each z in Ω. With respect to the notation employed for (5.29), we introduce

$$c_{i_1, i_2, \ldots, i_n}(z) \equiv d_{\chi(i_1, i_2, \ldots, i_n)}(z), \quad \text{on } \Omega, \text{ for } 0 \leq i_1 \leq \cdots \leq i_n \leq m \text{ and } i_n \geq 1.$$

We set $c_{0,0,\ldots,0}(z) \equiv 1$ on Ω and, for any integers $0 \leq j_1, j_2, \ldots, j_n \leq m$ that do not satisfy $0 \leq j_1 \leq j_2 \leq \cdots \leq j_n \leq m$, we set $c_{j_1, j_2, \ldots, j_n}(z) \equiv c_{i_1, i_2, \ldots, i_n}(z)$, where $(i_1, i_2, \ldots, i_n)$ is an arrangement of $(j_1, j_2, \ldots, j_n)$ having $i_1 \leq i_2 \leq \cdots \leq i_n$. The (4.1) on Ω having these coefficients satisfies (4.2) and the function obtained by replacing $\boldsymbol{w}_{i_1, i_2, \ldots, i_n}^{(k)}$ in $\boldsymbol{D}$ with $c_{i_1, i_2, \ldots, i_n}^{(k)}(z)$ from this (4.1) is equal to the function $D(z)$ obtained by replacing each $\mathfrak{s}_h^{(k)}$ in $\boldsymbol{D}$ with the corresponding $d_h^{(k)}(z)$. Since we have $D(z) \neq 0$ for each z in Ω, this completes the proof. □

5.5. Relative invariants in terms of basic ones and $\boldsymbol{a}_{m,n}$

For $m \geq 2$ and $n \geq 1$, Corollary 5.3 shows that the polynomial

$$(5.31) \quad \boldsymbol{a}_{m,n} \equiv \frac{1}{\binom{m+1}{3}}\left[\boldsymbol{w}_{\underbrace{0, \ldots, 0}_{n-1}, 2} - \frac{m-1}{2}\boldsymbol{w}_{\underbrace{0, \ldots, 0}_{n-1}, 1}^{(1)} - \frac{m-1}{2m}(\boldsymbol{w}_{\underbrace{0, \ldots, 0}_{n-1}, 1})^2\right]$$

of $\mathcal{R}_{m,n}$ is an isobaric semi-invariant of the first kind for $\mathcal{C}_{m,n}$ having weight 2. We note that the use of $\boldsymbol{a}_{m,n}$ rather than $\boldsymbol{G}_{m,n}$ of (5.2) has advantages in the formulations of (4.22) and (4.48). That is our key reason for selecting $\boldsymbol{a}_{m,n}$ rather than $\boldsymbol{G}_{m,n}$ in the formulation of the next result.

THEOREM 5.14. *For any $m \geq 2$, each relative invariant in $\mathcal{R}_{m,n}$ for $\mathcal{C}_{m,n}$ is expressible as a differential-polynomial combination over $\mathbb{Q}$ of $\boldsymbol{a}_{m,n}$ and the basic relative invariants in $\mathcal{R}_{m,n}$ for $\mathcal{C}_{m,n}$.*

PROOF. For $m \geq 2$, suppose that $\boldsymbol{F}$ is a relative invariant in $\mathcal{R}_{m,n}$ for $\mathcal{C}_{m,n}$. Then, $\boldsymbol{F}$ is expressible as a differential-polynomial combination over $\mathbb{Q}$ of those variables from (4.9) given by

$$(5.32) \quad \boldsymbol{w}_{i_1, i_2, \ldots, i_n}, \quad \text{for } 0 \leq i_1 \leq i_2 \leq \cdots \leq i_n \leq m \text{ and } i_n \geq 1.$$

The variables of (5.32) are totally ordered by the rule that $\boldsymbol{w}_{i_1, i_2, \ldots, i_n}$ precedes $\boldsymbol{w}_{j_1, j_2, \ldots, j_n}$ if and only if either $i_1 < j_1$ or else there is an integer r satisfying $2 \leq r \leq n$ such that $i_k = j_k$ for $1 \leq k \leq r-1$ and $i_r < j_r$.

Proposition 5.11 shows that $M = \binom{m+n}{n} - 1$ is the number of variables in (5.32). Let the total ordering for the variables in (5.32) be represented by

$$(5.33) \quad \mathfrak{s}_1, \mathfrak{s}_2, \ldots, \mathfrak{s}_M, \quad \text{with } \mathfrak{s}_1 \equiv \boldsymbol{w}_{\underbrace{0, \ldots, 0}_{n-1}, 1}, \mathfrak{s}_2 \equiv \boldsymbol{w}_{\underbrace{0, \ldots, 0}_{n-1}, 2}, \ldots.$$

When $n = 1$, we observe that $M = m$ and (5.33) consists of $\boldsymbol{w}_1, \boldsymbol{w}_2, \boldsymbol{w}_3, \ldots, \boldsymbol{w}_m$. When $n = 2$, we find that the variables of (5.33) are respectively given by

$$\boldsymbol{w}_{0,1}, \boldsymbol{w}_{0,2}, \ldots, \boldsymbol{w}_{0,m}, \boldsymbol{w}_{1,1}, \boldsymbol{w}_{1,2}, \ldots, \boldsymbol{w}_{1,m}, \boldsymbol{w}_{2,2}, \ldots, \boldsymbol{w}_{2,m}, \ldots, \boldsymbol{w}_{m,m}$$

We use (5.31) to obtain

$$(5.34) \quad \boldsymbol{w}_{\underbrace{0, \ldots, 0}_{n-1}, 2} \equiv \binom{m+1}{3}\boldsymbol{a}_{m,n} + \frac{m-1}{2}\boldsymbol{w}_{\underbrace{0, \ldots, 0}_{n-1}, 1}^{(1)} + \frac{m-1}{2m}(\boldsymbol{w}_{\underbrace{0, \ldots, 0}_{n-1}, 1})^2$$

$$\equiv \binom{m+1}{3}\boldsymbol{a}_{m,n} + \boldsymbol{R}_{m,n;\underbrace{0, \ldots, 0}_{n-1}, 2; 0},$$

where $\boldsymbol{R}_{m,n;\,0,\,\ldots,\,0,\,2;\,0}$ is a differential-polynomial combination over $\mathbb{Q}$ of the only variable in (5.33) that precede $\boldsymbol{w}_{0,\,\ldots,\,0,\,2}$. The structure of each basic relative invariant $\boldsymbol{\mathcal{I}}_{m,n;\,i_1,\,i_2,\,\ldots,\,i_n}$ in $\mathcal{R}_{m,n}$ for $\mathcal{C}_{m,n}$ yields

$$(5.35) \qquad \boldsymbol{w}_{i_1,\,i_2,\,\ldots,\,i_n} \equiv \boldsymbol{\mathcal{I}}_{m,n;\,i_1,\,i_2,\,\ldots,\,i_n} + \boldsymbol{R}_{m,n;\,i_1,\,i_2,\,\ldots,\,i_n;\,0},$$

for $i_n \geq 3$ or $n \geq 2$ and $i_{n-1} \geq 1$,

where $\boldsymbol{R}_{m,n;\,i_1,\,i_2,\,\ldots,\,i_n;\,0}$ is a differential-polynomial combination over $\mathbb{Q}$ of variables in (5.33) that precede $\boldsymbol{w}_{i_1,\,i_2,\,\ldots,\,i_n}$. By repeatedly applying the derivation $'$ for $\mathcal{R}_{m,n}$ to (5.34) and (5.35), we deduce for $k \geq 0$ that

$$(5.36) \qquad \boldsymbol{w}^{(k)}_{\underbrace{0,\,\ldots,\,0}_{n-1},\,2} \equiv \binom{m+1}{3} \boldsymbol{a}^{(k)}_{m,n} + \boldsymbol{R}_{m,n;\,\underbrace{0,\,\ldots,\,0}_{n-1},\,2;\,k}$$

and

$$(5.37) \qquad \boldsymbol{w}^{(k)}_{i_1,\,i_2,\,\ldots,\,i_n} \equiv \boldsymbol{\mathcal{I}}^{(k)}_{m,n;\,i_1,\,i_2,\,\ldots,\,i_n} + \boldsymbol{R}_{m,n;\,i_1,\,i_2,\,\ldots,\,i_n;\,k},$$

when $i_n \geq 3$ or $n \geq 2$ and $i_{n-1} \geq 1$,

where $\boldsymbol{R}_{m,n;\,i_1,\,i_2,\,\ldots,\,i_n;\,k}$ in (5.37) or (5.36) is a differential-polynomial combination over $\mathbb{Q}$ of variables in (5.33) that precede $\boldsymbol{w}_{i_1,\,i_2,\,\ldots,\,i_n}$.

Starting with $\boldsymbol{F}$ as a differential-polynomial combination over $\mathbb{Q}$ of the variables in (5.33), we use (5.37) to eliminate from $\boldsymbol{F}$ the variables $\boldsymbol{\mathfrak{s}}^{(k)}_{M}$, for each $k \geq 0$. Then, from the resulting expression, we use (5.37) or (5.36) to eliminate $\boldsymbol{\mathfrak{s}}^{(k)}_{M-1}$, for each $k \geq 0$. We continue in this manner using either (5.37) or (5.36) until finally we eliminate the variables $\boldsymbol{\mathfrak{s}}^{(k)}_{2}$, for each $k \geq 0$. The resulting expression is then a differential-polynomial combination over $\mathbb{Q}$ of the basic relative invariants in $\mathcal{R}_{m,n}$ for $\mathcal{C}_{m,n}$ as well as $\boldsymbol{a}_{m,n}$ and $\boldsymbol{\mathfrak{s}}_1$. With $\boldsymbol{\mathfrak{s}}_1 \equiv \boldsymbol{w}_{0,\,\ldots,\,0,\,1}$, we therefore write $\boldsymbol{F}$ as

$$(5.38) \qquad \boldsymbol{F} \equiv \boldsymbol{F}_0 + \boldsymbol{Z}_0,$$

where $\boldsymbol{F}_0$ is the sum of the terms of $\boldsymbol{F}$ that do not involve $\boldsymbol{w}^{(k)}_{0,\,\ldots,\,0,\,1}$, for any $k \geq 0$, and $\boldsymbol{Z}_0$ is the sum of any remaining terms. Thus, $\boldsymbol{F}_0$ is a differential-polynomial combination over $\mathbb{Q}$ of $\boldsymbol{a}_{m,n}$ and basic relative invariants in $\mathcal{R}_{m,n}$ for $\mathcal{C}_{m,n}$.

For any (4.1)–(4.2) on some region Ω, there is a meromorphic function $\rho_0(z) \not\equiv 0$ on a subregion $\mathcal{U}$ of Ω such that

$$(5.39) \qquad \rho_0^{(1)}(z) + \frac{c_{0,\,\ldots,\,0,\,1}(z)}{m} \rho_0(z) \equiv 0, \quad \text{on } \mathcal{U}.$$

Due to (5.39), an application of Theorem 3.1 on page 19 shows that the substitution $y(z) = \rho_0(z)\, v(z)$ transforms the restriction of (4.1)–(4.2) to $\mathcal{U}$ into a corresponding equation (4.4)–(4.5) on $\mathcal{U}$ having

$$(5.40) \qquad c^*_{0,\,\ldots,\,0,\,1}(z) \equiv c_{0,\,\ldots,\,0,\,1}(z) + m \frac{\rho_0^{(1)}(z)}{\rho_0(z)} \equiv 0.$$

Since $\boldsymbol{F}_0$ is a differential-polynomial combination over $\mathbb{Q}$ of semi-invariants of the first kind, Proposition 5.1 shows that the corresponding functions $F_0^*(z)$ and $F_0(z)$ obtained from (4.4) and (4.1) via Notation 1.0 satisfy $F_0^*(z) \equiv F_0(z)$, on $\mathcal{U}$. Because any nonzero term of $\boldsymbol{Z}_0$ would effectively involve $\boldsymbol{w}^{(k)}_{0,\,\ldots,\,0,\,1}$ for some $k \geq 0$, we use (5.40) to see that the substitution of $c^{*(k)}_{i_1,\,i_2,\,\ldots,\,i_n}(z)$ from this (4.4) for $\boldsymbol{w}^{(k)}_{i_1,\,i_2,\,\ldots,\,i_n}$

in $\boldsymbol{Z}_0$ yields $Z_0^*(z) \equiv 0$, on $\mathcal{U}$. Thus, we obtain
$$F(z) \equiv F^*(z) \equiv F_0^*(z) + Z_0^*(z) \equiv F_0^*(z) \equiv F_0(z), \quad \text{on } \mathcal{U}.$$
Hence, we have $F(z) \equiv F_0(z)$ on Ω, for any (4.1)–(4.2). Thus, the replacement of each $\boldsymbol{w}_{i_1, i_2, \ldots, i_n}^{(k)}$ in $\boldsymbol{F} - \boldsymbol{F}_0$ with the corresponding $c_{i_1, i_2, \ldots, i_n}^{(k)}(z)$ from any (4.1)–(4.2) yields a function that is identically zero. In view of Corollary 5.13, we conclude that $\boldsymbol{F} - \boldsymbol{F}_0 \equiv 0$. This yields $\boldsymbol{F} \equiv \boldsymbol{F}_0$ and completes the proof. $\square$

REMARK 5.15. The algorithmic substitutions employed in the preceding proof are well suited for computer algebra. Section 7.2 gives examples.

That technique can be applied to any nonconstant element $\boldsymbol{F}$ in $\mathcal{R}_{m,n}$ without knowing ahead of time whether $\boldsymbol{F}$ is a relative invariant. If that process fails to express $\boldsymbol{F}$ as a differential-polynomial combination over $\mathbb{Q}$ of $\boldsymbol{a}_{m,n}$ and the basic relative invariants in $\mathcal{R}_{m,n}$ for $\mathcal{C}_{m,n}$, then $\boldsymbol{F}$ is not a relative invariant for $\mathcal{C}_{m,n}$. However, when the process does express $\boldsymbol{F}$ in that manner, the representation may be quite useful for the further study needed to decide whether $\boldsymbol{F}$ is a relative invariant. Illustrations of that situation are provided by the representations (7.18), (7.19), and (7.20).

CHAPTER 6

Use of Computer Algebra

Programs to exhibit the basic relative invariants in $\mathcal{R}_{m,1}$ for $\mathcal{C}_{m,1}$ and in $\mathcal{R}_{m,2}$ for $\mathcal{C}_{m,2}$ are given in Sections 6.1 and 6.3. Independent verifications are provided in Sections 6.2 and 6.4. With minor alterations, these programs can be rewritten for $\mathcal{R}_{m,3}$, $\mathcal{R}_{m,4}$, ... Our separate focus on $n = 1$ or $n = 2$ or ... enables us to use standard commands that can be easily modified for other systems.

6.1. $\mathcal{I}_{m,1;e_1}$ in $\mathcal{R}_{m,1}$ for $\mathcal{C}_{m,1}$ when $m \geq 3$

We select a version of *Mathematica* from [55, 56, 57, 58, 59] as the system of computer algebra. We represent the derivation $'$ for $\mathcal{R}_{m,1}$ as differentiation with respect to a fictitious independent variable denoted by z. Then, the evaluation of D[w[i1][z],{z,k}] represents the corresponding variable $w_{i_1}^{(k)}$ of (4.9). We use (4.18)–(4.29) and (4.17) as our guide for the following fourteen input statements

```
a[m_,1][z_] := (1/Binomial[m+1,3])( w[2][z]
  -((m-1)/2)w[1]'[z]-((m-1)/(2m))w[1][z]^2 )

d[m_,1][z_] := (1/(m(m-1)))w[1][z]

K[m_,1,i_,j_][z_] := 0 /; i <= -1

K[m_,1,0,j_][z_] := 1

K[m_,1,i_,j_][z_] :=
  ( Sum[( D[K[m,1,i-1,k][z],z]
  -(m-1)*d[m,1][z]*K[m,1,i-1,k][z]
  +(m+2-i-k)(2-i-k)a[m,1][z]*
  K[m,1,i-2,k][z]),{k,j+1,m}] ) /; i >= 1

w[0][z_] = 1;    X[k_][z_] := w[k][z]

L[m_,1,i_][z_] :=
  Sum[ K[m,1,i-j,j][z]*X[j][z], {j, 0, i}]

M[m_,1,e1_,i_][z_] :=
  FunctionExpand[Binomial[m-i,e1-i]]*
  Product[(e1-r), {r,1,e1-i}]*L[m,1,i][z]

A[e1_,i_] := -1/(e1+i-1) /; i >= 1
```

```
B[e1_,i_] := (e1-i)/(e1+i-2) /; i >= 1

inv[m_,1,e1_,0][z_] := 0

inv[m_,1,e1_,1][z_] := 0

inv[m_,1,e1_,i_][z_] := ( M[m,1,e1,i][z]
  +A[e1,i-1]*D[ inv[m,1,e1,i-1][z], z]
  +B[e1,i-1]*a[m,1][z]*
  inv[m,1,e1,i-2][z] ) /; i >= 2

basicInv[m_,1,e1_][z_] := inv[m,1,e1,e1][z]
```

that are to be evaluated by the selected version of *Mathematica*. The additional evaluation of

```
Expand[ basicInv[3,1,3][z] ]
```

yields the expression for $\mathcal{I}_{3,1;3}$ given in (1.6) on page 2. After evaluating

```
list[e1_] := Table[ D[w[j][z], {z,k}],
  {j,1,e1},{k,0,e1-j}] // Flatten
```

to help simplify later output, we find that the evaluation of

```
Collect[ basicInv[m,1,3][z], list[3], Factor]
```

yields the expression for $\mathcal{I}_{m,1;3}$ in (1.13) of page 3. In particular, setting $m = 4$ in the preceding expression or making an evaluation of

```
Collect[ basicInv[4,1,3][z], list[3], Factor]
```

gives the expression for $\mathcal{I}_{4,1;3}$ in (1.15) on page 4. The evaluation of

```
Collect[ basicInv[4,1,4][z], list[4], Factor]
```

yields the expression for $\mathcal{I}_{4,1;4}$ in (1.17) on page 4. The preceding nineteen input statements can be evaluated in several seconds on a typical personal computer of current capabilities. When m is merely a symbol for an integer assumed to satisfy $3 \leq e_1 \leq m$, the times required to obtain analogous representations of $\mathcal{I}_{m,1;e_1}$ are approximately equal to: 1 second for $e_1 = 4$; 5 seconds for $e_1 = 5$; 15 seconds for $e_1 = 6$; 65 seconds for $e_1 = 7$; 4 minutes for $e_1 = 8$; 30 minutes for $e_1 = 9$; 4 hours for $e_1 = 10$; and 19 hours for $e_1 = 11$.

6.2. Alternative computation for $\mathcal{I}_{m,1;e_1}$ when $m \geq 3$

Here, we show how *Mathematica* can be employed to obtain a computer representation for $\mathcal{J}_{m,1;e_1}$ in (4.31). Since we have $\mathcal{I}_{m,1;e_1} \equiv \mathcal{J}_{m,1;e_1}$, an excellent check on our computer representations for $\mathcal{I}_{m,1;e_1}$ and $\mathcal{J}_{m,1;e_1}$ is provided when the subtraction of one from the other is identically zero. Based on (4.32)–(4.43) and (4.31), the following fourteen input statements

6.2. ALTERNATIVE COMPUTATION FOR $\mathcal{I}_{m,1;e_1}$ WHEN $m \geq 3$

```
b[m_,1][z_] := (1/Binomial[m+1,3])( w[2][z]
  -((m-2)/3)w[1]'[z]
  -((3m-1)(m-2)/(6m(m-1)))w[1][z]^2 )

d[m_,1][z_] := (1/(m(m-1)))w[1][z]

U[m_,1,i_,j_][z_] := 0 /; i <= -1

U[m_,1,0,j_][z_] := 1

U[m_,1,i_,j_][z_] := ( Sum[ D[U[m,1,i-1,k][z],z]
  +2(i-1+k-m)d[m,1][z]*U[m,1,i-1,k][z]
  +(m-i-k+2)(2-i-k)b[m,1][z]*
  U[m,1,i-2,k][z], {k,j+1,m}] ) /; i >= 1

w[0][z_] = 1;   X[k_][z_] := w[k][z]

V[m_,1,i_][z_] :=
  Sum[ U[m,1,i-j,j][z]*X[j][z], {j,0,i}]

W[m_,1,e1_,i_][z_] :=
  FunctionExpand[Binomial[m-i,e1-i]]*
  Product[(e1-r), {r,1,e1-i}]*V[m,1,i][z]

A[e1_,i_] := -1/(e1+i-1) /; i >= 1

B[e1_,i_] := (e1-i)/(e1+i-2) /; i >= 1

jnv[m_,1,e1_,0][z_] := 0; jnv[m_,1,e1_,1][z_] := 0

jnv[m_,1,e1_,i_][z_] := ( W[m,1,e1,i][z]
  +A[e1,i-1]*(D[ jnv[m,1,e1,i-1][z], z]
  +2(i-1)*d[m,1][z]*jnv[m,1,e1,i-1][z])
  +B[e1,i-1]*b[m,1][z]*
  jnv[m,1,e1,i-2][z] ) /; i >= 2

basicJ[m_,1,e1_][z_] := jnv[m,1,e1,e1][z]

list[e1_] := Table[ D[w[j][z],{z, k}],
  {j,1,e1},{k,0,e1-j}] // Flatten
```

are to be evaluated by the same version of *Mathematica* employed for Section 6.1. The additional evaluation of

```
Collect[basicJ[m,1,3][z], list[3], Factor]
```

yields a computer representation for $\mathcal{J}_{m,1;3}$; an evaluation of

```
Collect[basicJ[m,1,4][z], list[4], Factor]
```

gives a computer representation for $\mathcal{J}_{m,1;4}$; etc. When not set equal to a specific integer, the symbol m for the order of the differential equation is a free variable that represents any integer greater than or equal to the number assigned to e1.

To check the accuracy of the the Latex file for pages 53–55 of this monograph, we copied the input and pasted it into a *Mathematica* notebook. After evaluating each item in the notebook, we then inserted the input commands

```
diff[e1_] := Together[
   basicInv[m,1,e1][z] - basicJ[m,1,e1][z]]

Do[Print["diff[",e1,"] = ", diff[e1]], {e1,3,9}]
```

and found that their evaluation yielded 0 for each difference as it should.

6.3. $\mathcal{I}_{m,2;e_1,e_2}$ in $\mathcal{R}_{m,2}$ for $\mathcal{C}_{m,2}$ when $m \geq 2$

To present a *Mathematica* notebook based on Theorem 4.6 that is suitable for making explicit the structure of various relative invariants $\mathcal{I}_{m,2;e_1,e_2}$ in $\mathcal{R}_{m,2}$ for $\mathcal{C}_{m,2}$, we clear computer memory of previous usage and successively evaluate each of the following sixteen input statements

```
a[m_,2][z_] := (1/Binomial[m+1,3])( w[0,2][z]
  -((m-1)/2)w[0,1]'[z]
  -((m-1)/(2m))w[0,1][z]^2 )

d[m_,2][z_] := (1/(m(m-1)))w[0,1][z]

K[m_,2,i_,j_][z_] := 0 /; i <= -1

K[m_,2,0,j_][z_] := 1

K[m_,2,i_,j_][z_] :=
  ( Sum[( D[K[m,2,i-1,k][z], z]
  -(m-1)*d[m,2][z]*K[m,2,i-1,k][z]
  +(m+2-i-k)(2-i-k)a[m,2][z]*
  K[m,2,i-2,k][z]), {k,j+1,m}] ) /; i >= 1

w[0, 0][z_] = 1;

X[j1_,j2_][z_] := Apply[w, Sort[{j1,j2}]][z]

L[m_,2,i1_,i2_][z_] :=
  Sum[ K[m,2,i1-j1,j1][z]*K[m,2,i2-j2,j2][z]*
  X[j1,j2][z], {j1,0,i1}, {j2,0,i2} ]

H[m_,2,e1_,e2_,h1_,h2_] := ( Binomial[m-h1,e1-h1]
  *Product[e1-r, {r,1,e1-h1}])(Binomial[m-h2,e2-h2]
  *Product[e2-r, {r,1,e2-h2}] )
```

```
M[m_,2,e1_,e2_,h1_,i_][z_] :=
  H[m,2,e1,e2,h1,i-h1]*L[m,2,h1,i-h1][z]

A[e1_,e2_,i_] := -1/(e1+e2+i-1) /; i >= 1

B[e1_,e2_,i_] := (e1+e2-i)/(e1+e2+i-2) /; i >= 1

inv[m_,2,e1_,e2_,h1_,0][z_] := 0

inv[m_,2,e1_,e2_,h1_,1][z_] := 0

inv[m_,2,e1_,e2_,h1_,i_][z_] :=
  ( M[m,2,e1,e2,h1,i][z] +
  A[e1,e2,i-1]*D[ inv[m,2,e1,e2,h1,i-1][z], z]
  + B[e1,e2,i-1]*a[m,2][z]
  *inv[m,2,e1,e2,h1,i-2][z] ) /; i >= 2

basicInv[m_,2,e1_,e2_][z_] :=
  Sum[ inv[m,2,e1,e2,h1,e1+e2][z], {h1,0,e1} ]
```

with a selected version of *Mathematica*. Then, evaluations of

```
list[e1_,e2_] := Table[ D[w[j1,j2][z],{z,k}],
  {j2,1,e2},{j1,0,Min[j2,e1]},{k,0,e1+e2-j1-j2}] // Flatten

Collect[ basicInv[m,2,1,1][z], list[1,1], Factor]
```

give

(6.1) $$\mathcal{I}_{m,2;1,1} \equiv w_{1,1} - (w_{0,1})^2, \quad \text{with } m \geq 2,$$

as the basic relative invariant of index $(1,1)$ in $\mathcal{R}_{m,2}$ for $\mathcal{C}_{m,2}$. We evaluate

```
Collect[ basicInv[m,2,1,2][z], list[1,2], Factor]
```

to obtain

(6.2) $$\mathcal{I}_{m,2;1,2} \equiv w_{1,2} - \frac{m-1}{m} w_{0,1} w_{1,1} - w_{0,1} w_{0,2} + \frac{m-1}{m}(w_{0,1})^3$$
$$- \frac{m-1}{4} w_{1,1}^{(1)} + \frac{m-1}{2} w_{0,1} w_{0,1}^{(1)}, \quad \text{with } m \geq 2.$$

as the basic relative invariant of index $(1,2)$ in $\mathcal{R}_{m,2}$ for $\mathcal{C}_{m,2}$. The evaluation of

```
Collect[ basicInv[m,2,2,2][z], list[2,2], Factor]
```

yields

$$\text{(6.3)} \quad \mathcal{I}_{m,2;2,2} \equiv w_{2,2} - \frac{2(m-1)}{m} w_{0,1} w_{1,2} + \frac{6(m-1)}{5(m+1)m} w_{0,2} w_{1,1} - (w_{0,2})^2$$
$$+ \frac{(5m+2)(m-1)^2}{5(m+1)m^2} (w_{0,1})^2 w_{1,1} + \frac{2(5m+2)(m-1)}{5(m+1)m} (w_{0,1})^2 w_{0,2}$$
$$- \frac{(5m+2)(m-1)^2}{5(m+1)m^2} (w_{0,1})^4 - \frac{m-1}{3} w^{(1)}_{1,2} + \frac{(m-1)^2}{30} w^{(2)}_{1,1}$$
$$+ \frac{(m-1)^2}{3m} w_{0,1} w^{(1)}_{1,1} + \frac{(5m-4)(m-1)^2}{15(m+1)m} w^{(1)}_{0,1} w_{1,1}$$
$$+ \frac{m-1}{3} w_{0,1} w^{(1)}_{0,2} + \frac{m-1}{3} w^{(1)}_{0,1} w_{0,2} - \frac{(m-1)^2}{15} w_{0,1} w^{(2)}_{0,1}$$
$$- \frac{(m-1)^2}{15} (w^{(1)}_{0,1})^2 - \frac{(5m+2)(m-1)^2}{5(m+1)m} (w_{0,1})^2 w^{(1)}_{0,1}, \quad \text{with } m \geq 2,$$

as the basic relative invariant of index (2, 2) in $\mathcal{R}_{m,2}$ for $\mathcal{C}_{m,2}$.

6.4. Alternative computation for $\mathcal{I}_{m,2;e_1,e_2}$ when $m \geq 2$

The program presented here is based on Theorem 4.8 of page 34 for $\mathcal{J}_{m,2;e_1,e_2}$. It provides an independent check on the computations of Section 6.3 for $\mathcal{I}_{m,2;e_1,e_2}$. Namely, after the first sixteen input commands of Section 6.3 and the following thirteen input commands

```
b[m_,2][z_] := ( (1/(Binomial[m+1,3])) ( w[0,2][z]
   -((m-2)/3)w[0, 1]'[z]
   -((3m-1)(m-2)/(6m(m-1)))w[0,1][z]^2) )

d[m_,2][z_] := (1/(m(m-1)))w[0,1][z]

U[m_,2,i_,j_][z_] := 0 /; i <= -1

U[m_,2,0,j_][z_] := 1

U[m_,2,i_,j_][z_] :=
   ( Sum[ D[U[m,2,i-1,k][z], z]
   +2(i-1+k-m)d[m, 2][z]*U[m,2,i-1,k][z]
   +(m-i-k+2)(2-i-k)b[m,2][z]*U[m,2,i-2,k][z],
   {k,j+1,m}] ) /; i >= 1

w[0, 0][z_] = 1;

X[j1_,j2_][z_] := Apply[w, Sort[{j1,j2}]][z]

V[m_,2,i1_,i2_][z_] :=
   Sum[ U[m,2,i1-j1,j1][z]*U[m,2,i2-j2,j2][z]*
   X[j1,j2][z], {j1,0,i1}, {j2,0,i2} ]

W[m_,2,e1_,e2_,h1_,i_][z_] :=
   H[m,2,e1,e2,h1,i-h1]*V[m,2,h1,i-h1][z]
```

```
jnv[m_,2,e1_,e2_,h1_,0][z_] := 0

jnv[m_,2,e1_,e2_,h1_,1][z_] := 0

jnv[m_,2,e1_,e2_,h1_,i_][z_] :=
  ( W[m,2,e1,e2,h1,i][z]
  +A[e1,e2,i-1]*(D[jnv[m,2,e1,e2,h1,i-1][z],z]
  +2(i-1)*d[m,2][z]*jnv[m,2,e1,e2,h1,i-1][z])
  +B[e1,e2,i-1]*b[m,2][z]
  *jnv[m,2,e1,e2,h1,i-2][z] ) /; i >= 2

basicJnv[m_,2,e1_,e2_][z_] :=
  Sum[ jnv[m,2,e1,e2,h1,e1+e2][z], {h1,0,e1} ]
```

are evaluated by the selected version of *Mathematica* with m as a mere symbol representing any integer greater than or equal to e_2, the evaluation of

```
diff[e1_,e2_] := Together[
  basicInv[m,2,e1,e2][z] - basicJnv[m,2,e1,e2][z]]

Do[Print["diff[",e1,", ",e2,"] = ", diff[e1,e2]],
  {e2,1,3},{e1,0,e2}]
```

yields 0 for each difference.

With m as a variable for any integer ≥ 2, a program to explicitly specify the basic relative invariants in any particular one of $\mathcal{R}_{m,3}$, $\mathcal{R}_{m,4}$, ... can be written by making minor modifications to the input commands of Sections 6.3 and 6.4.

6.5. $C_{p,q,r}(P, Q)$ as a differential-polynomial combination of P, Q, and $a_{m,n}$ over $\mathbb{Q}$

Let P and Q denote relative invariants in $\mathcal{R}_{m,n}$ for $\mathcal{C}_{m,n}$ having respective weights p and q. For $r=0$ and $r=1$, we note that (4.52) and (4.51) give

$$C_{p,q,0}(P, Q) \equiv PQ$$

and

$$C_{p,q,1}(P, Q) \equiv PQ^{(1)} - \frac{q}{p}P^{(1)}Q$$

as explicit differential-polynomial combinations of P and Q over $\mathbb{Q}$. Here, for $r \geq 2$ and $m \geq 2$, we present a program that yields $C_{p,q,r}(P, Q)$ as an explicit differential-polynomial combination of P, Q, and $a_{m,n}$ over $\mathbb{Q}$.

For $r \geq 2$, $m \geq 2$, and a version of *Mathematica* from [55, 56, 57, 58, 59] as the system of computer algebra, we shall use the evaluations of

```
D[P[z],{z,i}],  D[Q[z],{z,j}],  and  D[a[m,n][z],{z,k}]
```

to represent $P^{(i)}$, $Q^{(j)}$, and $a_{m,n}^{(k)}$. To obtain a representation of $C_{p,q,r}(P, Q)$, we clear the computer memory and then have *Mathematica* evaluate the six input statements

```
B[h_,i_,j_][z_] := 0   /; i <= -1

B[h_,0, j_][z_] := 1;

B[h_,i_,j_][z_] := Sum[ D[B[h,i-1,k][z], z]
  -k(2h+k-1)a[m,n][z]*B[h,i-2,k-1][z],
  {k,i-1,j-1}] /; i >= 1

Ce[p_,q_,r_,mu_] := FunctionExpand[
  (-1)^mu *Binomial[r,mu]*
  Binomial[2q+r-1,mu]/Binomial[2p+mu-1,mu] ]

A[p_,q_,r_,s_,t_][z_] := Sum[Ce[p,q,r,t+k]*
  B[p,k,t+k][z]*B[q,s-k,r-t-k][z],{k,0,s}]

poly[p_,q_,r_,P_,Q_][z_]:= Module[{expr, mon},
  mon = Join[Table[D[a[m,n][z],{z,i}], {i,0,r-2}],
  Table[D[P[z],{z,j}], {j,0,r}],
  Table[D[Q[z],{z,k}], {k,0,r}] ];
  expr = Sum[ A[p,q,r,s,t][z]*Derivative[t][P][z]*
  Derivative[r-s-t][Q][z], {s,0,r}, {t,0,r-s}];
  Collect[ expr, mon, Factor] ]
```

that are based on formulas (4.46)–(4.50) and (4.44).

For $r = 2$, we evaluate `poly[p,q,2,P,Q][z]` and find that

$$(6.4) \quad C_{p,q,2}(P, Q) \equiv P Q^{(2)} + \alpha\, P^{(1)} Q^{(1)} + \beta\, P^{(2)} Q + \gamma\, a_{m,n}\, P Q,$$

where

$$(6.5) \quad \alpha = \frac{-(2q+1)}{p}, \quad \beta = \frac{q(2q+1)}{p(2p+1)}, \quad \text{and} \quad \gamma = \frac{-4q(p+q+1)}{2p+1}.$$

By Theorem 4.10 on page 36, (6.4) is a relative invariant in $\mathcal{R}_{m,n}$ for $\mathcal{C}_{m,n}$ of weight $p + q + 2$. Computer algebra shows that α, β, γ in (6.5) are the only coefficients for which the right member of (6.4) is a relative invariant.

For $r = 3$, we evaluate `poly[p,q,3,P,Q][z]` and find that

$$(6.6) \quad C_{p,q,3}(P, Q) \equiv P Q^{(3)} + x_1\, P^{(1)} Q^{(2)} + x_2\, P^{(2)} Q^{(1)} + x_3\, P^{(3)} Q$$
$$+ x_4\, a_{m,n}\, P Q^{(1)} + x_5\, a_{m,n}\, P^{(1)} Q + x_6\, a^{(1)}_{m,n}\, P Q,$$

where

$$(6.7) \quad x_1 = \frac{-3(q+1)}{p}, \qquad x_4 = \frac{-4(3pq+3q^2+p+6q+2)}{2p+1},$$

$$(6.8) \quad x_2 = \frac{3(q+1)(2q+1)}{p(2p+1)}, \qquad x_5 = \frac{4q(q+1)(3p^2+3pq+6p+q+2)}{p(p+1)(2p+1)},$$

$$(6.9) \quad x_3 = \frac{-q(q+1)(2q+1)}{p(p+1)(2p+1)}, \qquad x_6 = \frac{2q(q-p)(2p+2q+3)}{(p+1)(2p+1)}.$$

6.5. $C_{p,q,r}(P, Q)$ AS A DIFFERENTIAL COMBINATION OF P, Q, AND $a_{m,n}$

When P and Q are linearly independent over $\mathcal{Q}$, Theorem 4.10 establishes that (6.6) is a relative invariant in $\mathcal{R}_{m,n}$ for $\mathcal{C}_{m,n}$ of weight $p+q+3$. The coefficients (6.7)–(6.9) first appeared for the context $n=2$ in [**20**, page 150]. They are the only ones for which the right member of (6.6) is a relative invariant.

For $r=4$, we evaluate `poly[p,q,4,P,Q][z]` and find that

$$(6.10) \quad C_{p,q,4}(P, Q)$$
$$\equiv P Q^{(4)} + y_1 P^{(1)} Q^{(3)} + y_2 P^{(2)} Q^{(2)} + y_3 P^{(3)} Q^{(1)} + y_4 P^{(4)} Q$$
$$+ a_{m,n}\left[y_5 P Q^{(2)} + y_6 P^{(1)} Q^{(1)} + y_7 P^{(2)} Q\right]$$
$$+ a_{m,n}^{(1)}\left[y_8 P Q^{(1)} + y_9 P^{(1)} Q\right] + \left[y_{10} a_{m,n}^{(2)} + y_{11} (a_{m,n})^2\right] P Q,$$

where

$$(6.11) \quad y_1 = \frac{-2(2q+3)}{p},$$

$$(6.12) \quad y_2 = \frac{6(q+1)(2q+3)}{p(2p+1)},$$

$$(6.13) \quad y_3 = \frac{-2(q+1)(2q+1)(2q+3)}{p(p+1)(2p+1)},$$

$$(6.14) \quad y_4 = \frac{q(q+1)(2q+1)(2q+3)}{p(p+1)(2p+1)(2p+3)},$$

$$(6.15) \quad y_5 = \frac{-4(6pq + 6q^2 + 4p + 18q + 11)}{2p+1},$$

$$(6.16) \quad y_6 = \frac{8(2q+3)(3p^2q + 3pq^2 + p^2 + 9pq + q^2 + 3p + 3q + 1)}{p(p+1)(2p+1)},$$

$$(6.17) \quad y_7 = \frac{-4q(q+1)(2q+3)(6p^2 + 6pq + 18p + 4q + 11)}{p(p+1)(2p+1)(2p+3)},$$

$$(6.18) \quad y_8 = \frac{-2(8p^2q - 8q^3 + 2p^2 + 12pq - 24q^2 + 3p - 18q - 5)}{(p+1)(2p+1)},$$

$$(6.19) \quad y_9 = \frac{2q(2q+3)(8p^3 - 8pq^2 + 24p^2 - 12pq - 2q^2 + 18p - 3q + 5)}{p(p+1)(2p+1)(2p+3)},$$

$$(6.20) \quad y_{10} = \frac{-2q(p+q+2)(4p^2 - 4pq + 4q^2 + 4p + 4q + 3)}{(p+1)(2p+1)(2p+3)},$$

and

$$(6.21) \quad y_{11} = \frac{48q(q+1)(p+q+2)(p+q+3)}{(2p+1)(2p+3)}.$$

Theorem 4.10 of page 36, shows that (6.10) is a relative invariant in $\mathcal{R}_{m,n}$ for $\mathcal{C}_{m,n}$ of weight $p+q+4$. The coefficients (6.11)–(6.21) first appeared for the context $n=2$ in [**20**, page 151]. They are the only ones for which the right member of (6.10) is a relative invariant.

Analogous results can be made explicit for larger values of r.

6.6. Several identities

PROPOSITION 6.1. *Suppose that* P, Q, *and* R *are relative invariants in* $\mathcal{R}_{m,n}$ *for* $\mathcal{C}_{m,n}$ *having respective weights* p, q, *and* r. *Then,* P, Q, *and* R *satisfy*

(6.22) $$C_{r,p+q,2}(R, PQ) \equiv -\left(\frac{2p}{2q+1}\right) R \, C_{p,q,2}(P, Q)$$
$$+ \left(\frac{2p+2q+1}{2q+1}\right) P \, C_{r,q,2}(R, Q)$$
$$+ \left(\frac{2p+2q+1}{2p+1}\right) Q \, C_{r,p,2}(R, P)$$

and

(6.23) $$R \, C_{p,q,2}(P, Q)$$
$$\equiv -\left(\frac{r(2q+1)(2r+1)}{2p(p+q+r+1)(2p+2q+2r+1)}\right) C_{r,p+q,2}(R, PQ)$$
$$+ \left(\frac{(2q+1)(2p+2q+1)}{2(p+q+r+1)(2p+2q+2r+1)}\right) C_{p,q+r,2}(P, QR)$$
$$+ \left(\frac{q(2q+1)(2p+2q+1)}{2p(p+q+r+1)(2p+2q+2r+1)}\right) C_{q,p+r,2}(Q, PR).$$

PROOF. While pencil-and-paper computations yield the desired results, the computer algebra of pages 59–60 enables us to check that the evaluation of

```
Together[ poly[r,p+q,2,R,Function[x,P[x]*Q[x]]][z]
 +(2p/(2q+1))R[z]*poly[p,q,2,P,Q][z]
 -((2p+2q+1)/(2q+1))P[z]*poly[r,q,2,R,Q][z]
 -((2p+2q+1)/(2p+1))Q[z]*poly[r,p,2,R,P][z] ]
```

has 0 as its output. Thus, (6.22) is valid. The evaluation of

```
Together[ R[z]*poly[p,q,2,P,Q][z]
 +(r(2q+1)(2r+1)/(2p(p+q+r+1)(2p+2q+2r+1)))*
  poly[r,p+q,2,R,Function[x,P[x]*Q[x]]][z]
 -((2q+1)(2p+2q+1)/(2(p+q+r+1)(2p+2q+2r+1)))*
  poly[p,q+r,2,P,Function[x,Q[x]*R[x]]][z]
 -(q(2q+1)(2p+2q+1)/(2p(p+q+r+1)(2p+2q+2r+1)))*
  poly[q,p+r,2,Q,Function[x,P[x]*R[x]]][z] ]
```

yields the output 0 and establishes (6.23). This completes the proof. □

By setting $R \equiv Q \equiv P$ and $r = q = p$ in either (6.22) or (6.23), we obtain

(6.24) $$C_{p,2p,2}(P, P^2) \equiv \frac{2(3p+1)}{2p+1} P \, C_{p,p,2}(P, P).$$

The definition of $C_{p,q,r}(P, Q)$ in (4.44)–(4.52) is applicable to any isobaric polynomials P and Q in $\mathcal{R}_{m,n}$ having respective weights p and q. Each of (6.22), (6.23), and (6.24) as well as (8.16), (8.17), and (8.18) remains valid as an identity in that context.

6.7. Observations about computations

Versions 3.0, 7.0.1, 8.0.1, 9.0.1, 10.1, and 11.2 of *Mathematica* can be used to evaluate correctly all of the notebooks for [19] and [20]. We specifically found that [54] and [55] performed flawlessly. Supplementary input not needed for Version 7.0.1 was supplied to the notebooks of [19] and [20] in order to enable Version 3.0 to efficiently evaluate symbolic sums like those for $\boldsymbol{K}_{m,n;i,j}$ of (4.22) and $\boldsymbol{U}_{m,n;i,j}$ of (4.36); in this regard, see [19, pages 24–28]. With such supplements, Version 3.0 could be employed here.

For computer representations of $\boldsymbol{C}_{p,q,r}(\boldsymbol{P}, \boldsymbol{Q})$ when $\boldsymbol{P}$ and $\boldsymbol{Q}$ have machine representations, see Sections 13.3 and 14.2.

Correct evaluations for each of the *Mathematica* notebooks in this monograph can be downloaded by using the Google browser *Chrome* to visit the web page
    `http://homepages.uc.edu/~chalklr/Notebooks.htm`
and then making a selection of a particular notebook labeled according to the section in which it appears.

CHAPTER 7

Principal Theorems Applied to Paul Appell's Study of $\mathcal{C}_{2,2}$

Theorem 4.6 and Theorem 4.10 will be employed in this chapter to show that relative invariants can be used to precisely characterize the equations of $\mathcal{C}_{2,2}$ for which solution procedures studied by Paul Appell in [4] are applicable.

7.1. Solution procedures for two special kinds of equations in $\mathcal{C}_{2,2}$

We use (4.1) with $m = n = 2$, $c_{1,0} \equiv c_{0,1}$, $c_{2,0} \equiv c_{0,2}$, and $c_{2,1} \equiv c_{1,2}$ to see that: for each region Ω, the differential equations of $\mathcal{C}_{2,2}$ on Ω are specified by

$$(7.1) \quad \bigl(y''(z)\bigr)^2 + 2c_{0,1}(z)\,y''(z)\,y'(z) + 2c_{0,2}(z)\,y''(z)\,y(z)$$
$$+ c_{1,1}(z)\bigl(y'(z)\bigr)^2 + 2c_{1,2}(z)\,y'(z)\,y(z) + c_{2,2}(z)\bigl(y(z)\bigr)^2 = 0,$$

where the $c_{i,j}(z)$ are meromorphic functions on Ω.

To characterize the equations (7.1) that have a nontrivial factorization when restricted to various subregion of Ω, we introduce

$$(7.2) \quad \boldsymbol{D}_2 \equiv \begin{vmatrix} 1 & \boldsymbol{w}_{0,1} & \boldsymbol{w}_{0,2} \\ \boldsymbol{w}_{0,1} & \boldsymbol{w}_{1,1} & \boldsymbol{w}_{1,2} \\ \boldsymbol{w}_{0,2} & \boldsymbol{w}_{1,2} & \boldsymbol{w}_{2,2} \end{vmatrix}$$

in $\mathcal{R}_{2,2}$ and we let $D_2(z)$ denote the function on Ω obtained by replacing each $\boldsymbol{w}_{i,j}$ in $\boldsymbol{D}_2$ with the corresponding $c_{i,j}(z)$ from (7.1). In particular, the situation $k = 2$ of [20, Theorem 15.10] shows that $\boldsymbol{D}_2$ is a relative invariant of weight 6 for $\mathcal{C}_{2,2}$. We know from [20, Example 15.6] that: (i) if $D_2(z) \equiv 0$, then each subregion $\mathcal{U}$ of Ω contains a subregion $\mathcal{V}$ on which there are meromorphic functions α, β, γ, δ such that (7.1) is expressible on $\mathcal{V}$ as

$$(7.3) \quad \bigl(y''(z) + \alpha(z)\,y'(z) + \beta(z)\,y(z)\bigr)\bigl(y''(z) + \gamma(z)\,y'(z) + \delta(z)\,y(z)\bigr) = 0;$$

and, (ii) if (7.1) is expressible as (7.3) on a subregion $\mathcal{V}$ of Ω, then $D_2(z) \equiv 0$.

Paul Appell presented the main ideas of the preceding paragraph in [4] of 1889. Thus, if (7.3) exists for (7.1) and is abbreviated as $L_1 L_2 = 0$ on $\mathcal{V}$, then the solutions (7.1) on subregions of $\mathcal{V}$ are the solutions of $L_1 = 0$ or $L_2 = 0$.

To describe another kind of (7.1) that is solvable by a technique Paul Appell presented in [4], let $Q(z)$ denote the left member of (7.1), let $Q^{(1)}(z)$ denote the formal derivative of $Q(z)$ with respect to z, and let $\mathfrak{F}_\Omega$ denote the field of meromorphic functions on Ω. In [4], Paul Appell studied the equations (7.1) for which

$$(7.4) \quad \text{there is a } \lambda(z) \text{ in } \mathfrak{F}_\Omega \text{ such that } Q^{(1)}(z) + \lambda(z)\,Q(z) \equiv L_2(z)\,L_3(z),$$

where $L_2(z)$ and $L_3(z)$ are respectively linear forms in $y''(z)$, $y'(z)$, $y(z)$ and $y'''(z)$, $y''(z)$, $y'(z)$, $y(z)$ over $\mathfrak{F}_\Omega$.

When (7.1) satisfies $D_2(z) \not\equiv 0$ and (7.4), each subregion $\mathcal{U}$ of Ω contains a subregion $\mathcal{V}$ on which there are linearly independent meromorphic functions $v_0(z)$, $v_1(z)$, $v_2(z)$ with respect to the field $\mathbb{C}$ of complex numbers such that

(7.5) $$\bigl(v_1(z)\bigr)^2 - 4v_0(z)\,v_2(z) \not\equiv 0, \quad \text{on } \mathcal{V},$$

and each of the functions

(7.6) $$y(z) \equiv C^2 v_0(z) + CK v_1(z) + K^2 v_2(z), \quad \text{for any } C, K \text{ in } \mathbb{C},$$

is a solution of (7.1) on $\mathcal{V}$. Moreover, if there is a subregion $\mathcal{V}$ of Ω on which are defined linearly independent meromorphic functions $v_0(z)$, $v_1(z)$, $v_2(z)$ over $\mathbb{C}$ such that (7.5) is satisfied and each function of (7.6) is a solution of (7.1), then (7.1) satisfies $D_2(z) \not\equiv 0$ and (7.4). For complete details, see [**20**, pages 193–208].

In [**4**], Paul Appell sought relative invariants with which to characterize the equations (7.1) that satisfy $D_2(z) \not\equiv 0$ and (7.4). He was motivated by the way that the relative invariant $\boldsymbol{D}_2$ specifies the condition $D_2(z) \equiv 0$ for (7.3).

The differential polynomials $\boldsymbol{E}_6$ and $\boldsymbol{E}_7$ defined in $\mathcal{R}_{2,2}$ by (7.10) and (7.12) have the property that: *the functions $D_2(z)$, $E_6(z)$, and $E_7(z)$ obtained by replacing each $w_{i,j}^{(k)}$ in $\boldsymbol{D}_2$, $\boldsymbol{E}_6$, and $\boldsymbol{E}_7$ with the corresponding $c_{i,j}^{(k)}(z)$ from (7.1) satisfy $D_2(z) \not\equiv 0 \equiv E_6(z) \equiv E_7(z)$ if and only if (7.1) satisfies $D_2(z) \not\equiv 0$ and (7.4)*. This was established in [**20**, Theorem 1.10] with [**13**, pages 84–87] and it was independently verified for [**20**, Remark 19.3].

Next, the constructive techniques of Theorem 5.14 and Theorem 4.10 will be employed to express $\boldsymbol{E}_6$ and $\boldsymbol{E}_7$ as sums of relative invariants and thereby verify that $\boldsymbol{E}_6$ and $\boldsymbol{E}_7$ are relative invariants in $\mathcal{R}_{2,2}$ for $\mathcal{C}_{2,2}$ of respective weights 6 and 7.

7.2. Representations for $\boldsymbol{E}_6$, $\boldsymbol{E}_7$, and $\boldsymbol{D}_2$

In $\mathcal{R}_{2,2}$, we set

(7.7) $$\boldsymbol{B}_{1,1} \equiv \boldsymbol{w}_{1,1} - (\boldsymbol{w}_{0,1})^2,$$

(7.8) $$\boldsymbol{B}_{1,2} \equiv \boldsymbol{w}_{1,2} - \boldsymbol{w}_{0,1}\,\boldsymbol{w}_{0,2},$$

(7.9) $$\boldsymbol{B}_{2,2} \equiv \boldsymbol{w}_{2,2} - (\boldsymbol{w}_{0,2})^2,$$

(7.10) $$\boldsymbol{E}_6 \equiv \boldsymbol{B}_{1,1}\boldsymbol{B}_{2,2} - 2(\boldsymbol{B}_{1,2})^2 + \boldsymbol{w}_{0,1}\boldsymbol{B}_{1,1}\boldsymbol{B}_{1,2} \\ - \boldsymbol{w}_{0,2}(\boldsymbol{B}_{1,1})^2 + \boldsymbol{B}_{1,1}\boldsymbol{B}_{1,2}^{(1)} - \boldsymbol{B}_{1,1}^{(1)}\boldsymbol{B}_{1,2},$$

(7.11) $$\boldsymbol{H}_7 \equiv \boldsymbol{B}_{1,1}\boldsymbol{B}_{2,2}^{(1)} - \boldsymbol{B}_{1,1}^{(1)}\boldsymbol{B}_{2,2} - 2\boldsymbol{B}_{1,2}\boldsymbol{B}_{2,2} \\ + 2\boldsymbol{w}_{0,1}\boldsymbol{B}_{1,1}\boldsymbol{B}_{2,2} - 2\boldsymbol{w}_{0,2}\boldsymbol{B}_{1,1}\boldsymbol{B}_{1,2},$$

and

(7.12) $$\boldsymbol{E}_7 \equiv \boldsymbol{H}_7 - \frac{1}{6}\Bigl(\boldsymbol{E}_6^{(1)} + 6\,\boldsymbol{w}_{0,1}\,\boldsymbol{E}_6\Bigr).$$

The polynomial $\boldsymbol{D}_2$ of (7.2) can also be written as

(7.13) $$\boldsymbol{D}_2 \equiv \boldsymbol{B}_{1,1}\boldsymbol{B}_{2,2} - (\boldsymbol{B}_{1,2})^2.$$

7.2. REPRESENTATIONS FOR E_6, E_7, AND D_2

By substituting $m = 2$ in (6.1), (6.2), and (6.3), we see that the three basic relative invariants for $C_{2,2}$ are given by

(7.14) $\quad \mathcal{I}_{2,2;1,1} \equiv w_{1,1} - (w_{0,1})^2,$

(7.15) $\quad \mathcal{I}_{2,2;1,2} \equiv w_{1,2} - \frac{1}{2} w_{0,1} w_{1,1} - w_{0,1} w_{0,2}$
$\qquad + \frac{1}{2}(w_{0,1})^3 - \frac{1}{4} w_{1,1}^{(1)} + \frac{1}{2} w_{0,1} w_{0,1}^{(1)},$

and

(7.16) $\quad \mathcal{I}_{2,2;2,2} \equiv w_{2,2} - w_{0,1} w_{1,2} + \frac{1}{5} w_{0,2} w_{1,1} - (w_{0,2})^2 + \frac{1}{5}(w_{0,1})^2 w_{1,1}$
$\qquad + \frac{4}{5}(w_{0,1})^2 w_{0,2} - \frac{1}{5}(w_{0,1})^4 - \frac{1}{3} w_{1,2}^{(1)} + \frac{1}{30} w_{1,1}^{(2)}$
$\qquad + \frac{1}{6} w_{0,1} w_{1,1}^{(1)} + \frac{1}{15} w_{0,1}^{(1)} w_{1,1} + \frac{1}{3} w_{0,1} w_{0,2}^{(1)} + \frac{1}{3} w_{0,1}^{(1)} w_{0,2}$
$\qquad - \frac{1}{15} w_{0,1} w_{0,1}^{(2)} - \frac{1}{15} (w_{0,1}^{(1)})^2 - \frac{2}{5}(w_{0,1})^2 w_{0,1}^{(1)}.$

From (5.31), we also have

(7.17) $\quad a_{2,2} \equiv w_{0,2} - \frac{1}{2} w_{0,1}^{(1)} - \frac{1}{4}(w_{0,1})^2.$

For any F in $\mathcal{R}_{2,2}$, the algorithmic technique used to prove Theorem 5.14 is applicable to express F as a differential-polynomial combination over $\mathbb{Q}$ of $\mathcal{I}_{2,2;2,2}$, $\mathcal{I}_{2,2;1,2}$, $\mathcal{I}_{2,2;1,1}$, $a_{2,2}$, and $w_{0,1}$. Namely, after transposing all of the terms other than $w_{2,2}$ in the right member of (7.16) to its left member and designating the resulting left member by $x_{2,2}$, we obtain $w_{2,2} \equiv x_{2,2}$. Similarly, we rewrite (7.15), (7.14), and (7.17) to obtain $x_{1,2}$, $x_{1,1}$, and $x_{0,2}$ such that $w_{1,2} \equiv x_{1,2}$, $w_{1,1} \equiv x_{1,1}$, and $w_{0,2} \equiv x_{0,2}$. We use $w_{2,2}^{(k)} \equiv x_{2,2}^{(k)}$ to eliminate from F the various $w_{2,2}^{(k)}$; in the resulting expression, we use $w_{1,2}^{(k)} \equiv x_{1,2}^{(k)}$ to eliminate the various $w_{1,2}^{(k)}$; then, we use $w_{1,1}^{(k)} \equiv x_{1,1}^{(k)}$ to eliminate the various $w_{1,1}^{(k)}$; and, we finally use $w_{0,2}^{(k)} \equiv x_{0,2}^{(k)}$ to eliminate the various $w_{0,2}^{(k)}$.

To perform such computations with a selected version of *Mathematica* from [55, 56, 57, 58, 59] in terms of (7.7)–(7.17) when F is replaced with E_6 or E_7 or D_2, let the derivation $'$ for $\mathcal{R}_{2,2}$ be represented as differentiation with respect to a fictitious independent variable denoted by z and successively evaluate

```
B[1,1][z_] = w[1,1][z] - (w[0,1][z])^2;

B[1,2][z_] = w[1,2][z] - w[0,1][z]*w[0,2][z];

B[2,2][z_] = w[2,2][z] - (w[0,2][z])^2;

D2[z_] = B[1,1][z]*B[2,2][z] - (B[1,2][z])^2;

E6[z_] = ( B[1,1][z]*B[2,2][z] - 2(B[1,2][z])^2
        + w[0,1][z]*B[1,1][z]*B[1,2][z]
        - w[0,2][z](B[1,1][z])^2
        + B[1,1][z]*D[B[1,2][z],z]
        - D[B[1,1][z],z]*B[1,2][z] );
```

```
H7[z_] = ( B[1,1][z]*D[B[2,2][z],z]
            - D[B[1,1][z],z]*B[2,2][z]
            - 2B[1,2][z]*B[2,2][z]
            + 2w[0,1][z]*B[1,1][z]*B[2,2][z]
            - 2w[0,2][z]*B[1,1][z]*B[1,2][z] );

E7[z_] = ( H7[z] - (1/6)(D[E6[z],z] + 6w[0,1][z]*E6[z]) );

x[2,2][z_] = ( inv[2,2][z]
              + w[0,1][z]*w[1,2][z]
              - (1/5)w[0,2][z]*w[1,1][z]
              + (w[0,2][z])^2
              - (1/5)(w[0,1][z])^2*w[1,1][z]
              - (4/5)(w[0,1][z])^2*w[0,2][z]
              + (1/5)(w[0,1][z])^4
              + (1/3)D[w[1,2][z],z]
              - (1/30)D[w[1,1][z],{z,2}]
              - (1/6)w[0,1][z]*D[w[1,1][z],z]
              - (1/15)D[w[0,1][z],z]*w[1,1][z]
              - (1/3)w[0,1][z]*D[w[0,2][z],z]
              - (1/3)D[w[0,1][z],z]*w[0,2][z]
              + (1/15)w[0,1][z]*D[w[0,1][z],{z,2}]
              + (1/15)(D[w[0,1][z],z])^2
              + (2/5)(w[0,1][z])^2*D[w[0,1][z],z] );

x[1,2][z_] = ( inv[1,2][z]
              + (1/2)w[0,1][z]*w[1,1][z]
              + w[0,1][z]*w[0,2][z]
              - (1/2)(w[0,1][z])^3
              + (1/4)D[w[1,1][z],z]
              - (1/2)w[0,1][z]*D[w[0,1][z],z] );

x[1,1][z_] = ( inv[1,1][z] + (w[0,1][z])^2 );

x[0,2][z_] = ( a[2,2][z] + (1/2)D[w[0,1][z],z]
              + (1/4)(w[0,1][z])^2 );

{a1,b1,c1} = Expand[ {E6[z],E7[z],D2[z]} /.
                    w[2,2]'[z]->x[2,2]'[z] ];

{a2,b2,c2} = Expand[ {a1,b1,c1} /.
                    w[2,2][z]->x[2,2][z] ];

{a3,b3,c3} = Expand[ {a2,b2,c2} /.
                    w[1,2]''[z]->x[1,2]''[z] ];

{a4,b4,c4} = Expand[ {a3,b3,c3} /.
                    w[1,2]'[z]->x[1,2]'[z] ];
```

```
{a5,b5,c5} = Expand[ {a4,b4,c4} /.
                     w[1,2][z]->x[1,2][z] ];

{a6,b6,c6} = Expand[ {a5,b5,c5} /.
                     w[1,1]''[z]->x[1,1]''[z] ];

{a7,b7,c7} = Expand[ {a6,b6,c6} /.
                     w[1,1]'[z]->x[1,1]'[z] ];

{a8,b8,c8} = Expand[ {a7,b7,c7} /.
                     w[1,1][z]->x[1,1][z] ];

{a9,b9,c9} = Expand[ {a8,b8,c8} /.
                     w[0,2][z]->x[0,2][z] ]
```

with that selected version. The *Mathematica* output for a9 is a representation of $\boldsymbol{E}_6$ and shows that $\boldsymbol{E}_6$ is expressible as

$$\boldsymbol{E}_6 \equiv \boldsymbol{\mathcal{I}}_{2,2;1,1}\boldsymbol{\mathcal{I}}_{2,2;2,2} - 2\left(\boldsymbol{\mathcal{I}}_{2,2;1,2}\right)^2$$
$$+ \frac{4}{3}\left[\boldsymbol{\mathcal{I}}_{2,2;1,1}\boldsymbol{\mathcal{I}}^{(1)}_{2,2;1,2} - \frac{3}{2}\boldsymbol{\mathcal{I}}^{(1)}_{2,2;1,1}\boldsymbol{\mathcal{I}}_{2,2;1,2}\right]$$
$$+ \frac{3}{20}\left[2\boldsymbol{\mathcal{I}}_{2,2;1,1}\boldsymbol{\mathcal{I}}^{(2)}_{2,2;1,1} - \frac{5}{2}\left(\boldsymbol{\mathcal{I}}^{(1)}_{2,2;1,1}\right)^2 - 8\,a_{2,2}\left(\boldsymbol{\mathcal{I}}_{2,2;1,1}\right)^2\right].$$

By applying (4.52), (4.51), and (4.53)-(4.54) to the preceding formula, we obtain

(7.18) $\quad \boldsymbol{E}_6 \equiv C_{2,4,0}(\boldsymbol{\mathcal{I}}_{2,2;1,1},\,\boldsymbol{\mathcal{I}}_{2,2;2,2}) - 2C_{3,3,0}(\boldsymbol{\mathcal{I}}_{2,2;1,2},\,\boldsymbol{\mathcal{I}}_{2,2;1,2})$
$$+ \frac{4}{3}C_{2,3,1}(\boldsymbol{\mathcal{I}}_{2,2;1,1},\,\boldsymbol{\mathcal{I}}_{2,2;1,2}) + \frac{3}{20}C_{2,2,2}(\boldsymbol{\mathcal{I}}_{2,2\,1,1},\,\boldsymbol{\mathcal{I}}_{2,2\,1,1}).$$

Moreover, the output b9 for $\boldsymbol{E}_7$ gives

$$\boldsymbol{E}_7 \equiv -2\boldsymbol{\mathcal{I}}_{2,2;1,2}\boldsymbol{\mathcal{I}}_{2,2;2,2} + \frac{5}{6}\left[\boldsymbol{\mathcal{I}}_{2,2;1,1}\boldsymbol{\mathcal{I}}^{(1)}_{2,2;2,2} - 2\boldsymbol{\mathcal{I}}^{(1)}_{2,2;1,1}\boldsymbol{\mathcal{I}}_{2,2;2,2}\right]$$
$$+ \frac{1}{9}\left[\begin{array}{l}\boldsymbol{\mathcal{I}}_{2,2;1,1}\boldsymbol{\mathcal{I}}^{(2)}_{2,2;1,2} - \frac{7}{2}\boldsymbol{\mathcal{I}}^{(1)}_{2,2;1,1}\boldsymbol{\mathcal{I}}^{(1)}_{2,2;1,2}\\ + \frac{21}{10}\boldsymbol{\mathcal{I}}^{(2)}_{2,2;1,1}\boldsymbol{\mathcal{I}}_{2,2;1,2} - \frac{72}{5}a_{2,2}\boldsymbol{\mathcal{I}}_{2,2;1,1}\boldsymbol{\mathcal{I}}_{2,2;1,2}\end{array}\right].$$

In view of (4.52), (4.51), and (4.53)-(4.54), it yields

(7.19) $\quad \boldsymbol{E}_7 \equiv -2C_{3,4,0}(\boldsymbol{\mathcal{I}}_{2,2;1,2},\,\boldsymbol{\mathcal{I}}_{2,2;2,2}) + \frac{5}{6}C_{2,4,1}(\boldsymbol{\mathcal{I}}_{2,2;1,1},\,\boldsymbol{\mathcal{I}}_{2,2;2,2})$
$$+ \frac{1}{9}C_{2,3,2}(\boldsymbol{\mathcal{I}}_{2,2,\,1,1},\,\boldsymbol{\mathcal{I}}_{2,2\,1,2}).$$

After observing that the output c9 for D_2 yields
$$D_2 \equiv \mathcal{I}_{2,2;\,1,1}\mathcal{I}_{2,2;\,2,2} - \left(\mathcal{I}_{2,2;\,1,2}\right)^2$$
$$+ \frac{1}{3}\left[\mathcal{I}_{2,2;\,1,1}\mathcal{I}^{(1)}_{2,2;\,1,2} - \frac{3}{2}\mathcal{I}^{(1)}_{2,2;\,1,1}\mathcal{I}_{2,2;\,1,2}\right]$$
$$+ \frac{1}{40}\left[2\mathcal{I}_{2,2;\,1,1}\mathcal{I}^{(2)}_{2,2;\,1,1} - \frac{5}{2}(\mathcal{I}^{(1)}_{2,2;\,1,1})^2 - 8a_{2,2}\left(\mathcal{I}_{2,2;\,1,1}\right)^2\right],$$

we employ (4.52), (4.51), and (4.53)-(4.54) to deduce

(7.20) $\quad D_2 \equiv C_{2,4,0}(\mathcal{I}_{2,2;\,1,1},\,\mathcal{I}_{2,2;\,2,2}) - C_{3,3,0}(\mathcal{I}_{2,2;\,1,2},\,\mathcal{I}_{2,2;\,1,2})$
$$+ \frac{1}{3}C_{2,3,1}(\mathcal{I}_{2,2;\,1,1},\,\mathcal{I}_{2,2;\,1,2}) + \frac{1}{40}C_{2,2,2}(\mathcal{I}_{2,2;\,1,1},\,\mathcal{I}_{2,2;\,1,1}).$$

Since (4.52), (4.51), and the consequence (4.53)–(4.54) of Theorem 4.10 show that each term in the right members of (7.20), (7.18), and (7.19) is a relative invariant for $\mathcal{C}_{2,2}$ we conclude that D_2, E_6, and E_7 are relative invariants in $\mathcal{R}_{2,2}$ for $\mathcal{C}_{2,2}$ of respective weights 6, 6, and 7.

The subscript of D_2 is employed to signify that D_2 is the special case $k = 2$ of a relative invariant D_k in $\mathcal{R}_{m,2}$ for $\mathcal{C}_{m,2}$ of weight $k(k+1)$ that was introduced and examined in [20, pages 162–168] for any integer k satisfying $1 \leq k \leq m$.

Each relative invariant in $\mathcal{R}_{2,2}$ for $\mathcal{C}_{2,2}$ of weight ≤ 12 has a representation analogous to the ones of (7.18), (7.19), and (7.20). We present details about that in Section 12.3 and there is no evidence that 12 is a necessary limitation.

CHAPTER 8

Separate Examination of $C_{p,q,1}(P, Q)$

Given $m, n \geq 1$, suppose that P and Q are relative invariants in $\mathcal{R}_{m,n}$ for $\mathcal{C}_{m,n}$ of respective weights p and q. We see that $C_{p,q,1}(P, Q)$ is defined in $\mathcal{R}_{m,n}$ by

$$(8.1) \qquad C_{p,q,1}(P, Q) \equiv PQ^{(1)} - \frac{q}{p}P^{(1)}Q.$$

In particular, this does not require the restriction $m \geq 2$ of Theorem 4.10.

8.1. Properties of $C_{p,q,1}(P, Q)$

PROPOSITION 8.1. *Let γ denote a rational number and set*

$$(8.2) \qquad F \equiv PQ^{(1)} + \gamma P^{(1)}Q, \quad \text{in } \mathcal{R}_{m,n}.$$

Then, with respect to Notation 1.0 of page 30, the conditions

$$(8.3) \qquad F^*(z) \equiv F(z), \text{ on } \Omega, \text{ and } F^{**}(\zeta) \equiv \big(f'(\zeta)\big)^{p+q+1} F\big(f(\zeta)\big), \text{ on } \Omega^{**},$$

are satisfied for any (4.1)–(4.2), (4.3), and (4.6) if and only if $\gamma = -q/p$ and F is therefore $C_{p,q,1}(P, Q)$ in (8.1). Moreover, $C_{p,q,1}(P, Q)$ is a relative invariant in $\mathcal{R}_{m,n}$ for $\mathcal{C}_{m,n}$ of weight $p+q+1$ if and only if $C_{p,q,1}(P, Q) \not\equiv 0$ and that occurs if and only if P^q and Q^p are linearly independent over $\mathbb{Q}$.

PROOF. Since P and Q are semi-invariants of the first kind in $\mathcal{R}_{m,n}$ for $\mathcal{C}_{m,n}$ while F is a differential-polynomial combination over $\mathbb{Q}$ of P and Q, Proposition 5.1 yields $F^*(z) \equiv F(z)$.

For any (4.1)–(4.2) and (4.6) that transforms (4.1)–(4.2) into a corresponding (4.7)–(4.8), the properties of P and Q as semi-invariants of the second kind having respective weights p and q yield

$$(8.4) \qquad P^{**}(\zeta) \equiv \big(f'(\zeta)\big)^p P\big(f(\zeta)\big), \quad \text{on } \Omega^{**},$$

and

$$(8.5) \qquad Q^{**}(\zeta) \equiv \big(f'(\zeta)\big)^q Q\big(f(\zeta)\big), \quad \text{on } \Omega^{**}.$$

For each ζ in Ω^{**}, we use (8.4) and (8.5) to obtain

$$(8.6) \qquad P^{**(1)}(\zeta) \equiv \big(f'(\zeta)\big)^{p+1} P^{(1)}\big(f(\zeta)\big) + p\big(f'(\zeta)\big)^{p-1} f''(\zeta) P\big(f(\zeta)\big)$$

and

$$(8.7) \qquad Q^{**(1)}(\zeta) \equiv \big(f'(\zeta)\big)^{q+1} Q^{(1)}\big(f(\zeta)\big) + q\big(f'(\zeta)\big)^{q-1} f''(\zeta) Q\big(f(\zeta)\big).$$

For γ in $\mathbb{Q}$, we employ (8.2) as well as (8.4)–(8.7) to deduce
$$F^{**}(\zeta) \equiv P^{**}(\zeta)\,Q^{**(1)}(\zeta) + \gamma\,P^{**(1)}(\zeta)\,Q^{**}(\zeta)$$
$$\equiv (f'(\zeta))^{p+q+1}\Big(P\big(f(\zeta)\big)\,Q^{(1)}\big(f(\zeta)\big) + \gamma P^{(1)}\big(f(\zeta)\big)\,Q\big(f(\zeta)\big)\Big)$$
$$+ (q+\gamma p)(f'(\zeta))^{p+q-1} f''(\zeta)\,P\big(f(\zeta)\big)\,Q\big(f(\zeta)\big)$$

and

(8.8) $$F^{**}(\zeta) - (f'(\zeta))^{p+q+1} F\big(f(\zeta)\big)$$
$$\equiv (q+\gamma p)(f'(\zeta))^{p+q-1} f''(\zeta)\,P\big(f(\zeta)\big)\,Q\big(f(\zeta)\big), \quad \text{on } \Omega^{**}.$$

Corollary 5.13 on page 49 shows that there is an equation (4.1)–(4.2) on some region Ω such that the function $P(z)\,Q(z)$ obtained by replacing each $w_{i_1,i_2,\ldots,i_n}^{(k)}$ in $\boldsymbol{P}\boldsymbol{Q}$ with the corresponding $c_{i_1,i_2,\ldots,i_n}^{(k)}(z)$ from (4.1) yields $P(z)\,Q(z) \not\equiv 0$ on Ω. For this (4.1)–(4.2) and a transformation (4.6) with $f''(\zeta) \not\equiv 0$, the coefficient of $(q+\gamma p)$ in (8.8) is not identically zero. Thus, (8.3) is satisfied for any (4.1)–(4.2) and compatible transformations (4.3), (4.6) if and only if $q+\gamma p = 0$, $\gamma = -q/p$, and $\boldsymbol{F}$ equals $\boldsymbol{C}_{p,q,1}(\boldsymbol{P},\boldsymbol{Q})$. Since $\boldsymbol{P}$ and $\boldsymbol{Q}$ are isobaric polynomials, we see that $\boldsymbol{F}$ in (8.2) is not a nonzero constant. Thus, $\boldsymbol{C}_{p,q,1}(\boldsymbol{P},\boldsymbol{Q})$ is a relative invariant in $\mathcal{R}_{m,n}$ for $\mathcal{C}_{m,n}$ if and only if it is nonzero.

To investigate the condition $\boldsymbol{C}_{p,q,1}(\boldsymbol{P},\boldsymbol{Q}) \not\equiv 0$, we let $\mathcal{Q}_{m,n}$ denote the quotient field for $\mathcal{R}_{m,n}$. Then, an element of $\mathcal{Q}_{m,n}$ has a representation $\boldsymbol{A}/\boldsymbol{B}$, where $\boldsymbol{A}$ and $\boldsymbol{B}$ are polynomials in $\mathcal{R}_{m,n}$ and $\boldsymbol{B} \not\equiv 0$. The derivation $'$ for $\mathcal{R}_{m,n}$ has a unique extension to a derivation $'$ for $\mathcal{Q}_{m,n}$; e.g., see [5, page 45, Proposition 11]. For any $\boldsymbol{A}$ and $\boldsymbol{B} \not\equiv 0$ in $\mathcal{R}_{m,n}$, it specifies

(8.9) $$\left(\frac{\boldsymbol{A}}{\boldsymbol{B}}\right)' \equiv \frac{\boldsymbol{B}\boldsymbol{A}' - \boldsymbol{A}\boldsymbol{B}'}{\boldsymbol{B}^2}, \quad \text{in } \mathcal{Q}_{m,n}.$$

When $\boldsymbol{A}$ and $\boldsymbol{B} \not\equiv 0$ are relatively prime and $(\boldsymbol{A}/\boldsymbol{B})' \equiv 0$, we use $\boldsymbol{B}\boldsymbol{A}' \equiv \boldsymbol{A}\boldsymbol{B}'$ to deduce that $\boldsymbol{B}$ divides $\boldsymbol{B}'$ in $\mathcal{R}_{m,n}$. This requires $\boldsymbol{B}' \equiv 0$ as well as $\boldsymbol{A}' \equiv 0$. Thus, the constants of $\mathcal{Q}_{m,n}$ are the constants of $\mathcal{R}_{m,n}$, namely the elements of $\mathbb{Q}$.

We use the derivation $'$ for $\mathcal{Q}_{m,n}$ as well as (8.1) to obtain
$$\left(\frac{\boldsymbol{P}^q}{\boldsymbol{Q}^p}\right)' \equiv \frac{q\boldsymbol{Q}^p\,\boldsymbol{P}^{q-1}\boldsymbol{P}^{(1)} - p\boldsymbol{P}^q\,\boldsymbol{Q}^{p-1}\boldsymbol{Q}^{(1)}}{\boldsymbol{Q}^{2p}}$$
$$\equiv -p\left(\frac{\boldsymbol{P}^{q-1}\boldsymbol{Q}^{p-1}}{\boldsymbol{Q}^{2p}}\right)\boldsymbol{C}_{p,q,1}(\boldsymbol{P},\boldsymbol{Q}).$$

This yields $\boldsymbol{C}_{p,q,1}(\boldsymbol{P},\boldsymbol{Q}) \equiv 0$ if and only if $(\boldsymbol{P}^q/\boldsymbol{Q}^p)' \equiv 0$. We observe that the condition $(\boldsymbol{P}^q/\boldsymbol{Q}^p)' \equiv 0$ is satisfied if and only if there is a rational number α such that $\boldsymbol{P}^q \equiv \alpha \boldsymbol{Q}^p$. Hence, $\boldsymbol{C}_{p,q,1}(\boldsymbol{P},\boldsymbol{Q}) \not\equiv 0$ is valid if and only if $\boldsymbol{P}^q$ and $\boldsymbol{Q}^p$ are linearly independent over $\mathbb{Q}$. This completes the proof. $\square$

When $p = q$, (8.1) yields
$$\left(\frac{\boldsymbol{P}}{\boldsymbol{Q}}\right)' \equiv \frac{\boldsymbol{Q}\boldsymbol{P}^{(1)} - \boldsymbol{P}\boldsymbol{Q}^{(1)}}{\boldsymbol{Q}^2} \equiv -\left(\frac{1}{\boldsymbol{Q}^2}\right)\boldsymbol{C}_{p,q,1}(\boldsymbol{P},\boldsymbol{Q}).$$

Thus, for the situation $p = q$, the condition $\boldsymbol{C}_{p,q,1}(\boldsymbol{P},\boldsymbol{Q}) \not\equiv 0$ is satisfied if and only if $\boldsymbol{P}$ and $\boldsymbol{Q}$ are linearly independent over $\mathbb{Q}$.

8.2. The condition that P^q and Q^p are linearly independent over $\mathbb{Q}$

For integers $m, n \geq 1$, $\mathcal{R}_{m,n}$ is defined on page 30 as the ring of polynomials in the variables of (4.9) over the field $\mathbb{Q}$ of rational numbers. For each $r \geq 1$, let $\mathcal{R}_{m,n,r}$ be the polynomial ring over $\mathbb{Q}$ in the variables

(8.10) $\quad w^{(k)}_{i_1, i_2, \ldots, i_n}, \quad$ for $\quad 1 \leq i_n \leq m, \ 0 \leq i_1 \leq i_2 \leq \cdots \leq i_n, \ $ and $\ 0 \leq k \leq r$.

Since (8.10) consists of finitely many variables and $\mathbb{Q}$ is a field, it follows that each $\mathcal{R}_{m,n,r}$ is a unique-factorization ring; e.g., see [**39**, page 199]. Because each element of $\mathcal{R}_{m,n}$ belongs to $\mathcal{R}_{m,n,r}$ for some r, we conclude that $\mathcal{R}_{m,n}$ is a unique-factorization ring. Consequently, for P and Q as relative invariants in $\mathcal{R}_{m,n}$ of respective weights p and q, there are pairwise inequivalent irreducible polynomials $R_1, R_2, \ldots, R_s$ in $\mathcal{R}_{m,n}$ such that P and Q have unique representations

(8.11) $\qquad P = u_1 R_1^{\alpha_1} R_2^{\alpha_2} \cdots R_s^{\alpha_s} \quad$ and $\quad Q = u_2 R_1^{\beta_1} R_2^{\beta_2} \cdots R_s^{\beta_s}$.

where $\alpha_1, \alpha_2, \ldots, \alpha_s$ and $\beta_1, \beta_2, \ldots, \beta_s$ are nonnegative integers while u_1, u_2 are nonzero rational numbers. We note that the corresponding sequences

(8.12) $\qquad v_P = (\alpha_1, \alpha_2, \ldots, \alpha_s) \quad$ and $\quad v_Q = (\beta_1, \beta_2, \ldots, \beta_s)$

are elements of the vector space $\mathbb{Q}^{(s)}$ over $\mathbb{Q}$.

PROPOSITION 8.2. *The polynomials P^q and Q^p are linearly independent over $\mathbb{Q}$ if and only if $v_P \neq (p/q) v_Q$.*

PROOF. Instead, we shall establish the equivalent statement that: P^q and Q^p are linearly dependent over $\mathbb{Q}$ if and only if $v_P = (p/q) v_Q$.

Suppose that P^q and Q^p are linearly dependent over $\mathbb{Q}$. Then, there is a nonzero rational number u_3 such that

(8.13) $\qquad\qquad\qquad\qquad P^q = u_3 Q^p$.

In view of (8.11), we use (8.13) to obtain

(8.14) $\qquad u_1^q R_1^{q\alpha_1} R_2^{q\alpha_2} \cdots R_s^{q\alpha_s} = u_2^p u_3 R_1^{p\beta_1} R_2^{p\beta_2} \cdots R_s^{p\beta_s}$

Due to uniqueness of factorization in $\mathcal{R}_{m,n}$, we find that (8.14) requires

(8.15) $\qquad\qquad u_3 = u_1^q/u_2^p \quad$ and $\quad q\alpha_i = p\beta_i \quad$ for $1 \leq i \leq s$.

Thus, the condition $v_P = (p/q) v_Q$ is satisfied.

Suppose the condition $v_P = (p/q) v_Q$ is satisfied. Then, we use (8.11) to define u_3 by $u_3 = u_1^q/u_2^p$. Thus, (8.15) is satisfied and it yields (8.14). Since (8.14) and (8.11) give (8.13), we conclude that P^q and Q^p are linearly dependent over $\mathbb{Q}$. This completes the proof. $\square$

EXAMPLE 8.3. For $m = 4$ and $n = 1$, let I_3 denote the relative invariant $\mathcal{I}_{4,1;3}$ of weight 3 in $\mathcal{R}_{4,1}$ given by (1.15) and let I_4 denote the relative invariant $\mathcal{I}_{4,1;4}$ of weight 4 in $\mathcal{R}_{4,1}$ given by (1.17). Since $\mathcal{I}_{4,1;3}$ in (1.15) has the term w_3 and $\mathcal{I}_{4,1;4}$ in (1.17) has the term w_4, both I_3 and I_4 are irreducible polynomials in $\mathcal{R}_{4,1}$.

For the situation $P \equiv I_3 I_4$ with $p = 7$ and $Q \equiv (I_3)^2 (I_4)^2$ with $q = 14$, we note that $v_P = (1, 1)$, $v_Q = (2, 2)$, $v_P = (p/q) v_Q$, and $C_{7,14,1}(P, Q) \equiv 0$.

For the situation $P \equiv I_3 I_4$ with $p = 7$ and $Q \equiv (I_3)^2 (I_4)$ with $q = 10$, we observe that $v_P = (1, 1)$, $v_Q = (2, 1)$, $v_P \neq (p/q) v_Q$, and a relative invariant in $\mathcal{R}_{4,1}$ of weight $p + q + 1 = 18$ is specified by $C_{7,10,1}(I_3 I_4, (I_3)^2 (I_4))$.

8.3. Several identities

PROPOSITION 8.4. *If P, Q, and R are relative invariants in $\mathcal{R}_{m,n}$ for $\mathcal{C}_{m,n}$ having respective weights p, q, and r, then*

(8.16) $\quad p\,C_{p,q+r,1}(P,\,QR) + q\,C_{q,p+r,1}(Q,\,PR) + r\,C_{r,p+q,1}(R,\,PQ) \equiv 0.$

PROOF. Using (8.1), we obtain

$p\,C_{p,q+r,1}(P,\,QR) + q\,C_{q,p+r,1}(Q,\,PR) + r\,C_{r,p+q,1}(R,\,PQ)$

$$\equiv \begin{bmatrix} p\left(P(QR)^{(1)} - \dfrac{q+r}{p}P^{(1)}(QR)\right) \\ +q\left(Q(PR)^{(1)} - \dfrac{p+r}{q}Q^{(1)}(PR)\right) \\ +r\left(R(PQ)^{(1)} - \dfrac{p+q}{r}R^{(1)}(PQ)\right) \end{bmatrix} \equiv \begin{bmatrix} pP\left((QR)^{(1)} - (QR)^{(1)}\right) \\ +qQ\left((PR)^{(1)} - (PR)^{(1)}\right) \\ +rR\left((PQ)^{(1)} - (PQ)^{(1)}\right) \end{bmatrix}$$

and (8.16). This completes the proof. $\square$

PROPOSITION 8.5. *If P and Q are relative invariants of respective weights p and q in $\mathcal{R}_{m,n}$ for $\mathcal{C}_{m,n}$, then*

(8.17) $\quad C_{q,p,1}(Q,\,P) \equiv -\dfrac{p}{q} C_{p,q,1}(P,\,Q)$

and

(8.18) $\quad C_{p,p+q,1}(P,\,PQ) \equiv P\,C_{p,q,1}(P,\,Q) \equiv C_{2p,q,1}(P^2,\,Q).$

PROOF. We apply (8.1) to deduce

$$C_{q,p,1}(Q,\,P) \equiv -\dfrac{p}{q}\left(PQ^{(1)} - \dfrac{q}{p}P^{(1)}Q\right) \equiv -\dfrac{p}{q}C_{p,q,1}(P,\,Q)$$

and see that (8.17) is valid. Using (8.1) and (8.17), we find that

$$C_{p,p+q,1}(P,\,PQ) \equiv P(PQ)^{(1)} - \dfrac{p+q}{p}P^{(1)}(PQ)$$

$$\equiv PPQ^{(1)} - \dfrac{q}{p}PP^{(1)}Q \equiv P\,C_{p,q,1}(P,\,Q)$$

$$\equiv -\dfrac{q}{2p}\left(Q(2PP^{(1)}) - \dfrac{2p}{q}Q^{(1)}P^2\right)$$

$$\equiv -\dfrac{q}{2p}C_{q,2p,1}(Q,\,P^2) \equiv C_{2p,q,1}(P^2,\,Q).$$

This yields (8.18) and completes the proof. $\square$

Part 2

Proof of Theorem 4.10

CHAPTER 9

Invariant Character of $C_{p,q,r}(P, Q)$ when $m, r \geq 2$

The principal part of Theorem 4.10 is established in this chapter by means of arguments developed during the discovery of (4.44)–(4.50).

9.1. Introduction of R and $\phi_{h,i,j}(z)$

For $m, r \geq 2$ and $(m, n) \neq (2, 1)$, let P and Q denote relative invariants in $\mathcal{R}_{m,n}$ for $\mathcal{C}_{m,n}$ of respective weights p and q. We use (4.44)–(4.50) to introduce

(9.1) $$R \equiv C_{p,q,r}(P, Q).$$

Because $a_{m,n}$ of (4.45) as well as P and Q are semi-invariants of the first kind in $\mathcal{R}_{m,n}$, formulas (9.1) and (4.44)–(4.50) show that R is a differential-polynomial combination over $\mathbb{Q}$ of semi-invariants of the first kind. Thus, for any (4.1)-(4.2) defined on a region Ω and any transformation (4.3) of that (4.1)-(4.2) into a corresponding (4.4)–(4.5), Proposition 5.1 shows that the function $R(z)$ on Ω obtained by replacing $w_{i_1, i_2, \ldots, i_n}^{(k)}$ in R with $c_{i_1, i_2, \ldots, i_n}^{(k)}(z)$ from (4.1) and the function $R^*(z)$ on Ω obtained by replacing $w_{i_1, i_2, \ldots, i_n}^{(k)}$ in R with $c_{i_1, i_2, \ldots, i_n}^{*(k)}(z)$ from (4.4) satisfy

(9.2) $$R^*(z) \equiv R(z), \quad \text{for } z \text{ in } \Omega.$$

We use the context of Section 4.1 where: (4.1)–(4.2) on Ω is a given differential equation of $\mathcal{C}_{m,n}$; $\zeta = g(z)$ is a univalent analytic function on Ω and therefore satisfies $g'(z) \neq 0$, for each z in Ω; $z = f(\zeta)$ on $\Omega^{**} = g(\Omega)$ is the univalent analytic inverse function of $\zeta = g(z)$ on Ω; and, (4.7)–(4.8) on Ω^{**} is the unique differential equation of $\mathcal{C}_{m,n}$ into which the substitution $z = f(\zeta)$ transforms (4.1)–(4.2). We shall establish in Theorem 9.14 on page 95 that the function $R(z)$ on Ω obtained by replacing $w_{i_1, i_2, \ldots, i_n}^{(k)}$ in R with $c_{i_1, i_2, \ldots, i_n}^{(k)}(z)$ from (4.1) and the function $R^{**}(\zeta)$ on Ω^{**} obtained by replacing $w_{i_1, i_2, \ldots, i_n}^{(k)}$ in R with $c_{i_1, i_2, \ldots, i_n}^{**(k)}(\zeta)$ from (4.7) satisfy

(9.3) $$R^{**}(\zeta) \equiv \bigl(f'(\zeta)\bigr)^{p+q+r} R\bigl(f(\zeta)\bigr), \quad \text{for } \zeta \text{ in } \Omega^{**}.$$

We begin by introducing the analytic functions $\phi_{h,i,j}(z)$ on Ω defined by

(9.4) $$\phi_{h,i,j}(z) \equiv 0, \quad \text{for } i \leq -1 \text{ and any } h, j,$$

(9.5) $$\phi_{h,0,j}(z) \equiv 1, \quad \text{for any } h, j,$$

and

(9.6) $$\phi_{h,i+1,j}(z) \equiv \sum_{k=i}^{j-1} \left[\phi_{h,i,k}^{(1)}(z) - (h+k)\frac{g''(z)}{g'(z)} \phi_{h,i,k}(z) \right],$$
$$\text{for } i \geq 0 \text{ and any } h, j.$$

LEMMA 9.1. *Formulas (9.4)–(9.6) yield*

(9.7) $$\phi_{h,i,j}^{(1)}(z) \equiv \phi_{h,i+1,j+1}(z) - \phi_{h,i+1,j}(z) + (h+j)\frac{g''(z)}{g'(z)}\phi_{h,i,j}(z),$$

for $i \leq j$ and any h.

PROOF. For h, i, j satisfying $0 \leq i \leq j$, we use (9.6) to deduce

(9.8) $$\phi_{h,i+1,j+1}(z) - \phi_{h,i+1,j}(z) \equiv \phi_{h,i,j}^{(1)}(z) - (h+j)\frac{g''(z)}{g'(z)}\phi_{h,i,j}(z).$$

For $i = -1$ and any h, j, we employ (9.5) and (9.4) to obtain $1 - 1$ and $0 - 0$ as the left and right members of (9.8). For $i \leq -2$ and any h, j, we note that (9.4) yields $0 - 0$ and $0 - 0$ as the corresponding left and right members of (9.8). We rewrite (9.8) to obtain (9.7) and complete the proof. □

PROPOSITION 9.2. *Let $\Psi(z)$ be a meromorphic function defined on Ω and set*

(9.9) $$\Phi(\zeta) \equiv \bigl(f'(\zeta)\bigr)^h \Psi\bigl(f(\zeta)\bigr), \quad \text{on } \Omega^{**},$$

where h is a fixed nonnegative integer. Then, the derivatives of $\Phi(\zeta)$ with respect to ζ are given in terms of (9.4)–(9.6) by

(9.10) $$\Phi^{(j)}(\zeta) \equiv \bigl(f'(\zeta)\bigr)^{h+j} \sum_{i=0}^{j} \phi_{h,i,j}\bigl(f(\zeta)\bigr) \Psi^{(j-i)}\bigl(f(\zeta)\bigr),$$

on Ω^{**} for $j \geq 0$.

PROOF. Due to (9.9) and (9.5), (9.10) is correct for $j = 0$. Suppose that j is a fixed nonnegative integer for which (9.10) is valid. Then, we obtain

$$\Phi^{(j+1)}(\zeta) \equiv \sum_{i=0}^{j} \left\{ \begin{array}{l} (h+j)\bigl(f'(\zeta)\bigr)^{h+j-1} f''(\zeta)\, \phi_{h,i,j}\bigl(f(\zeta)\bigr)\, \Psi^{(j-i)}\bigl(f(\zeta)\bigr) \\ + \bigl(f'(\zeta)\bigr)^{h+j+1} \phi_{h,i,j}^{(1)}\bigl(f(\zeta)\bigr)\, \Psi^{(j-i)}\bigl(f(\zeta)\bigr) \\ + \bigl(f'(\zeta)\bigr)^{h+j+1} \phi_{h,i,j}\bigl(f(\zeta)\bigr)\, \Psi^{(j+1-i)}\bigl(f(\zeta)\bigr) \end{array} \right\}$$

$$\equiv \bigl(f'(\zeta)\bigr)^{h+j+1} \left\{ \begin{array}{l} \displaystyle\sum_{i=1}^{j+1} \left[(h+j)\dfrac{f''(\zeta)}{\bigl(f'(\zeta)\bigr)^2} \phi_{h,i-1,j}\bigl(f(\zeta)\bigr) \\ + \phi_{h,i-1,j}^{(1)}\bigl(f(\zeta)\bigr) \end{array} \right] \Psi^{(j+1-i)}\bigl(f(\zeta)\bigr) \\ + \displaystyle\sum_{i=0}^{j} \phi_{h,i,j}\bigl(f(\zeta)\bigr)\, \Psi^{(j+1-i)}\bigl(f(\zeta)\bigr) \end{array} \right\}.$$

After replacing i with $i-1$ and z with $f(\zeta)$ in (9.7), we use the identity

$$\frac{g''\bigl(f(\zeta)\bigr)}{g'\bigl(f(\zeta)\bigr)} \equiv -\frac{f''(\zeta)}{\bigl(f'(\zeta)\bigr)^2}, \quad \text{for } \zeta \text{ in } \Omega^{**},$$

derived from $g\bigl(f(\zeta)\bigr) \equiv \zeta$ to deduce

$$\left[(h+j)\frac{f''(\zeta)}{\bigl(f'(\zeta)\bigr)^2}\phi_{h,i-1,j}\bigl(f(\zeta)\bigr) + \phi_{h,i-1,j}^{(1)}\bigl(f(\zeta)\bigr) + \phi_{h,i,j}\bigl(f(\zeta)\bigr)\right] \equiv \phi_{h,i,j+1}\bigl(f(\zeta)\bigr),$$

on Ω^{**}, for $i \leq j+1$ and any h.

Since (9.6) gives $\phi_{h,i+1,\nu}(z) \equiv 0$, for $i \geq 0$ and $\nu \leq i$, we have $\phi_{h,j+1,j}(z) \equiv 0$. Moreover, (9.5) yields $\phi_{h,0,j}\bigl(f(\zeta)\bigr) \equiv 1 \equiv \phi_{h,0,j+1}\bigl(f(\zeta)\bigr)$. By combining these observations with the expression for $\Phi^{(j+1)}(\zeta)$, we find that

$$\frac{\Phi^{(j+1)}(\zeta)}{(f'(\zeta))^{h+j+1}} \equiv \left\{ \sum_{i=1}^{j+1} \begin{bmatrix} (h+j)\dfrac{f''(\zeta)}{(f'(\zeta))^2}\,\phi_{h,i-1,j}(f(\zeta)) \\ +\,\phi^{(1)}_{h,i-1,j}(f(\zeta)) + \phi_{h,i,j}(f(\zeta)) \end{bmatrix} \Psi^{(j+1-i)}(f(\zeta)) \right\}$$
$$+\,\phi_{h,0,j}(f(\zeta))\,\Psi^{(j+1)}(f(\zeta))$$

$$\equiv \left\{ \sum_{i=1}^{j+1} \phi_{h,i,j+1}(f(\zeta))\,\Psi^{(j+1-i)}(f(\zeta)) \atop +\,\phi_{h,0,j+1}(f(\zeta))\,\Psi^{(j+1)}(f(\zeta)) \right\}$$

$$\equiv \sum_{i=0}^{j+1} \phi_{h,i,j+1}(f(\zeta))\,\Psi^{(j+1-i)}(f(\zeta)).$$

Thus, (9.10) is valid with j replaced by $j+1$. This completes the proof. □

9.2. Formula for $R^{**}(\zeta)$ that involves $\mathfrak{A}_{p,q,r,s,t}(\zeta)$ of (9.11)

With respect to relative invariants $\boldsymbol{P}$ and $\boldsymbol{Q}$ in $\mathcal{R}_{m,n}$ for $\mathcal{C}_{m,n}$, functions $P(z)$ and $Q(z)$ on Ω are obtained by replacing $w^{(k)}_{i_1,i_2,\ldots,i_n}$ in $\boldsymbol{P}$ and in $\boldsymbol{Q}$ with $c^{(k)}_{i_1,i_2,\ldots,i_n}(z)$ from (4.1). For a transformation (4.6) of (4.1)–(4.2) into (4.7)–(4.8), functions $P^{**}(\zeta)$, $Q^{**}(\zeta)$, $A^{**}_{p,q,r,s,t}(\zeta)$, and $R^{**}(\zeta)$ on Ω^{**} are obtained by replacing each $w^{(k)}_{i_1,i_2,\ldots,i_n}$ in $\boldsymbol{P}$, $\boldsymbol{Q}$, $\boldsymbol{A}_{p,q,r,s,t}$ of (4.50), and $\boldsymbol{R}$ of (9.1) with the corresponding $c^{**(k)}_{i_1,i_2,\ldots,i_n}(\zeta)$ from (4.7). We use (9.4)–(9.6) to define $\mathfrak{A}_{p,q,r,s,t}(\zeta)$ on Ω^{**} by

$$(9.11)\quad \mathfrak{A}_{p,q,r,s,t}(\zeta) \equiv \sum_{u=0}^{s}\sum_{v=0}^{s-u} \frac{A^{**}_{p,q,r,u,t+v}(\zeta)}{(f'(\zeta))^u}\,\phi_{p,v,t+v}(f(\zeta))\,\phi_{q,s-u-v,r-t-u-v}(f(\zeta)),$$
$$\text{for } 0 \leq s \leq r \text{ and } 0 \leq t \leq r-s.$$

THEOREM 9.3. *For each ζ in Ω^{**}, $R^{**}(\zeta)$ is given by*

$$(9.12)\quad R^{**}(\zeta) \equiv (f'(\zeta))^{p+q+r} \sum_{s=0}^{r}\sum_{t=0}^{r-s} \mathfrak{A}_{p,q,r,s,t}(\zeta)\,P^{(t)}(f(\zeta))\,Q^{(r-s-t)}(f(\zeta)).$$

PROOF. We use (9.1) and (4.44) to obtain

$$(9.13)\quad R^{**}(\zeta) \equiv \sum_{s=0}^{r}\sum_{t=0}^{r-s} A^{**}_{p,q,r,s,t}(\zeta)\,P^{**(t)}(\zeta)\,Q^{**(r-s-t)}(\zeta).$$

In view of $P^{**}(\zeta) \equiv (f'(\zeta))^p P(f(\zeta))$, Proposition 9.2 yields

$$(9.14)\quad P^{**(t)}(\zeta) \equiv (f'(\zeta))^{p+t} \sum_{\mu=0}^{t} \phi_{p,\mu,t}(f(\zeta))\,P^{(t-\mu)}(f(\zeta)).$$

With respect to $Q^{**}(\zeta) \equiv (f'(\zeta))^q Q(f(\zeta))$, Proposition 9.2 gives

$$(9.15)\quad Q^{**(r-s-t)}(\zeta) \equiv (f'(\zeta))^{q+r-s-t} \sum_{\nu=0}^{r-s-t} \begin{bmatrix} \phi_{q,\nu,r-s-t}(f(\zeta)) \\ \times\,Q^{(r-s-t-\nu)}(f(\zeta)) \end{bmatrix}.$$

We introduce

(9.16) $$X_{p,q,r,s,t,\mu,\nu}(\zeta) \equiv \frac{A^{**}_{p,q,r,s,t}(\zeta)}{(f'(\zeta))^s} \left[\begin{array}{l} \phi_{p,\mu,t}(f(\zeta)) \, \phi_{q,\nu,r-s-t}(f(\zeta)) \\ \times P^{(t-\mu)}(f(\zeta)) \, Q^{(r-s-t-\nu)}(f(\zeta)) \end{array} \right]$$

and then combine (9.13)–(9.16) to deduce

(9.17) $$R^{**}(\zeta) \equiv (f'(\zeta))^{p+q+r} \sum_{s=0}^{r} \sum_{t=0}^{r-s} \sum_{\mu=0}^{t} \sum_{\nu=0}^{r-s-t} X_{p,q,r,s,t,\mu,\nu}(\zeta)$$

$$\equiv (f'(\zeta))^{p+q+r} \sum_{s=0}^{r} \sum_{\mu=0}^{r-s} \sum_{t=\mu}^{r-s} \sum_{\nu=0}^{r-s-t} X_{p,q,r,s,t,\mu,\nu}(\zeta)$$

$$\equiv (f'(\zeta))^{p+q+r} \sum_{\mu=0}^{r} \sum_{s=0}^{r-\mu} \sum_{t=\mu}^{r-s} \sum_{\nu=0}^{r-s-t} X_{p,q,r,s,t,\mu,\nu}(\zeta)$$

$$\equiv (f'(\zeta))^{p+q+r} \sum_{\mu=0}^{r} \sum_{s=0}^{r-\mu} \sum_{\nu=0}^{r-s-\mu} \sum_{t=\mu}^{r-s-\nu} X_{p,q,r,s,t,\mu,\nu}(\zeta)$$

$$\equiv (f'(\zeta))^{p+q+r} \sum_{\mu=0}^{r} \sum_{\nu=0}^{r-\mu} \sum_{s=0}^{r-\mu-\nu} \sum_{t=\mu}^{r-s-\nu} X_{p,q,r,s,t,\mu,\nu}(\zeta)$$

$$\equiv (f'(\zeta))^{p+q+r} \sum_{\mu=0}^{r} \sum_{\nu=0}^{r-\mu} \sum_{s=0}^{r-\mu-\nu} \sum_{t=0}^{r-s-\mu-\nu} X_{p,q,r,s,t+\mu,\mu,\nu}(\zeta)$$

$$\equiv (f'(\zeta))^{p+q+r} \sum_{\mu=0}^{r} \sum_{\nu=0}^{r-\mu} \sum_{s=\mu+\nu}^{r} \sum_{t=0}^{r-s} X_{p,q,r,s-\mu-\nu,t+\mu,\mu,\nu}(\zeta)$$

$$\equiv (f'(\zeta))^{p+q+r} \sum_{\mu=0}^{r} \sum_{s=\mu}^{r} \sum_{\nu=0}^{s-\mu} \sum_{t=0}^{r-s} X_{p,q,r,s-\mu-\nu,t+\mu,\mu,\nu}(\zeta)$$

$$\equiv (f'(\zeta))^{p+q+r} \sum_{s=0}^{r} \sum_{\mu=0}^{s} \sum_{\nu=0}^{s-\mu} \sum_{t=0}^{r-s} X_{p,q,r,s-\mu-\nu,t+\mu,\mu,\nu}(\zeta)$$

$$\equiv (f'(\zeta))^{p+q+r} \sum_{s=0}^{r} \sum_{\mu=0}^{s} \sum_{t=0}^{r-s} \sum_{\nu=0}^{s-\mu} X_{p,q,r,s-\mu-\nu,t+\mu,\mu,\nu}(\zeta)$$

$$\equiv (f'(\zeta))^{p+q+r} \sum_{s=0}^{r} \sum_{t=0}^{r-s} \sum_{\mu=0}^{s} \sum_{\nu=0}^{s-\mu} X_{p,q,r,s-\mu-\nu,t+\mu,\mu,\nu}(\zeta)$$

$$\equiv (f'(\zeta))^{p+q+r} \sum_{s=0}^{r} \sum_{t=0}^{r-s} \sum_{\lambda=0}^{s} \sum_{\mu=0}^{\lambda} X_{p,q,r,s-\lambda,t+\mu,\mu,\lambda-\mu}(\zeta)$$

$$\equiv (f'(\zeta))^{p+q+r} \sum_{s=0}^{r} \sum_{t=0}^{r-s} \sum_{u=0}^{s} \sum_{v=0}^{s-u} X_{p,q,r,u,t+v,v,s-u-v}(\zeta).$$

We use (9.16) and (9.11) to verify that

$$\sum_{u=0}^{s} \sum_{v=0}^{s-u} X_{p,q,r,u,t+v,v,s-u-v}(\zeta) \equiv \mathfrak{A}_{p,q,r,s,t}(\zeta) \, P^{(t)}(f(\zeta)) \, Q^{(r-s-t)}(f(\zeta)).$$

By combining this with (9.17), we obtain (9.12) and complete the proof. $\square$

9.3. Reformulation for $\mathfrak{A}_{p,q,r,s,t}(g(z))$

With $f(g(z)) \equiv z$ and $f'(g(z)) \equiv 1/g'(z)$ on Ω, we note that (9.11) yields

$$(9.18) \quad \mathfrak{A}_{p,q,r,s,t}(g(z)) \equiv \sum_{u=0}^{s} \sum_{v=0}^{s-u} \frac{A^{**}_{p,q,r,u,t+v}(g(z))}{(g'(z))^{-u}} \phi_{p,v,t+v}(z) \, \phi_{q,s-u-v,r-t-u-v}(z),$$

on Ω, for $0 \le s \le r$ and $0 \le t \le r-s$.

Our immediate goal is to simplify the expression for $\mathfrak{A}_{p,q,r,s,t}(g(z))$ in (9.18).

With respect to the $\phi_{h,i,j}(z)$ on Ω defined by (9.4)–(9.6) and the $B^{**}_{h,i,j}(\zeta)$ on Ω^{**} obtained by substituting $c^{**(k)}_{i_1,i_2,\ldots,i_n}(\zeta)$ from (4.7) for $w^{(k)}_{i_1,i_2,\ldots,i_n}$ in $\boldsymbol{B}_{h,i,j}$ of (4.46)–(4.48) on page 36, we introduce

$$(9.19) \quad E_{h,i,j}(z) \equiv \sum_{k=0}^{i} (g'(z))^k B^{**}_{h,k,j}(g(z)) \, \phi_{h,i-k,j-k}(z),$$

on Ω for any h, i, j.

PROPOSITION 9.4. *The functions* $\mathfrak{A}_{p,q,r,s,t}(g(z))$ *of (9.18) are specified in terms of (9.19) and the rational numbers* $\mathfrak{C}_{p,q,r,\mu}$ *of (4.49) by*

$$(9.20) \quad \mathfrak{A}_{p,q,r,s,t}(g(z)) \equiv \sum_{\lambda=0}^{s} \mathfrak{C}_{p,q,r,t+\lambda} \, E_{p,\lambda,t+\lambda}(z) \, E_{q,s-\lambda,r-t-\lambda}(z),$$

on Ω, for $0 \le s \le r$ and $0 \le t \le r-s$.

PROOF. We use (4.50) on page 36 to see that

$$(9.21) \quad \boldsymbol{A}_{p,q,r,u,t+v} \equiv \sum_{k=0}^{u} \mathfrak{C}_{p,q,r,t+v+k} \, \boldsymbol{B}_{p,k,t+v+k} \, \boldsymbol{B}_{q,u-k,r-t-v-k}.$$

By substituting $c^{**(k)}_{i_1,i_2,\ldots,i_n}(\zeta)$ from (4.7) for $w^{(k)}_{i_1,i_2,\ldots,i_n}$ in (9.21) and writing $g(z)$ for ζ, we find that

$$(9.22) \quad A^{**}_{p,q,r,u,t+v}(g(z)) \equiv \sum_{k=0}^{u} \left[\begin{array}{c} \mathfrak{C}_{p,q,r,t+v+k} \, B^{**}_{p,k,t+v+k}(g(z)) \\ \times B^{**}_{q,u-k,r-t-v-k}(g(z)) \end{array} \right].$$

After introducing

$$(9.23) \quad Y_{p,t,v,k}(z) \equiv (g'(z))^k B^{**}_{p,k,t+v+k}(g(z)) \, \phi_{p,v,t+v}(z)$$

and

$$(9.24) \quad Z_{q,r,s,t,u,v,k}(z) \equiv (g'(z))^{u-k} \left[\begin{array}{c} B^{**}_{q,u-k,r-t-v-k}(g(z)) \\ \times \phi_{q,s-u-v,r-t-u-v}(z) \end{array} \right],$$

we employ (9.18) and (9.22)–(9.24) to verify that

$$
(9.25) \quad \mathfrak{A}_{p,q,r,s,t}\big(g(z)\big) \equiv \sum_{u=0}^{s} \sum_{v=0}^{s-u} \sum_{k=0}^{u} \mathfrak{C}_{p,q,r,t+v+k}\, Y_{p,t,v,k}(z)\, Z_{q,r,s,t,u,v,k}(z)
$$

$$
\equiv \sum_{v=0}^{s} \sum_{u=0}^{s-v} \sum_{k=0}^{u} \mathfrak{C}_{p,q,r,t+v+k}\, Y_{p,t,v,k}(z)\, Z_{q,r,s,t,u,v,k}(z)
$$

$$
\equiv \sum_{v=0}^{s} \sum_{k=0}^{s-v} \sum_{u=k}^{s-v} \mathfrak{C}_{p,q,r,t+v+k}\, Y_{p,t,v,k}(z)\, Z_{q,r,s,t,u,v,k}(z)
$$

$$
\equiv \sum_{\lambda=0}^{s} \sum_{v=0}^{\lambda} \sum_{u=\lambda-v}^{s-v} \mathfrak{C}_{p,q,r,t+\lambda}\, Y_{p,t,v,\lambda-v}(z)\, Z_{q,r,s,t,u,v,\lambda-v}(z)
$$

$$
\equiv \sum_{\lambda=0}^{s} \mathfrak{C}_{p,q,r,t+\lambda} \sum_{v=0}^{\lambda} \sum_{u=\lambda}^{s} Y_{p,t,v,\lambda-v}(z)\, Z_{q,r,s,t,u-v,v,\lambda-v}(z)
$$

$$
\equiv \sum_{\lambda=0}^{s} \mathfrak{C}_{p,q,r,t+\lambda} \sum_{v=0}^{\lambda} Y_{p,t,v,\lambda-v}(z) \sum_{u=\lambda}^{s} Z_{q,r,s,t,u-v,v,\lambda-v}(z).
$$

We apply (9.24) and (9.19) to deduce

$$
\sum_{u=\lambda}^{s} Z_{q,r,s,t,u-v,v,\lambda-v}(z) \equiv \sum_{u=\lambda}^{s} \big(g'(z)\big)^{u-\lambda} B^{**}_{q,u-\lambda,r-t-\lambda}\big(g(z)\big)\, \phi_{q,s-u,r-t-u}(z)
$$

$$
\equiv \sum_{k=0}^{s-\lambda} \big(g'(z)\big)^{k} B^{**}_{q,k,r-t-\lambda}\big(g(z)\big)\, \phi_{q,s-\lambda-k,r-t-\lambda-k}(z)
$$

$$
\equiv E_{q,s-\lambda,r-t-\lambda}(z).
$$

Moreover, (9.23) and (9.19) yield

$$
\sum_{v=0}^{\lambda} Y_{p,t,v,\lambda-v}(z) \equiv \sum_{v=0}^{\lambda} \big(g'(z)\big)^{\lambda-v} B^{**}_{p,\lambda-v,t+\lambda}\big(g(z)\big)\, \phi_{p,v,t+v}(z)
$$

$$
\equiv \sum_{k=0}^{\lambda} \big(g'(z)\big)^{k} B^{**}_{p,k,t+\lambda}\big(g(z)\big)\, \phi_{p,\lambda-k,t+\lambda-k}(z)
$$

$$
\equiv E_{p,\lambda,t+\lambda}(z).
$$

By combining the two preceding formulas with (9.25), we obtain (9.20) and complete the proof. $\square$

9.4. Initial simplification for $E_{h,i,j}(z)$

To obtain information about $\mathfrak{A}_{p,q,r,s,t}\big(g(z)\big)$ from the involvement of $E_{h,i,j}(z)$ in (9.20), we deduce properties of $E_{h,i,j}(z)$ via its definition in (9.19). The results about $E_{h,i,j}(z)$ in Proposition 9.9 depend on four lemmas for which we introduce

$$
(9.26) \quad \mathcal{D}(z) \equiv -\frac{g'''(z)}{2g'(z)} + \frac{3}{4}\left[\frac{g''(z)}{g'(z)}\right]^{2}, \quad \text{on } \Omega.
$$

9.4. INITIAL SIMPLIFICATION FOR $E_{h,i,j}(z)$

LEMMA 9.5. *The function $a_{m,n}(z)$ on Ω that is obtained by replacing each $\boldsymbol{w}_{i_1,i_2,\ldots,i_n}^{(k)}$ in $\boldsymbol{a}_{m,n}$ of (4.45) with the corresponding $c_{i_1,i_2,\ldots,i_n}^{(k)}(z)$ on Ω from (4.1) and the function $a_{m,n}^{**}(\zeta)$ on Ω^{**} that is obtained by replacing each $\boldsymbol{w}_{i_1,i_2,\ldots,i_n}^{(k)}$ in $\boldsymbol{a}_{m,n}$ with the corresponding $c_{i_1,i_2,\ldots,i_n}^{**(k)}(\zeta)$ on Ω^{**} from (4.7) satisfy*

$$(9.27) \quad a_{m,n}^{**}(g(z)) \equiv (g'(z))^{-2}(a_{m,n}(z) + \mathcal{D}(z)), \quad \text{on } \Omega,$$

and

$$(9.28) \quad a_{m,n}^{**}(\zeta) \equiv (f'(\zeta))^2 a_{m,n}(f(\zeta)) + \frac{f'''(\zeta)}{2f'(\zeta)} - \frac{3}{4}\left[\frac{f''(\zeta)}{f'(\zeta)}\right]^2, \quad \text{on } \Omega^{**}.$$

PROOF. Using the abbreviations of (5.20) on page 45 as well as (4.45), (5.23), and (5.22), we obtain

$$a_{m,n}^{**}(\zeta) \equiv \frac{1}{\binom{m+1}{3}}\left[\mathfrak{c}_2^{**}(\zeta) - \frac{m-1}{2}\mathfrak{c}_1^{**(1)}(\zeta) - \frac{m-1}{2m}(\mathfrak{c}_1^{**}(\zeta))^2\right]$$

$$\equiv \frac{1}{\binom{m+1}{3}}\begin{bmatrix}(f'(\zeta))^2\mathfrak{c}_2(f(\zeta)) - \binom{m-1}{2}f''(\zeta)\mathfrak{c}_1(f(\zeta))\\ -\binom{m}{3}\frac{f'''(\zeta)}{f'(\zeta)} + 3\binom{m+1}{4}\left[\frac{f''(\zeta)}{f'(\zeta)}\right]^2\\ -\frac{m-1}{2}(f'(\zeta))^2\mathfrak{c}_1^{(1)}(f(\zeta)) - \frac{m-1}{2}f''(\zeta)\mathfrak{c}_1(f(\zeta))\\ +\frac{(m-1)}{2}\binom{m}{2}\frac{f'''(\zeta)}{f'(\zeta)} - \frac{(m-1)}{2}\binom{m}{2}\left[\frac{f''(\zeta)}{f'(\zeta)}\right]^2\\ -\frac{m-1}{2m}(f'(\zeta))^2(\mathfrak{c}_1(f(\zeta)))^2 - \frac{m-1}{2m}\left[\binom{m}{2}\frac{f''(\zeta)}{f'(\zeta)}\right]^2\\ +\frac{(m-1)}{m}\binom{m}{2}f''(\zeta)\mathfrak{c}_1(f(\zeta))\end{bmatrix}$$

$$\equiv \frac{(f'(\zeta))^2}{\binom{m+1}{3}}\left[\mathfrak{c}_2(f(\zeta)) - \frac{m-1}{2}\mathfrak{c}_1^{(1)}(f(\zeta)) - \frac{m-1}{2m}(\mathfrak{c}_1(f(\zeta)))^2\right]$$
$$+ \frac{f'''(\zeta)}{2f'(\zeta)} - \frac{3}{4}\left[\frac{f''(\zeta)}{f'(\zeta)}\right]^2$$

$$\equiv (f'(\zeta))^2 a_{m,n}(f(\zeta)) + \frac{f'''(\zeta)}{2f'(\zeta)} - \frac{3}{4}\left[\frac{f''(\zeta)}{f'(\zeta)}\right]^2.$$

Thus, (9.28) is valid. We substitute $\zeta = g(z)$ in (9.28) and use $f(g(z)) = z$ as well as $f'(g(z)) \equiv 1/g'(z)$,

$$f''(g(z)) \equiv \frac{-g''(z)}{(g'(z))^3}, \quad f'''(g(z)) \equiv \frac{-g'''(z)}{(g'(z))^4} + 3\frac{(g''(z))^2}{(g'(z))^5},$$

$$\frac{f'''(g(z))}{2f'(g(z))} - \frac{3}{4}\left[\frac{f''(g(z))}{f'(g(z))}\right]^2 \equiv \frac{1}{(g'(z))^2}\left[-\frac{g'''(z)}{2g'(z)} + \frac{3}{4}\left[\frac{g''(z)}{g'(z)}\right]^2\right],$$

and (9.26) to deduce (9.27), This completes the proof. $\square$

LEMMA 9.6. *The* $\boldsymbol{B}_{h,i,j}$ *defined in* $\mathcal{R}_{m,n}$ *by* (4.46)–(4.48) *satisfy*

$$(9.29) \quad \boldsymbol{B}^{(1)}_{h,i,j} \equiv \boldsymbol{B}_{h,i+1,j+1} - \boldsymbol{B}_{h,i+1,j} + j(2h+j-1)\boldsymbol{a}_{m,n}\boldsymbol{B}_{h,i-1,j-1},$$
$$\text{for } i \leq j \text{ and any } h.$$

PROOF. For h, i, j such that $0 \leq i \leq j$, we employ (4.48) to deduce

$$(9.30) \quad \boldsymbol{B}_{h,i+1,j+1} - \boldsymbol{B}_{h,i+1,j} \equiv \boldsymbol{B}^{(1)}_{h,i,j} - j(2h+j-1)\boldsymbol{a}_{m,n}\boldsymbol{B}_{h,i-1,j-1}.$$

For $i = -1$ and any h, j, we use (4.47) and (4.46) to obtain $1 - 1$ and $0 - 0$ as the left and right members of (9.30). For $i \leq -2$ and any h, j, we note that (4.46) yields $0 - 0$ and $0 - 0$ as the left and right members of (9.30). By rewriting (9.30), we find that (9.29) is valid. This completes the proof. □

The two members of (9.29) are not equal when $i = 1$ and $1 - 2h \neq j \leq -1$. The two members of (9.7) are not equal when $i = 0$ and $-h \neq j \leq -1$. Thus, care is needed when using (9.7) and (9.29).

LEMMA 9.7. *In terms of* $a_{m,n}(z)$ *and* $\mathcal{D}(z)$ *as presented for* Lemma 9.5, *the functions* $E_{h,i,j}(z)$ *defined by* (9.19) *satisfy*

$$(9.31) \quad E_{h,i+1,j+1}(z) - E_{h,i+1,j}(z) \equiv E^{(1)}_{h,i,j}(z) - (h+j)\frac{g''(z)}{g'(z)}E_{h,i,j}(z)$$
$$- j(2h+j-1)\big(a_{m,n}(z) + \mathcal{D}(z)\big)E_{h,i-1,j-1}(z),$$
$$\text{on } \Omega, \text{ for } 0 \leq i \leq j \text{ and any } h.$$

PROOF. For h, i, j subject to $0 \leq i \leq j$, (9.19) and differentiation yield

$$(9.32) \quad E^{(1)}_{h,i,j}(z) \equiv \sum_{k=0}^{i} \left[\begin{array}{l} k\big(g'(z)\big)^{k-1} g''(z)\, B^{**}_{h,k,j}\big(g(z)\big)\, \phi_{h,i-k,j-k}(z) \\ + \big(g'(z)\big)^{k+1} B^{**(1)}_{h,k,j}\big(g(z)\big)\, \phi_{h,i-k,j-k}(z) \\ + \big(g'(z)\big)^{k} B^{**}_{h,k,j}\big(g(z)\big)\, \phi^{(1)}_{h,i-k,j-k}(z) \end{array} \right].$$

We replace each $\boldsymbol{w}^{(k)}_{i_1,i_2,\ldots,i_n}$ in (9.29) with the corresponding $c^{**(k)}_{i_1,i_2,\ldots,i_n}(\zeta)$ from (4.7) and use the resulting expression to verify that: for $0 \leq k \leq i \leq j$,

$$(9.33) \quad B^{**(1)}_{h,k,j}\big(g(z)\big) \equiv B^{**}_{h,k+1,j+1}\big(g(z)\big) - B^{**}_{h,k+1,j}\big(g(z)\big)$$
$$+ j(2h+j-1)\, a^{**}_{m,n}\big(g(z)\big)\, B^{**}_{h,k-1,j-1}\big(g(z)\big).$$

With $i - k \leq j - k$, we employ (9.7) to deduce

$$(9.34) \quad \phi^{(1)}_{h,i-k,j-k}(z) \equiv \phi_{h,i+1-k,j+1-k}(z) - \phi_{h,i+1-k,j-k}(z)$$
$$+ (h+j-k)\frac{g''(z)}{g'(z)}\phi_{h,i-k,j-k}(z).$$

After replacing $B^{**(1)}_{h,k,j}\big(g(z)\big)$ and $\phi^{(1)}_{h,i-k,j-k}(z)$ in (9.32) with the right sides of (9.33) and (9.34), we find that the resulting expression gives

$$(9.35) \quad E^{(1)}_{h,i,j}(z) \equiv \mathcal{E}_1 + \mathcal{E}_2 + \mathcal{E}_3 + \mathcal{E}_4 + \mathcal{E}_5 + \mathcal{E}_6 + \mathcal{E}_7,$$

where

$$\mathcal{E}_1 \equiv \sum_{k=0}^{i} k\big(g'(z)\big)^k \frac{g''(z)}{g'(z)} B^{**}_{h,k,j}\big(g(z)\big)\, \phi_{h,i-k,j-k}(z),$$

$$\mathcal{E}_2 \equiv \sum_{k=0}^{i} \big(g'(z)\big)^{k+1} B^{**}_{h,k+1,j+1}\big(g(z)\big)\, \phi_{h,i-k,j-k}(z),$$

$$\mathcal{E}_3 \equiv \sum_{k=0}^{i} -\big(g'(z)\big)^{k+1} B^{**}_{h,k+1,j}\big(g(z)\big)\, \phi_{h,i-k,j-k}(z),$$

$$\mathcal{E}_4 \equiv \sum_{k=0}^{i} \big(g'(z)\big)^{k+1} j(2h+j-1)\, a^{**}_{m,n}\big(g(z)\big)\, B^{**}_{h,k-1,j-1}\big(g(z)\big)\, \phi_{h,i-k,j-k}(z),$$

$$\mathcal{E}_5 \equiv \sum_{k=0}^{i} \big(g'(z)\big)^{k} B^{**}_{h,k,j}\big(g(z)\big)\, \phi_{h,i+1-k,j+1-k}(z),$$

$$\mathcal{E}_6 \equiv \sum_{k=0}^{i} -\big(g'(z)\big)^{k} B^{**}_{h,k,j}\big(g(z)\big)\, \phi_{h,i+1-k,j-k}(z),$$

and

$$\mathcal{E}_7 \equiv \sum_{k=0}^{i} \big(g'(z)\big)^{k} \frac{g''(z)}{g'(z)} (h+j-k) B^{**}_{h,k,j}\big(g(z)\big)\, \phi_{h,i-k,j-k}(z).$$

We apply (9.19) to check that

$$(9.36) \qquad \mathcal{E}_1 + \mathcal{E}_7 \equiv \sum_{k=0}^{i} \big(g'(z)\big)^{k} \frac{g''(z)}{g'(z)} (h+j) B^{**}_{h,k,j}\big(g(z)\big)\, \phi_{h,i-k,j-k}(z)$$

$$\equiv (h+j)\frac{g''(z)}{g'(z)} E_{h,i,j}(z).$$

With (9.19), (4.47), and (9.5), we find that

$$\mathcal{E}_2 + \mathcal{E}_6 \equiv \sum_{k=1}^{i+1} \big(g'(z)\big)^{k} B^{**}_{h,k,j+1}\big(g(z)\big)\, \phi_{h,i+1-k,j+1-k}(z)$$

$$- \sum_{k=0}^{i} \big(g'(z)\big)^{k} B^{**}_{h,k,j}\big(g(z)\big)\, \phi_{h,i+1-k,j-k}(z)$$

$$\equiv E_{h,i+1,j+1}(z) - \phi_{h,i+1j+1}(z)$$

$$- E_{h,i+1,j}(z) + \big(g'(z)\big)^{i+1} B^{**}_{h,i+1,j}\big(g(z)\big),$$

$$\mathcal{E}_5 + \mathcal{E}_3 \equiv \sum_{k=0}^{i} \big(g'(z)\big)^{k} B^{**}_{h,k,j}\big(g(z)\big)\, \phi_{h,i+1-k,j+1-k}(z)$$

$$- \sum_{k=1}^{i+1} \big(g'(z)\big)^{k} B^{**}_{h,k,j}\big(g(z)\big)\, \phi_{h,i+1-k,j+1-k}(z)$$

$$\equiv \phi_{h,i+1,j+1}(z) - \big(g'(z)\big)^{i+1} B^{**}_{h,i+1,j}\big(g(z)\big),$$

and

$$\mathcal{E}_2 + \mathcal{E}_3 + \mathcal{E}_5 + \mathcal{E}_6 \equiv E_{h,i+1,j+1}(z) - E_{h,i+1,j}(z). \tag{9.37}$$

We use (9.27), (4.46), and (9.19) to establish

$$\mathcal{E}_4 \equiv \sum_{k=-1}^{i-1} \left[\begin{array}{c} (g'(z))^{k+2} j(2h+j-1) \, a^{**}_{m,n}(g(z)) \\ \times B^{**}_{h,k,j-1}(g(z)) \, \phi_{h,i-1-k,j-1-k}(z) \end{array} \right] \tag{9.38}$$

$$\equiv j(2h+j-1)(a_{m,n}(z) + \mathcal{D}(z)) \sum_{k=0}^{i-1} \left[\begin{array}{c} (g'(z))^{k} B^{**}_{h,k,j-1}(g(z)) \\ \times \phi_{h,i-1-k,j-1-k}(z) \end{array} \right]$$

$$\equiv j(2h+j-1)(a_{m,n}(z) + \mathcal{D}(z)) E_{h,i-1,j-1}(z).$$

By combining (9.35), (9.36), (9.37), and (9.38), we deduce

$$E^{(1)}_{h,i,j}(z) \equiv E_{h,i+1,j+1}(z) - E_{h,i+1,j}(z) + (h+j)\frac{g''(z)}{g'(z)} E_{h,i,j}(z) \tag{9.39}$$
$$+ j(2h+j-1)(a_{m,n}(z) + \mathcal{D}(z)) E_{h,i-1,j-1}(z).$$

We rewrite (9.39) to obtain (9.31) and complete the proof. □

LEMMA 9.8. *The functions* $E_{h,i,j}(z)$ *defined on* Ω *by* (9.19) *satisfy*
$$E_{h,i+1,j}(z) \equiv 0, \quad \text{for } i \geq 0, \, j \leq i \text{ and any } h. \tag{9.40}$$

PROOF. For fixed integers h, i, j subject to $i \geq 0$ and $j \leq i$, (9.19) yields

$$E_{h,i+1,j}(z) \equiv \sum_{\nu=0}^{i+1} (g'(z))^\nu B^{**}_{h,\nu,j}(g(z)) \, \phi_{h,i+1-\nu,j-\nu}(z). \tag{9.41}$$

If $0 \leq \nu \leq i$, then $i-\nu \geq 0$, $j-\nu-1 < i+1-\nu$, and (9.6) gives $\phi_{h,i+1-\nu,j-\nu}(z) \equiv 0$. If $\nu = i+1$, then (4.48) and $j \leq i$ show that $B^{**}_{h,i+1,j}(g(z)) \equiv 0$. Thus, we conclude from (9.41) that $E_{h,i+1,j}(z) \equiv 0$. This completes the proof. □

PROPOSITION 9.9. *The functions* $E_{h,i,j}(z)$ *defined on* Ω *by* (9.19) *satisfy*

$$E_{h,i,j}(z) \equiv 0, \quad \text{for } i \leq -1 \text{ and any } h, \, j, \tag{9.42}$$

$$E_{h,0,j}(z) \equiv 1, \quad \text{for any } h, \, j, \tag{9.43}$$

and

$$E_{h,i+1,j}(z) \equiv \sum_{k=i}^{j-1} \left[\begin{array}{c} E^{(1)}_{h,i,k}(z) - (h+k)\frac{g''(z)}{g'(z)} E_{h,i,k}(z) \\ - k(2h+k-1)(a_{m,n}(z) + \mathcal{D}(z)) E_{h,i-1,k-1}(z) \end{array} \right], \tag{9.44}$$
$$\text{for } i \geq 0 \text{ and any } h, \, j.$$

PROOF. For $i \leq -1$ and any h, j, we see that (9.19) gives (9.42). For $i = 0$ and any h, j, we find that (9.19), (4.47) of page 36, and (9.5) of page 77 yield (9.43). If $i \geq 0$ and $j \leq i$, then (9.40) shows that (9.44) reduces to $0 \equiv 0$. Suppose that $i \geq 0$ and $j \geq i+1$. Then, in view of $E_{h,i+1,i}(z) \equiv 0$ from (9.40), we obtain

$$(9.45) \quad E_{h,i+1,j}(z) \equiv E_{h,i+1,j}(z) - E_{h,i+1,i}(z) \equiv \sum_{k=i}^{j-1} \bigl(E_{h,i+1,k+1}(z) - E_{h,i+1,k}(z)\bigr).$$

By setting $j = k$ in (9.31) and substituting its resulting right member for the summand $\bigl(E_{h,i+1,k+1}(z) - E_{h,i+1,k}(z)\bigr)$ in (9.45), we establish (9.44). This completes the proof. □

9.5. Properties of $E_{h,i,j}(z)$

We define functions $F_{h,i,j,\nu}(z)$ on Ω by means of

$$(9.46) \quad F_{h,i,j,\nu}(z) \equiv 0, \quad \text{for } h,\, i,\, j,\, \nu \text{ subject to } i < 0 \text{ or } \nu < 0 \text{ or } \nu > i,$$

$$(9.47) \quad F_{h,0,j,0}(z) \equiv 1, \quad \text{for any } h,\, j,$$

and

$$(9.48) \quad F_{h,i+1,j,\nu}(z) \equiv \sum_{k=i}^{j-1} \begin{bmatrix} F^{(1)}_{h,i,k,\nu}(z) - 2(\nu+1)\mathcal{D}(z)\, F_{h,i,k,\nu+1}(z) \\ + \tfrac{1}{2}(\nu - 1 - 2h - 2k) F_{h,i,k,\nu-1}(z) \\ - k(2h+k-1)\bigl(a_{m,n}(z) + \mathcal{D}(z)\bigr) F_{h,i-1,k-1,\nu}(z) \end{bmatrix},$$

for $i \geq 0$ and any $h,\, j,\, \nu$ having $0 \leq \nu \leq i+1$.

PROPOSITION 9.10. *The functions $E_{h,i,j}(z)$ defined on Ω by (9.19) and the functions $F_{s,i,j,\nu}(z)$ defined on Ω by (9.46)–(9.48) satisfy*

$$(9.49) \quad E_{h,i,j}(z) \equiv \sum_{\nu=0}^{i} F_{h,i,j,\nu}(z) \left[\frac{g''(z)}{g'(z)}\right]^{\nu}, \quad \text{on } \Omega,$$

for all integers $h,\, i,\, j$.

PROOF. In view of (9.42)–(9.43) and (9.46)–(9.47), we observe that (9.49) is true for any $h,\, i,\, j$ having $i \leq 0$.

Let i_0 be a nonnegative integer such that (9.49) is valid for all $h,\, i,\, j$ that satisfy $i \leq i_0$. Then, for any h and k, we use (9.49), (9.26), and (9.46) to obtain

$$(9.50) \quad E^{(1)}_{h,i_0,k}(z) \equiv \sum_{\nu=0}^{i_0} F^{(1)}_{h,i_0,k,\nu}(z) \left[\frac{g''(z)}{g'(z)}\right]^{\nu}$$
$$+ \sum_{\nu=0}^{i_0} F_{h,i_0,k,\nu}(z)\, \nu \left[\frac{g''(z)}{g'(z)}\right]^{\nu-1} \left[\frac{g'''(z)}{g'(z)} - \left[\frac{g''(z)}{g'(z)}\right]^2\right]$$
$$\equiv \sum_{\nu=0}^{i_0} F^{(1)}_{h,i_0,k,\nu}(z) \left[\frac{g''(z)}{g'(z)}\right]^{\nu}$$
$$+ \sum_{\nu=0}^{i_0} F_{h,i_0,k,\nu}(z)\, \nu \left[\frac{g''(z)}{g'(z)}\right]^{\nu-1} \left[-2\mathcal{D}(z) + \frac{1}{2}\left[\frac{g''(z)}{g'(z)}\right]^2\right]$$
$$\equiv \sum_{\nu=0}^{i_0+1} \begin{bmatrix} F^{(1)}_{h,i_0,k,\nu}(z) \\ - 2(\nu+1)\mathcal{D}(z)\, F_{h,i_0,k,\nu+1}(z) \\ + \bigl((\nu-1)/2\bigr) F_{h,i_0,k,\nu-1}(z) \end{bmatrix} \left[\frac{g''(z)}{g'(z)}\right]^{\nu}.$$

For any h and k, we employ (9.49) for $i \leq i_0$ as well as (9.46) to verify that

$$(9.51) \quad -(h+k)\frac{g''(z)}{g'(z)} E_{h,i_0,k}(z) \equiv \sum_{\nu=0}^{i_0} -(h+k) F_{h,i_0,k,\nu}(z) \left[\frac{g''(z)}{g'(z)}\right]^{\nu+1}$$

$$\equiv \sum_{\nu=0}^{i_0+1} -(h+k) F_{h,i_0,k,\nu-1}(z) \left[\frac{g''(z)}{g'(z)}\right]^{\nu}$$

and

$$(9.52) \quad -k(2h+k-1)\bigl(a_{m,n}(z) + \mathcal{D}(z)\bigr) E_{h,i_0-1,k-1}(z)$$

$$\equiv \sum_{\nu=0}^{i_0+1} -k(2h+k-1)\bigl(a_{m,n}(z) + \mathcal{D}(z)\bigr) F_{h,i_0-1,k-1,\nu}(z) \left[\frac{g''(z)}{g'(z)}\right]^{\nu}.$$

For any h and j, we find that (9.44), (9.50), (9.51), (9.52), and (9.48) yield

$$E_{h,i_0+1,j}(z) \equiv \sum_{k=i_0}^{j-1} \left[\begin{array}{l} E^{(1)}_{h,i_0,k}(z) - (h+k)\dfrac{g''(z)}{g'(z)} E_{h,i_0,k}(z) \\ - k(2h+k-1)\bigl(a_{m,n}(z) + \mathcal{D}(z)\bigr) E_{h,i_0-1,k-1}(z) \end{array} \right]$$

$$\equiv \sum_{\nu=0}^{i_0+1} \sum_{k=i_0}^{j-1} \left[\begin{array}{l} F^{(1)}_{h,i_0,k,\nu}(z) - 2(\nu+1)\mathcal{D}(z) F_{h,i_0,k,\nu+1}(z) \\ + \frac{1}{2}(\nu-1) F_{h,i_0,k,\nu-1}(z) - (h+k) F_{h,i_0,k,\nu-1}(z) \\ - k(2h+k-1)\bigl(a_{m,n}(z) + \mathcal{D}(z)\bigr) F_{h,i_0-1,k-1,\nu}(z) \end{array} \right] \left[\frac{g''(z)}{g'(z)}\right]^{\nu}$$

$$\equiv \sum_{\nu=0}^{i_0+1} F_{h,i_0+1,j,\nu}(z) \left[\frac{g''(z)}{g'(z)}\right]^{\nu}.$$

By induction on i, this shows that (9.49) is valid for all integers h, i, j and completes the proof. □

The proofs of Proposition 9.12 and Proposition 11.9 require the following result.

LEMMA 9.11. *Any integers κ, ν, ω subject to $\nu \geq 1$ satisfy*

$$\binom{\kappa}{\nu}\binom{\omega+\kappa-1}{\nu} + \frac{(\omega+2\kappa-\nu+1)}{\nu}\binom{\kappa}{\nu-1}\binom{\omega+\kappa-1}{\nu-1} \equiv \binom{\kappa+1}{\nu}\binom{\omega+\kappa}{\nu}.$$

PROOF. Let $\mathcal{L}_1$ and $\mathcal{L}_2$ denote the first and second terms in the left member of the preceding formula and let $\mathcal{R}$ denote its right member. We find that

$$\mathcal{L}_2 \equiv \left[\frac{\kappa+1}{\nu} + \frac{\omega+\kappa-\nu}{\nu}\right]\binom{\kappa}{\nu-1}\binom{\omega+\kappa-1}{\nu-1}$$

$$\equiv \binom{\kappa+1}{\nu}\binom{\omega+\kappa-1}{\nu-1} + \binom{\kappa}{\nu-1}\binom{\omega+\kappa-1}{\nu}$$

and

$$\mathcal{L}_1 + \mathcal{L}_2 \equiv \binom{\kappa+1}{\nu}\binom{\omega+\kappa-1}{\nu-1} + \left[\binom{\kappa}{\nu-1} + \binom{\kappa}{\nu}\right]\binom{\omega+\kappa-1}{\nu}$$
$$\equiv \binom{\kappa+1}{\nu}\binom{\omega+\kappa-1}{\nu-1} + \binom{\kappa+1}{\nu}\binom{\omega+\kappa-1}{\nu}$$
$$\equiv \binom{\kappa+1}{\nu}\binom{\omega+\kappa}{\nu} \equiv \mathcal{R}.$$

This establishes the validity of the formula and completes the proof. □

We introduce

(9.53) $$\mathfrak{P}_\nu = \prod_{k=1}^{\nu} k, \quad \text{for any integer } \nu.$$

In particular, this gives $\mathfrak{P}_\nu = \nu!$, when $\nu \geq 1$; and it yields $\mathfrak{P}_\nu = 1$, when $\nu \leq 0$.

PROPOSITION 9.12. *The functions $F_{h,i,j,\nu}(z)$ defined on Ω by (9.46)–(9.48) and the functions $B_{h,i,j}(z)$ obtained by replacing $\boldsymbol{w}^{(k)}_{i_1,i_2,...,i_n}$ in $\boldsymbol{B}_{h,i,j}$ of (4.46)–(4.48) with $c^{(k)}_{i_1,i_2,...,i_n}(z)$ from (4.1) are related by means of*

(9.54) $$F_{h,i,j,\nu}(z) \equiv \frac{(-1)^\nu \mathfrak{P}_\nu}{2^\nu}\binom{j}{\nu}\binom{2h+j-1}{\nu}B_{h,i-\nu,j-\nu}(z), \quad \text{on } \Omega,$$

for all h, i, j, ν that satisfy $j \geq 0$.

PROOF. If $i < 0$ or $\nu < 0$ or $\nu > i$, then (9.46), (4.46) of page 36, and the binomial coefficients in (9.54) show that (9.54) is valid as $0 \equiv 0$. In particular, if $i = 0$ and $\nu \neq 0$, then (9.54) is true.

If $i = 0$ and $\nu = 0$, then (9.47), (9.53), and (4.47) show that (9.54) is valid as $1 \equiv 1$. Thus, (9.54) is true for $i = 0$ and any h, j, ν.

Suppose that $i \geq 1$, $0 \leq \nu \leq i$, and $0 \leq j \leq i-1$. Then, (9.48) yields 0 for the left member of (9.54). When $0 \leq \nu \leq i-1$, we use (4.48) with the inequalities $i - \nu \geq 1$ and $j - \nu < i - \nu$ to obtain $B_{h,i-\nu,j-\nu}(z) \equiv 0$ in the right member of (9.54). When $\nu = i$, we have $\binom{j}{\nu} = 0$. Thus, (9.54) is valid as $0 \equiv 0$.

It remains to be proven that (9.54) is true for $i \geq 1$, $0 \leq \nu \leq i$, and $j \geq i$.

Let i_0 be a nonnegative integer such that (9.54) is valid for $i = i_0$ and any integers h, j, ν that satisfy $0 \leq \nu \leq i_0$ and $j \geq i_0$. Henceforth, in this argument, let h, j, ν denote fixed integers that satisfy $0 \leq \nu \leq i_0 + 1$ and $j \geq i_0 + 1$. We use (9.48) and (9.54) for $i = i_0$ to obtain

(9.55) $$F_{h,i_0+1,j,\nu}(z) \equiv \mathcal{F}_1 + \mathcal{F}_2 + \mathcal{F}_3 + \mathcal{F}_4 + \mathcal{F}_5,$$

where

$$\mathcal{F}_1 \equiv \sum_{k=i_0}^{j-1} F_{h,i_0,k,\nu}^{(1)}(z) \equiv \frac{(-1)^\nu \mathfrak{P}_\nu}{2^\nu} \sum_{k=i_0}^{j-1} \binom{k}{\nu}\binom{2h+k-1}{\nu} B_{h,i_0-\nu,k-\nu}^{(1)}(z),$$

$$\mathcal{F}_2 \equiv \sum_{k=i_0}^{j-1} -2(\nu+1)\mathcal{D}(z)\, F_{h,i_0,k,\nu+1}(z)$$

$$\equiv -2(\nu+1)\frac{(-1)^{\nu+1}\mathfrak{P}_{\nu+1}}{2^{\nu+1}} \mathcal{D}(z) \sum_{k=i_0}^{j-1} \left[\binom{k}{\nu+1}\binom{2h+k-1}{\nu+1} \right. \\ \left. \times B_{h,i_0-\nu-1,k-\nu-1}(z) \right]$$

$$\equiv \frac{(-1)^\nu \mathfrak{P}_\nu}{2^\nu} \mathcal{D}(z) \sum_{k=i_0}^{j-1} \left[k(2h+k-1)\binom{k-1}{\nu}\binom{2h+k-2}{\nu} \right. \\ \left. \times B_{h,i_0-\nu-1,k-\nu-1}(z) \right],$$

$$\mathcal{F}_3 \equiv \sum_{k=i_0}^{j-1} \tfrac{1}{2}(\nu-1-2h-2k)\, F_{h,i_0,k,\nu-1}(z)$$

$$\equiv \frac{(-1)^\nu \mathfrak{P}_{\nu-1}}{2^\nu} \sum_{k=i_0}^{j-1} \left[(2h+2k-\nu+1)\binom{k}{\nu-1}\binom{2h+k-1}{\nu-1} \right. \\ \left. \times B_{h,i_0+1-\nu,k+1-\nu}(z) \right],$$

$$\mathcal{F}_4 \equiv \sum_{k=i_0}^{j-1} -k(2h+k-1)a_{m,n}(z)\, F_{h,i_0-1,k-1,\nu}(z)$$

$$\equiv \frac{(-1)^\nu \mathfrak{P}_\nu}{2^\nu} \sum_{k=i_0}^{j-1} \left[-k(2h+k-1)\binom{k-1}{\nu}\binom{2h+k-2}{\nu} \right. \\ \left. \times a_{m,n}(z)\, B_{h,i_0-\nu-1,k-\nu-1}(z) \right]$$

$$\equiv \frac{(-1)^\nu \mathfrak{P}_\nu}{2^\nu} \sum_{k=i_0}^{j-1} \left[-(k-\nu)(2h+k-\nu-1)\binom{k}{\nu}\binom{2h+k-1}{\nu} \right. \\ \left. \times a_{m,n}(z)\, B_{h,i_0-\nu-1,k-\nu-1}(z) \right],$$

and

$$\mathcal{F}_5 \equiv \sum_{k=i_0}^{j-1} -k(2h+k-1)\mathcal{D}(z)\, F_{h,i_0-1,k-1,\nu}(z)$$

$$\equiv (-1)\frac{(-1)^\nu \mathfrak{P}_\nu}{2^\nu}\mathcal{D}(z) \sum_{k=i_0}^{j-1} \left[k(2h+k-1)\binom{k-1}{\nu}\binom{2h+k-2}{\nu} \right. \\ \left. \times B_{h,i_0-\nu-1,k-\nu-1}(z) \right].$$

For $i_0 \le k \le j-1$, we have $i_0 - \nu \le k - \nu$ and see that (9.29) yields

$$B_{h,i_0-\nu,k-\nu}^{(1)}(z) - (k-\nu)(2h+k-\nu-1)a_{m,n}(z)\, B_{h,i_0-\nu-1,k-\nu-1}(z)$$

$$\equiv B_{h,i_0+1-\nu,k+1-\nu}(z) - B_{h,i_0+1-\nu,k-\nu}(z).$$

Consequently, we find that

$$\mathcal{F}_1 + \mathcal{F}_4 \equiv \frac{(-1)^\nu \mathfrak{P}_\nu}{2^\nu} \sum_{k=i_0}^{j-1} \binom{k}{\nu}\binom{2h+k-1}{\nu} \begin{bmatrix} B_{h,i_0+1-\nu,k+1-\nu}(z) \\ - B_{h,i_0+1-\nu,k-\nu}(z) \end{bmatrix}.$$

We apply $\mathcal{F}_5 \equiv -\mathcal{F}_2$, (9.55), and the preceding formulas to verify that

(9.56) $\quad F_{h,i_0+1,j,\nu}(z) \equiv (\mathcal{F}_1 + \mathcal{F}_4) + \mathcal{F}_3$

$$\equiv \sum_{k=i_0}^{j-1} \frac{(-1)^\nu \mathfrak{P}_\nu}{2^\nu} \binom{k}{\nu}\binom{2h+k-1}{\nu} \begin{bmatrix} B_{h,i_0+1-\nu,k+1-\nu}(z) \\ - B_{h,i_0+1-\nu,k-\nu}(z) \end{bmatrix}$$

$$+ \sum_{k=i_0}^{j-1} \frac{(-1)^\nu \mathfrak{P}_{\nu-1}}{2^\nu} \begin{bmatrix} (2h+2k-\nu+1)\binom{k}{\nu-1}\binom{2h+k-1}{\nu-1} \\ \times B_{h,i_0+1-\nu,k+1-\nu}(z) \end{bmatrix}.$$

If $\nu = 0$, then we use $B_{h,i_0+1,i_0}(z) \equiv 0$ from (4.48) to see that (9.56) reduces to

(9.57) $\quad F_{h,i_0+1,j,0}(z) \equiv \sum_{k=i_0}^{j-1} (B_{h,i_0+1,k+1}(z) - B_{h,i_0+1,k}(z)) \equiv B_{h,i_0+1,j}(z).$

Suppose that $1 \leq \nu \leq i_0 + 1$. Then we obtain

$$\binom{k}{\nu}\binom{2h+k-1}{\nu} + \frac{(2h+2k-\nu+1)}{\nu}\binom{k}{\nu-1}\binom{2h+k-1}{\nu-1} \equiv \binom{k+1}{\nu}\binom{2h+k}{\nu}$$

by setting $\kappa = k$ and $\omega = 2h$ in the identity of Lemma 9.11 and we employ it to rewrite (9.56) as

$$F_{h,i_0+1,j,\nu}(z) \equiv \frac{(-1)^\nu \mathfrak{P}_\nu}{2^\nu} \sum_{k=i_0}^{j-1} \begin{bmatrix} \binom{k+1}{\nu}\binom{2h+k}{\nu} B_{h,i_0+1-\nu,k+1-\nu}(z) \\ - \binom{k}{\nu}\binom{2h+k-1}{\nu} B_{h,i_0+1-\nu,k-\nu}(z) \end{bmatrix}$$

$$\equiv \frac{(-1)^\nu \mathfrak{P}_\nu}{2^\nu} \begin{bmatrix} \binom{j}{\nu}\binom{2h+j-1}{\nu} B_{h,i_0+1-\nu,j-\nu}(z) \\ - \binom{i_0}{\nu}\binom{2h+i_0-1}{\nu} B_{h,i_0+1-\nu,i_0-\nu}(z) \end{bmatrix}.$$

Either $1 \leq \nu \leq i_0$ and $B_{h,i_0+1-\nu,i_0-\nu}(z) \equiv 0$ or $\nu = i_0 + 1$ and $\binom{i_0}{\nu} \equiv 0$. This gives

(9.58) $\quad F_{h,i_0+1,j,\nu}(z) \equiv \frac{(-1)^\nu \mathfrak{P}_\nu}{2^\nu} \binom{j}{\nu}\binom{2h+j-1}{\nu} B_{h,i_0+1-\nu,j-\nu}(z).$

In view of (9.57) and (9.58), we conclude that (9.54) is valid for $i = i_0 + 1$, $0 \leq \nu \leq i_0 + 1$, and $j \geq i_0 + 1$. This completes the induction and the proof. $\square$

9.6. Simplification for $\mathfrak{A}_{p,q,r,s,t}(\zeta)$

PROPOSITION 9.13. *The functions $\mathfrak{A}_{p,q,r,s,t}(\zeta)$ defined on Ω^{**} by (9.11) and the functions $A_{p,q,r,s,t}(z)$ on Ω obtained by replacing each $\boldsymbol{w}_{i_1,i_2,\ldots,i_n}^{(k)}$ in $\boldsymbol{A}_{p,q,r,s,t}$ of (4.50) with the corresponding $c_{i_1,i_2,\ldots,i_n}^{(k)}(z)$ from (4.1) satisfy*

$$(9.59) \quad \mathfrak{A}_{p,q,r,s,t}(\zeta) \equiv A_{p,q,r,s,t}(f(\zeta)), \quad \text{on } \Omega^{**} \text{ for } 0 \le s \le r \text{ and } 0 \le t \le s-r.$$

PROOF. Using (9.20) of page 81 and (9.49) of page 87, we find that

$$(9.60) \quad \mathfrak{A}_{p,q,r,s,t}(g(z)) \equiv \sum_{\nu=0}^{s} \mathfrak{C}_{p,q,r,t+\nu}\, E_{p,\nu,t+\nu}(z)\, E_{q,s-\nu,r-t-\nu}(z)$$

$$\equiv \sum_{\nu=0}^{s} \mathfrak{C}_{p,q,r,t+\nu} \sum_{\mu=0}^{\nu} F_{p,\nu,t+\nu,\mu}(z) \left[\frac{g''(z)}{g'(z)}\right]^{\mu} \sum_{\lambda=0}^{s-\nu} F_{q,s-\nu,r-t-\nu,\lambda}(z) \left[\frac{g''(z)}{g'(z)}\right]^{\lambda}$$

$$\equiv \sum_{\nu=0}^{s} \sum_{\mu=0}^{\nu} \sum_{\lambda=0}^{s-\nu} \mathfrak{C}_{p,q,r,t+\nu}\, F_{p,\nu,t+\nu,\mu}(z)\, F_{q,s-\nu,r-t-\nu,\lambda}(z) \left[\frac{g''(z)}{g'(z)}\right]^{\lambda+\mu}$$

$$\equiv \sum_{\mu=0}^{s} \sum_{\nu=\mu}^{s} \sum_{\lambda=0}^{s-\nu} \mathfrak{C}_{p,q,r,t+\nu}\, F_{p,\nu,t+\nu,\mu}(z)\, F_{q,s-\nu,r-t-\nu,\lambda}(z) \left[\frac{g''(z)}{g'(z)}\right]^{\lambda+\mu}$$

$$\equiv \sum_{\mu=0}^{s} \sum_{\lambda=0}^{s-\mu} \sum_{\nu=\mu}^{s-\lambda} \mathfrak{C}_{p,q,r,t+\nu}\, F_{p,\nu,t+\nu,\mu}(z)\, F_{q,s-\nu,r-t-\nu,\lambda}(z) \left[\frac{g''(z)}{g'(z)}\right]^{\lambda+\mu}$$

$$\equiv \sum_{k=0}^{s} \sum_{\mu=0}^{k} \sum_{\nu=\mu}^{s+\mu-k} \mathfrak{C}_{p,q,r,t+\nu}\, F_{p,\nu,t+\nu,\mu}(z)\, F_{q,s-\nu,r-t-\nu,k-\mu}(z) \left[\frac{g''(z)}{g'(z)}\right]^{k}$$

$$\equiv \sum_{k=0}^{s} G_{p,q,r,s,t,k}(z) \left[\frac{g''(z)}{g'(z)}\right]^{k}, \quad \text{on } \Omega \text{ for } 0 \le s \le r \text{ and } 0 \le t \le r-s,$$

where

$$(9.61) \quad G_{p,q,r,s,t,k}(z) \equiv \sum_{\mu=0}^{k} \sum_{\nu=\mu}^{s+\mu-k} \mathfrak{C}_{p,q,r,t+\nu}\, F_{p,\nu,t+\nu,\mu}(z)\, F_{q,s-\nu,r-t-\nu,k-\mu}(z)$$

$$\equiv \sum_{\mu=0}^{k} \sum_{\nu=0}^{s-k} \mathfrak{C}_{p,q,r,t+\mu+\nu}\, F_{p,\mu+\nu,t+\mu+\nu,\mu}(z)\, F_{q,s-\mu-\nu,r-t-\mu-\nu,k-\mu}(z)$$

$$\equiv \sum_{\nu=0}^{s-k} \sum_{\mu=0}^{k} \mathfrak{C}_{p,q,r,t+\mu+\nu}\, F_{p,\mu+\nu,t+\mu+\nu,\mu}(z)\, F_{q,s-\mu-\nu,r-t-\mu-\nu,k-\mu}(z).$$

Since (9.54) yields

$$F_{p,\mu+\nu,t+\mu+\nu,\mu}(z) \equiv \frac{(-1)^{\mu}\mu!}{2^{\mu}} \binom{t+\mu+\nu}{\mu}\binom{2p+t+\mu+\nu-1}{\mu} B_{p,\nu,t+\nu}(z)$$

and
$$F_{q,s-\mu-\nu,r-t-\mu-\nu,k-\mu}(z) \equiv \frac{(-1)^{k-\mu}(k-\mu)!}{2^{k-\mu}} \left[\begin{array}{c} \binom{r-t-\mu-\nu}{k-\mu} \\ \times \binom{2q+r-t-\mu-\nu-1}{k-\mu} \\ \times B_{q,s-\nu-k,r-t-\nu-k}(z) \end{array} \right],$$

we introduce the rational numbers $H_{p,q,r,t,k,\nu}$ defined by

(9.62) $H_{p,q,r,t,k,\nu}$
$$\equiv \sum_{\mu=0}^{k} \frac{(-1)^k \mu!\,(k-\mu)!)}{2^k} \left[\begin{array}{c} \mathfrak{C}_{p,q,r,t+\mu+\nu}\binom{t+\mu+\nu}{\mu} \\ \times \binom{2p+t+\mu+\nu-1}{\mu}\binom{r-t-\mu-\nu}{k-\mu} \\ \times \binom{2q+r-t-\mu-\nu-1}{k-\mu} \end{array} \right]$$

in order to rewrite (9.61) as

(9.63) $\qquad G_{p,q,r,s,t,k}(z) \equiv \sum_{\nu=0}^{s-k} H_{p,q,r,t,k,\nu}\, B_{p,\nu,t+\nu}(z)\, B_{q,s-\nu-k,r-t-\nu-k}(z).$

For p, q, $r \geq 2$ and t, μ, $\nu \geq 0$, we obtain

$$\mathfrak{C}_{p,q,r,t+\mu+\nu} \equiv (-1)^{t+\mu+\nu} \frac{\binom{r}{t+\mu+\nu}\binom{2q+r-1}{t+\mu+\nu}}{\binom{2p+t+\mu+\nu-1}{t+\mu+\nu}}$$

from (4.49) of page 36 and we check that

$$\binom{r}{t+\mu+\nu}\binom{r-t-\mu-\nu}{k-\mu} \equiv \binom{r}{t+k+\nu}\binom{t+k+\nu}{k-\mu},$$

$$\binom{2q+r-1}{t+\mu+\nu}\binom{2q+r-t-\mu-\nu-1}{k-\mu} \equiv \binom{2q+r-1}{t+k+\nu}\binom{t+k+\nu}{k-\mu},$$

and

$$\frac{\binom{2p+t+\mu+\nu-1}{\mu}}{\binom{2p+t+\mu+\nu-1}{t+\mu+\nu}} \equiv \frac{\binom{t+\mu+\nu}{\mu}}{\binom{2p+t+\nu-1}{t+\nu}}$$

are valid identities. Since the expression $L_{p,q,r,t,k,\nu}$ defined by

$$L_{p,q,r,t,k,\nu} \equiv \frac{(-1)^{t+k+\nu} k!}{2^k} \frac{\binom{r}{t+k+\nu}\binom{2q+r-1}{t+k+\nu}}{\binom{2p+t+\nu-1}{t+\nu}}$$

is independent of μ, we see that (9.62) yields

$$(9.64) \qquad H_{p,q,r,t,k,\nu} \equiv L_{p,q,r,t,k,\nu} \sum_{\mu=0}^{k} \frac{(-1)^{\mu}}{\binom{k}{\mu}} \left[\binom{t+k+\nu}{k-\mu} \binom{t+\mu+\nu}{\mu} \right]^{2}.$$

In view of the identity

$$\binom{t+k+\nu}{k-\mu} \binom{t+\mu+\nu}{\mu} \equiv \binom{t+k+\nu}{k} \binom{k}{\mu},$$

we rewrite (9.64) as

$$(9.65) \qquad H_{p,q,r,t,k,\nu} \equiv \left[\binom{t+k+\nu}{k} \right]^{2} L_{p,q,r,t,k,\nu} \sum_{\mu=0}^{k} (-1)^{\mu} \binom{k}{\mu}.$$

We have

$$\sum_{\mu=0}^{k} (-1)^{\mu} \binom{k}{\mu} \equiv (1+(-1))^{k} \equiv 0, \quad \text{for } 1 \leq k \leq s.$$

Thus, (9.65) gives

$$(9.66) \qquad H_{p,q,r,t,k,\nu} \equiv 0, \quad \text{for } 1 \leq k \leq s.$$

We use (9.63) and (9.66) to deduce

$$(9.67) \qquad G_{p,q,r,s,t,k}(z) \equiv 0, \quad \text{for } 1 \leq k \leq s.$$

Moreover, for $k = 0$, (9.62) gives

$$(9.68) \qquad H_{p,q,r,t,0,\nu} \equiv \mathfrak{C}_{p,q,r,t+\nu}.$$

As a consequence of (9.60), (9.67), (9.63), (9.68), and (4.50), we obtain

$$\mathfrak{A}_{p,q,r,s,t}(g(z)) \equiv \sum_{k=0}^{s} G_{p,q,r,s,t,k}(z) \left[\frac{g''(z)}{g'(z)} \right]^{k},$$

$$\equiv G_{p,q,r,s,t,0}(z)$$

$$\equiv \sum_{\nu=0}^{s} H_{p,q,r,t,0,\nu}\, B_{p,\nu,t+\nu}(z)\, B_{q,s-\nu,r-t-\nu}(z)$$

$$\equiv \sum_{\nu=0}^{s} \mathfrak{C}_{p,q,r,t+\nu}\, B_{p,\nu,t+\nu}(z)\, B_{q,s-\nu,r-t-\nu}(z)$$

$$\equiv A_{p,q,r,s,t}(z), \quad \text{on } \Omega.$$

With ζ in Ω^{**}, $z = f(\zeta)$, and $g(f(\zeta)) = \zeta$, this gives

$$\mathfrak{A}_{p,q,r,s,t}(\zeta) \equiv A_{p,q,r,s,t}(f(\zeta)), \quad \text{on } \Omega^{**}.$$

Hence, (9.59) is valid. This completes the proof of Proposition 9.13. $\square$

9.7. $C_{p,q,r}(P, Q)$ is a relative invariant when nonzero

We continue to write R as the abbreviation in (9.1) for $C_{p,q,r}(P, Q)$ of (4.44).

THEOREM 9.14. *The function $R(z)$ on Ω obtained by replacing $w^{(k)}_{i_1, i_2, \ldots, i_n}$ in R with $c^{(k)}_{i_1, i_2, \ldots, i_n}(z)$ from (4.1) and the function $R^{**}(\zeta)$ on Ω^{**} obtained by replacing $w^{(k)}_{i_1, i_2, \ldots, i_n}$ in R with $c^{**(k)}_{i_1, i_2, \ldots, i_n}(\zeta)$ from (4.7) are related by*

$$(9.69) \qquad R^{**}(\zeta) \equiv \left(f'(\zeta)\right)^{p+q+r} R(f(\zeta)), \quad \text{on } \Omega^{**}.$$

Moreover, R is a relative invariant in $\mathcal{R}_{m,n}$ for $\mathcal{C}_{m,n}$ of weight $p+q+r$ if and only if it is nonzero.

PROOF. Since (9.1) and (4.44) yield

$$R(z) = \sum_{s=0}^{r} \sum_{t=0}^{r-s} A_{p,q,r,s,t}(z) \, P^{(t)}(z) \, Q^{(r-s-t)}(z), \quad \text{on } \Omega,$$

we employ (9.12) of page 79 and (9.59) to deduce

$$R^{**}(\zeta) \equiv \left(f'(\zeta)\right)^{p+q+r} \sum_{s=0}^{r} \sum_{t=0}^{r-s} \mathfrak{A}_{p,q,r,s,t}(\zeta) \, P^{(t)}(f(\zeta)) \, Q^{(r-s-t)}(f(\zeta))$$

$$\equiv \left(f'(\zeta)\right)^{p+q+r} \sum_{s=0}^{r} \sum_{t=0}^{r-s} A_{p,q,r,s,t}(f(\zeta)) \, P^{(t)}(f(\zeta)) \, Q^{(r-s-t)}(f(\zeta))$$

$$\equiv \left(f'(\zeta)\right)^{p+q+r} R(f(\zeta)), \quad \text{on } \Omega^{**}.$$

Thus, (9.69) is valid.

Because R in (9.1) satisfies (9.2) and (9.3), R is a relative invariant in $\mathcal{R}_{m,n}$ for $\mathcal{C}_{m,n}$ of weight $p+q+r$ if and only if it is not a constant. Since R can not be a nonzero constant, this completes the proof. $\square$

CHAPTER 10

Conditions for $C_{p,q,r}(P, Q) \not\equiv 0$ when $m, r \geq 2$

For $m, r \geq 2$ and $(m, n) \neq (2, 1)$, let P and Q denote relative invariants of respective weights p and q in $\mathcal{R}_{m,n}$ for $\mathcal{C}_{m,n}$. We shall verify in Theorem 10.11 on page 103 that $C_{p,q,r}(P, Q)$ satisfies the condition $C_{p,q,r}(P, Q) \not\equiv 0$ if and only if either r is an even integer or P and Q are linearly independent over $\mathbb{Q}$. In those situations, $C_{p,q,r}(P, Q)$ is a relative invariant for $\mathcal{C}_{m,n}$ of weight $p + q + r$.

10.1. The dependence of $C_{q,p,r}(Q, P)$ on $C_{p,q,r}(P, Q)$

The following result is needed to establish dependence in Proposition 10.2.

PROPOSITION 10.1. *The $A_{p,q,r,s,t}$ defined in $\mathcal{R}_{m,n}$ by (4.50) of page 36 satisfy*

$$(10.1) \quad A_{q,p,r,s,r-s-t} \equiv (-1)^r \frac{\binom{2p+r-1}{r}}{\binom{2q+r-1}{r}} A_{p,q,r,s,t},$$

for $0 \leq s \leq r$, and $0 \leq t \leq r - s$.

PROOF. We use (4.50) to obtain

$$(10.2) \quad A_{q,p,r,s,r-s-t} \equiv \sum_{j=0}^{s} \mathfrak{C}_{q,p,r,r-s-t+j} B_{q,j,r-s-t+j} B_{p,s-j,s+t-j}$$

$$\equiv \sum_{k=0}^{s} \mathfrak{C}_{q,p,r,r-t-k} B_{p,k,t+k} B_{q,s-k,r-t-k}.$$

For $0 \leq s \leq r$, $0 \leq t \leq r - s$, and $0 \leq k \leq s \leq r - t$, we find that (4.49) yields

$$\frac{\mathfrak{C}_{q,p,r,r-t-k}}{\mathfrak{C}_{p,q,r,t+k}} \equiv \frac{(-1)^{r-t-k}\binom{r}{r-t-k}\binom{2p+r-1}{r-t-k}\binom{2p+t+k-1}{t+k}}{(-1)^{t+k}\binom{r}{t+k}\binom{2q+r-1}{t+k}\binom{2q+r-t-k-1}{r-t-k}}$$

from which we deduce

$$(10.3) \quad \frac{\mathfrak{C}_{q,p,r,r-t-k}}{\mathfrak{C}_{p,q,r,t+k}} \equiv (-1)^r \frac{\binom{2p+r-1}{r}}{\binom{2q+r-1}{r}}.$$

By rewriting (10.3), combining it with (10.2), and using (4.50), we find that

$$\boldsymbol{A}_{q,p,r,s,r-s-t} \equiv (-1)^r \frac{\binom{2p+r-1}{r}}{\binom{2q+r-1}{r}} \sum_{k=0}^{s} \mathfrak{C}_{p,q,r,t+k}\, \boldsymbol{B}_{p,k,t+k}\, \boldsymbol{B}_{q,s-k,r-t-k}$$

$$\equiv (-1)^r \frac{\binom{2p+r-1}{r}}{\binom{2q+r-1}{r}} \boldsymbol{A}_{p,q,r,s,t}.$$

This establishes (10.1) and completes the proof. □

PROPOSITION 10.2. *The polynomials $\boldsymbol{C}_{p,q,r}(\boldsymbol{P}, \boldsymbol{Q})$ and $\boldsymbol{C}_{q,p,r}(\boldsymbol{Q}, \boldsymbol{P})$ defined in $\mathcal{R}_{m,n}$ by (4.44) of page 36 are related by*

(10.4) $$\boldsymbol{C}_{q,p,r}(\boldsymbol{Q}, \boldsymbol{P}) \equiv (-1)^r \frac{\binom{2p+r-1}{r}}{\binom{2q+r-1}{r}} \boldsymbol{C}_{p,q,r}(\boldsymbol{P}, \boldsymbol{Q}).$$

PROOF. We use (4.44) and (10.1) to obtain

$$\boldsymbol{C}_{q,p,r}(\boldsymbol{Q}, \boldsymbol{P}) \equiv \sum_{s=0}^{r} \sum_{\tau=0}^{r-s} \boldsymbol{A}_{q,p,r,s,\tau}\, \boldsymbol{Q}^{(\tau)}\, \boldsymbol{P}^{(r-s-\tau)}$$

$$\equiv \sum_{s=0}^{r} \sum_{t=0}^{r-s} \boldsymbol{A}_{q,p,r,s,r-s-t}\, \boldsymbol{P}^{(t)}\, \boldsymbol{Q}^{(r-s-t)}$$

$$\equiv (-1)^r \frac{\binom{2p+r-1}{r}}{\binom{2q+r-1}{r}} \sum_{s=0}^{r} \sum_{t=0}^{r-s} \boldsymbol{A}_{p,q,r,s,t}\, \boldsymbol{P}^{(t)}\, \boldsymbol{Q}^{(r-s-t)}$$

$$\equiv (-1)^r \frac{\binom{2p+r-1}{r}}{\binom{2q+r-1}{r}} \boldsymbol{C}_{p,q,r}(\boldsymbol{P}, \boldsymbol{Q}).$$

This yields (10.4) and completes the proof. □

We shall apply the following result in the proof for Theorem 10.11.

COROLLARY 10.3. *If r is an odd integer, then $\boldsymbol{C}_{p,p,r}(\boldsymbol{P}, \boldsymbol{P}) \equiv 0$. Also, if r is odd and $\{\boldsymbol{P}, \boldsymbol{Q}\}$ is linearly dependent over $\mathbb{Q}$, then $\boldsymbol{C}_{p,q,r}(\boldsymbol{P}, \boldsymbol{Q}) \equiv 0$.*

PROOF. When r is an odd integer, we use (10.4) to obtain

$$\boldsymbol{C}_{p,p,r}(\boldsymbol{P}, \boldsymbol{P}) \equiv -\boldsymbol{C}_{p,p,r}(\boldsymbol{P}, \boldsymbol{P})$$

and thereby conclude that $\boldsymbol{C}_{p,p,r}(\boldsymbol{P}, \boldsymbol{P}) \equiv 0$. When r is odd and $\{\boldsymbol{P}, \boldsymbol{Q}\}$ is linearly dependent over $\mathbb{Q}$, we have: $\boldsymbol{Q} \equiv \gamma_0 \boldsymbol{P}$, for some γ_0 in $\mathbb{Q}$; $q = p$; and $\boldsymbol{C}_{p,q,r}(\boldsymbol{P}, \boldsymbol{Q}) \equiv \gamma_0\, \boldsymbol{C}_{p,p,r}(\boldsymbol{P}, \boldsymbol{P}) \equiv 0$. This completes the proof. □

10.2. Several lemmas for use in Section 10.3

In this section, the symbols $\boldsymbol{B}_{h,i,j}$ from (4.46)–(4.48) and $\boldsymbol{A}_{p,q,r,s,t}$ from (4.50) are to be regarded as differential-polynomial combinations over $\mathbb{Q}$ of $\boldsymbol{a}_{m,n}$. We first use (4.46)–(4.48) to observe that: for $h \geq 1$, $r \geq 1$, $k \geq 0$, as well as for any nonzero rational number γ and integer j,

(10.5) $\qquad\qquad \gamma(\boldsymbol{a}_{m,n})^r$ is not a term of $\boldsymbol{B}_{h,2k+1,j}$.

LEMMA 10.4. *For each $h \geq 1$, $i \geq 1$, and $j \geq 2i$, there is a corresponding positive integer $\gamma_{h,i,j}$ such that*

(10.6) $\qquad \varGamma_{h,i,j} \equiv (-1)^i \gamma_{h,i,j} \quad$ *is the coefficient of $(\boldsymbol{a}_{m,n})^i$ in $\boldsymbol{B}_{h,2i,j}$.*

PROOF. For $h \geq 1$, $i = 1$, and $j \geq 2$, we use (4.46)–(4.48) to verify that the coefficient of $\boldsymbol{a}_{m,n}$ in $\boldsymbol{B}_{h,2,j}$ is given by $\varGamma_{h,1,j} \equiv -\gamma_{h,1,j}$ for the positive integer

$$\gamma_{h,1,j} \equiv \sum_{k=1}^{j-1} k(2h+k-1).$$

Let i_0 be an integer subject to $i_0 \geq 1$ such that: for each $j \geq 2i_0$, there is a positive integer $\gamma_{h,i_0,j}$ for which (10.6) is valid when $i = i_0$ and $j \geq 2i_0$. We apply this with (4.48) to see that: for $j \geq 2(i_0+1)$, the coefficient of $(\boldsymbol{a}_{m,n})^{i_0+1}$ in $\boldsymbol{B}_{h,2(i_0+1),j}$ is

$$\varGamma_{h,i_0+1,j} \equiv \sum_{k=2i_0+1}^{j-1} (-k)(2h+k-1)\varGamma_{h,i_0,k-1}$$

$$\equiv \sum_{k=2i_0+1}^{j-1} (-k)(2h+k-1)(-1)^{i_0}\gamma_{h,i_0,k-1} \equiv (-1)^{i_0+1}\gamma_{h,i_0+1,j},$$

where $\gamma_{h,i_0+1,j}$ is the positive integer defined through

$$\gamma_{h,i_0+1,j} \equiv \sum_{k=2i_0+1}^{j-1} k(2h+k-1)\gamma_{h,i_0,k-1}, \quad \text{for } j \geq 2(i_0+1).$$

Thus, a positive integer $\gamma_{h,i,j}$ exists for (10.6) whenever $h \geq 1$, $i \geq 1$, and $j \geq 2i$. This completes the proof. $\square$

LEMMA 10.5. *For p, q, $i \geq 1$, there is a positive rational number $\alpha_{p,q,i}$ such that the coefficient $\mathcal{A}_{p,q,i}$ of $(\boldsymbol{a}_{m,n})^i$ in $\boldsymbol{A}_{p,q,2i,2i,0}$ is given by*

(10.7) $\qquad\qquad \mathcal{A}_{p,q,i} \equiv (-1)^i \alpha_{p,q,i}.$

PROOF. We employ (4.50) to obtain

(10.8) $\qquad \boldsymbol{A}_{p,q,2i,2i,0} \equiv \sum_{k=0}^{2i} \mathfrak{C}_{p,q,2i,k}\, \boldsymbol{B}_{p,k,k}\, \boldsymbol{B}_{q,2i-k,2i-k}.$

Using (10.8) and (10.5), we see that the coefficient $\mathcal{A}_{p,q,i}$ of $(\boldsymbol{a}_{m,n})^i$ in $\boldsymbol{A}_{p,q,2i,2i,0}$ is equal to the coefficient of $(\boldsymbol{a}_{m,n})^i$ in the expression

$$\sum_{\nu=0}^{i} \mathfrak{C}_{p,q,2i,2\nu}\, \boldsymbol{B}_{p,2\nu,2\nu}\, \boldsymbol{B}_{q,2i-2\nu,2i-2\nu}.$$

We apply Lemma 10.4 and the preceding observation to verify that

$$
(10.9) \qquad \mathcal{A}_{p,q,i} \equiv \sum_{\nu=0}^{i} \mathfrak{C}_{p,q,2i,2\nu}\, \Gamma_{p,\nu,2\nu}\, \Gamma_{q,i-\nu,2i-2\nu}
$$

$$
\equiv \sum_{\nu=0}^{i} \mathfrak{C}_{p,q,2i,2\nu}\, (-1)^{\nu}\gamma_{p,\nu,2\nu}\, (-1)^{i-\nu}\gamma_{q,i-\nu,2i-2\nu}
$$

$$
\equiv (-1)^{i}\alpha_{p,q,i},
$$

where

$$
(10.10) \qquad \alpha_{p,q,i} \equiv \sum_{\nu=0}^{i} \mathfrak{C}_{p,q,2i,2\nu}\, \gamma_{p,\nu,2\nu}\, \gamma_{q,i-\nu,2i-2\nu}.
$$

Since (4.49) of page 36 and Lemma 10.4 show that the three factors in each summand of (10.10) are positive rational numbers, $\alpha_{p,q,i}$ is a positive rational number for (10.9) and (10.7). This completes the proof. □

LEMMA 10.6. *For p, q, $i \geq 1$, there is a positive rational number $\beta_{p,q,i}$ such that the coefficient $\mathcal{B}_{p,q,i}$ of $(\boldsymbol{a}_{m,n})^{i}$ in $\boldsymbol{A}_{p,q,2i+1,2i,0}$ is given by*

$$
(10.11) \qquad \mathcal{B}_{p,q,i,} \equiv (-1)^{i}\beta_{p,q,i}.
$$

PROOF. We note that (4.50) yields

$$
(10.12) \qquad \boldsymbol{A}_{p,q,2i+1,2i,0} \equiv \sum_{k=0}^{2i} \mathfrak{C}_{p,q,2i+1,k}\, \boldsymbol{B}_{p,k,k}\, \boldsymbol{B}_{q,2i-k,2i+1-k}.
$$

In view of (10.12) and (10.5), the coefficient $\mathcal{B}_{p,q,i}$ of $(\boldsymbol{a}_{m,n})^{i}$ in $\boldsymbol{A}_{p,q,2i+1,2i,0}$ is equal to the coefficient of $(\boldsymbol{a}_{m,n})^{i}$ in the expression

$$
\sum_{\nu=0}^{i} \mathfrak{C}_{p,q,2i+1,2\nu}\, \boldsymbol{B}_{p,2\nu,2\nu}\, \boldsymbol{B}_{q,2i-2\nu,2i+1-2\nu}.
$$

We apply Lemma 10.4 and the preceding observation to obtain

$$
(10.13) \qquad \mathcal{B}_{p,q,i} \equiv \sum_{\nu=0}^{i} \mathfrak{C}_{p,q,2i+1,2\nu}\, \Gamma_{p,\nu,2\nu}\, \Gamma_{q,i-l,2i+1-2\nu}
$$

$$
\equiv \sum_{\nu=0}^{i} \mathfrak{C}_{p,q,2i+1,2\nu}\, (-1)^{\nu}\gamma_{p,\nu,2\nu}\, (-1)^{i-\nu}\gamma_{q,i-\nu,2i+1-2\nu}
$$

$$
\equiv (-1)^{i}\beta_{p,q,i},
$$

where

$$
(10.14) \qquad \beta_{p,q,i} \equiv \sum_{\nu=0}^{i} \mathfrak{C}_{p,q,2i+1,2\nu}\, \gamma_{p,\nu,2\nu}\, \gamma_{q,i-\nu,2i+1-2\nu}.
$$

Since (4.49) of page 36 and Lemma 10.4 show that the three factors in each summand of (10.14) are positive rational numbers, $\beta_{p,q,i}$ is a positive rational number for (10.13) and (10.11). This completes the proof. □

10.3. The situations where $C_{p,q,r}(P, Q) \not\equiv 0$

The proofs of Propositions 10.8–10.10 employ the following terminology.

DEFINITION 10.7. For any nonzero monomial M of $\mathcal{R}_{m,n}$ uniquely specified by

$$M \equiv \gamma \prod_{\lambda=1}^{v} w_{i(1,\lambda),\, i(2,\lambda),\, \ldots,\, i(n,\lambda)}^{(k_\lambda)}$$

in terms of a nonzero rational number γ, an integer $v \geq 0$, and variables from (4.9), the *order-sum* $\mathcal{O}(M)$ *of* M is the nonnegative integer $\mathcal{O}(M) = \sum_{\lambda=1}^{v} k_\lambda$.

PROPOSITION 10.8. *For $m \geq 2$, $(m, n) \neq (2, 1)$, and an even integer $r \geq 2$, let P and Q be relative invariants of respective weights p and q in $\mathcal{R}_{m,n}$ for $\mathcal{C}_{m,n}$. Then, (4.44) of page 36 satisfies $C_{p,q,r}(P, Q) \not\equiv 0$.*

PROOF. We use (4.44) to obtain

$$(10.15) \qquad C_{p,q,r}(P, Q) \equiv \sum_{s=0}^{r-1} \sum_{t=0}^{r-s} A_{p,q,r,s,t}\, P^{(t)} Q^{(r-s-t)} + A_{p,q,r,0}\, PQ.$$

Writing $r = 2i_0$ with $i_0 \geq 1$, we apply Lemma 10.5 and (4.45) of page 36 to see that $A_{p,q,r,0}$ has a nonzero term of the form $\gamma_0 (w_{0,0,\ldots,0,2})^{i_0}$, for some nonzero rational number γ_0. Thus, $A_{p,q,r,0}$ has a nonzero term whose order-sum is 0.

Let p_0 denote the least order-sum for the nonzero terms of P; let q_0 denote the least order-sum for the nonzero terms of Q; and let X_0 denote a nonzero term of $A_{p,q,r,0} PQ$ whose order-sum is $\mathcal{O}(X_0) = p_0 + q_0$. For $0 \leq s \leq r-1$ and $0 \leq t \leq r-s$, we observe that:

for any nonzero term M_1 of $P^{(t)}$, $\mathcal{O}(M_1) \geq o_1 = p_0 + t$;

for any nonzero term M_2 of $Q^{(r-s-t)}$, $\mathcal{O}(M_2) \geq o_2 = q_0 + r - s - t$, and

for any nonzero term M_3 of $A_{p,q,r,s,t}\, P^{(t)} Q^{(r-s-t)}$, $\mathcal{O}(M_3) \geq o_1 + o_2$.

In view of $o_1 + o_2 = p_0 + q_0 + r - s \geq p_0 + q_0 + 1 > p_0 + q_0 = \mathcal{O}(X_0)$ and (10.15), we conclude that X_0 is a nonzero term of $C_{p,p,r}(P, Q)$. This yields $C_{p,p,r}(P, Q) \not\equiv 0$ and completes the proof. $\square$

PROPOSITION 10.9. *For $m \geq 2$, $(m, n) \neq (2, 1)$, and an odd integer $r \geq 3$, let P and Q be relative invariants of respective weights p and q in $\mathcal{R}_{m,n}$ for $\mathcal{C}_{m,n}$ such that $p = q$ and $\{P, Q\}$ is linearly independent over $\mathbb{Q}$. Then, (4.44) of page 36 satisfies $C_{p,q,r}(P, Q) \not\equiv 0$.*

PROOF. We use (10.1) of page 97 with $(-1)^r = -1$ to deduce

$$(10.16) \qquad A_{p,p,r,0} \equiv -A_{p,p,r,0}$$

and

$$(10.17) \qquad A_{p,p,r,r-1,1} \equiv -A_{p,p,r,r-1,0}.$$

By combining (4.44) with (10.17) and $\boldsymbol{A}_{p,p,r,r,0} \equiv 0$ from (10.16), we obtain

$$(10.18) \qquad \boldsymbol{C}_{p,p,r}(\boldsymbol{P}, \boldsymbol{Q}) \equiv \sum_{s=0}^{r-2} \sum_{t=0}^{r-s} \boldsymbol{A}_{p,p,r,s,t} \boldsymbol{P}^{(t)} \boldsymbol{Q}^{(r-s-t)}$$
$$+ \boldsymbol{A}_{p,p,r,r-1,0}(\boldsymbol{P}\boldsymbol{Q}^{(1)} - \boldsymbol{P}^{(1)}\boldsymbol{Q}).$$

For $r = 2i_0 + 1$ with $i_0 \geq 1$, we apply Lemma 10.6 as well as (4.45) of page 36 to see that $\boldsymbol{A}_{p,p,r,r-1,0}$ has a nonzero term of the form $\gamma_0 (\boldsymbol{w}_{0,0,\ldots,0,2})^{i_0}$, where γ_0 is a nonzero rational number. Thus, $\boldsymbol{A}_{p,p,r,r-1,0}$ has at least one nonzero term whose order-sum is 0.

Since $\boldsymbol{P}$ and $\boldsymbol{Q}$ are linearly independent over $\mathbb{Q}$, the polynomial defined in $\mathcal{R}_{m,n}$ by $\boldsymbol{Y} \equiv \boldsymbol{P}\boldsymbol{Q}^{(1)} - \boldsymbol{P}^{(1)}\boldsymbol{Q}$ satisfies $\boldsymbol{Y} \not\equiv 0$. Let p_0 denote the least order-sum for the nonzero terms of $\boldsymbol{P}$; let q_0 denote the least order-sum for the nonzero terms of $\boldsymbol{Q}$; and let $\boldsymbol{Y}_0$ be a nonzero term of $\boldsymbol{A}_{p,p,r,r-1,0}(\boldsymbol{P}\boldsymbol{Q}^{(1)} - \boldsymbol{P}^{(1)}\boldsymbol{Q})$ whose order-sum is $\mathcal{O}(\boldsymbol{Y}_0) = p_0 + q_0 + 1$. For $0 \leq s \leq r - 2$ and $0 \leq t \leq r - s$, we observe that:

for any nonzero term $\boldsymbol{M}_1$ of $\boldsymbol{P}^{(t)}$, $\mathcal{O}(\boldsymbol{M}_1) \geq o_1 = p_0 + t$;

for any nonzero term $\boldsymbol{M}_2$ of $\boldsymbol{Q}^{(r-s-t)}$, $\mathcal{O}(\boldsymbol{M}_2) \geq o_2 = q_0 + r - s - t$, and

for any nonzero term $\boldsymbol{M}_3$ of $\boldsymbol{A}_{p,p,r,s,t} \boldsymbol{P}^{(t)} \boldsymbol{Q}^{(r-s-t)}$, $\mathcal{O}(\boldsymbol{M}_3) \geq o_1 + o_2$.

In view of $o_1 + o_2 = p_0 + q_0 + r - s \geq p_0 + q_0 + 2 > p_0 + q_0 + 1 = \mathcal{O}(\boldsymbol{Y}_0)$ and (10.18), we conclude that $\boldsymbol{Y}_0$ is a nonzero term of $\boldsymbol{C}_{p,p,r}(\boldsymbol{P}, \boldsymbol{Q})$. This yields $\boldsymbol{C}_{p,p,r}(\boldsymbol{P}, \boldsymbol{Q}) \not\equiv 0$ and completes the proof. □

PROPOSITION 10.10. *For $m \geq 2$, $(m, n) \neq (2, 1)$, and an odd integer $r \geq 3$, let $\boldsymbol{P}$ and $\boldsymbol{Q}$ be relative invariants of respective weights p and q in $\mathcal{R}_{m,n}$ for $\mathcal{C}_{m,n}$ such that $p \neq q$. Then, (4.44) of page 36 satisfies $\boldsymbol{C}_{p,q,r}(\boldsymbol{P}, \boldsymbol{Q}) \not\equiv 0$.*

PROOF. Let i_0 denote the positive integer such that $r = 2i_0 + 1$ and let γ be the coefficient of $(\boldsymbol{a}_{m,n})^{i_0-1}\boldsymbol{a}_{m,n}^{(1)}$ in $\boldsymbol{A}_{p,q,2i_0+1,2i_0+1,0}$. After using (4.46)–(4.50) to conclude that γ is a nonzero rational number, we employ (4.45) to see that there is a nonzero rational number γ_0 such that a nonzero term of $\boldsymbol{A}_{p,q,r,r,0}$ is given by $\boldsymbol{Z}_1 \equiv \gamma_0 (\boldsymbol{w}_{0,0,\ldots,0,2})^{i_0-1} \boldsymbol{w}_{0,0,\ldots,0,2}^{(1)}$. We note that (4.44) yields

$$(10.19) \qquad \boldsymbol{C}_{p,q,r}(\boldsymbol{P}, \boldsymbol{Q}) \equiv \sum_{s=0}^{r-2} \sum_{t=0}^{r-s} \boldsymbol{A}_{p,q,r,s,t} \boldsymbol{P}^{(t)} \boldsymbol{Q}^{(r-s-t)}$$
$$+ \boldsymbol{A}_{p,q,r,2i_0,0} \boldsymbol{P}\boldsymbol{Q}^{(1)} + \boldsymbol{A}_{p,q,r,2i_0,1} \boldsymbol{P}^{(1)}\boldsymbol{Q} + \boldsymbol{A}_{p,q,r,r,0} \boldsymbol{P}\boldsymbol{Q}.$$

Let p_0 denote the least order-sum for the nonzero terms of $\boldsymbol{P}$; let q_0 denote the least order-sum for the nonzero terms of $\boldsymbol{Q}$; and let $\boldsymbol{Z}_2$ denote a nonzero term of $\boldsymbol{P}\boldsymbol{Q}$ not involving either $\boldsymbol{w}_{0,0,\ldots,0,1}$ or $\boldsymbol{w}_{0,0,\ldots,0,2}$ as a factor and having order-sum $p_0 + q_0$. We set $\boldsymbol{Z}_0 \equiv \boldsymbol{Z}_1 \boldsymbol{Z}_2$. Thus, $\boldsymbol{Z}_0$ is a nonzero term of $\boldsymbol{A}_{p,q,r,r,0} \boldsymbol{P}\boldsymbol{Q}$ whose order-sum is $\mathcal{O}(\boldsymbol{Z}_0) = p_0 + q_0 + 1$.

When $0 \leq s \leq r - 2$ and $0 \leq t \leq r - s$, the order-sum of any nonzero term $\boldsymbol{T}$ of $\boldsymbol{A}_{p,q,r,s,t} \boldsymbol{P}^{(t)} \boldsymbol{Q}^{(r-s-t)}$ satisfies

$$(10.20) \qquad \mathcal{O}(\boldsymbol{T}) \geq 0 + (p_0 + t) + (q_0 + r - s - t) \geq p_0 + q_0 + 2.$$

Any nonzero term of $\boldsymbol{A}_{p,q,r,2i_0,0}$ or $\boldsymbol{A}_{p,q,r,2i_0,1}$ having order-sum 0 is expressible as the product of a nonzero rational number and

$$(10.21) \quad \left((\boldsymbol{w}_{0,0,\ldots,0,1})^2\right)^k \left((\boldsymbol{w}_{0,0,\ldots,0,2})\right)^{i_0-k}, \quad \text{for some } k \text{ satisfying } 0 \le k \le i_0.$$

Consequently, any nonzero term of $\boldsymbol{A}_{p,q,r,2i_0,0}\boldsymbol{P}\boldsymbol{Q}^{(1)}$ or $\boldsymbol{A}_{p,q,r,2i_0,1}\boldsymbol{P}^{(1)}\boldsymbol{Q}$ whose order-sum is p_0+q_0+1 must be divisible by a monomial of the type (10.21). Since $\boldsymbol{Z}_0$ is not divisible by any monomial of the type (10.21), we apply (10.19), (10.20), and the preceding observations to see that $\boldsymbol{Z}_0$ is a nonzero term of $\boldsymbol{C}_{p,q,r}(\boldsymbol{P},\boldsymbol{Q})$. This completes the proof. □

THEOREM 10.11. *For $m, r \ge 2$ and $(m, n) \ne (2, 1)$, suppose that $\boldsymbol{P}$ and $\boldsymbol{Q}$ are relative invariants of respective weights p and q in $\mathcal{R}_{m,n}$ for $\mathcal{C}_{m,n}$. Then, the polynomial $\boldsymbol{C}_{p,q,r}(\boldsymbol{P},\boldsymbol{Q})$ of (4.44) on page 36 satisfies*

$$(10.22) \qquad\qquad \boldsymbol{C}_{p,q,r}(\boldsymbol{P},\boldsymbol{Q}) \not\equiv 0$$

if and only if r is an even positive integer or $\{\boldsymbol{P},\boldsymbol{Q}\}$ is linearly independent over $\mathbb{Q}$.

PROOF. For the situation where r is an even positive integer, Proposition 10.8 establishes (10.22). When r is odd and $\{\boldsymbol{P},\boldsymbol{Q}\}$ is linearly independent over $\mathbb{Q}$ with $p=q$, (10.22) is a consequence of Proposition 10.9. If r is odd and $\{\boldsymbol{P},\boldsymbol{Q}\}$ is linearly independent over $\mathbb{Q}$ with $p \ne q$, then Proposition 10.10 yields (10.22). The remaining situation is that where r is odd and $\{\boldsymbol{P},\boldsymbol{Q}\}$ is linearly dependent over $\mathbb{Q}$; for it, we use Corollary 10.3 of page 98 to obtain $\boldsymbol{C}_{p,q,r}(\boldsymbol{P},\boldsymbol{Q}) \equiv 0$ and therefore see that (10.22) is not satisfied. Hence, we conclude that (10.22) is satisfied if and only if either r is even or $\{\boldsymbol{P},\boldsymbol{Q}\}$ is linearly independent over $\mathbb{Q}$. This completes the proof. □

Part 3

Independent Verification for Theorem 4.10

CHAPTER 11

Symmetry with Respect to Semi-Invariants

The basic relative invariants were developed symmetrically in [19, 20] with respect to semi-invariants of the first and second kinds. In particular, the definition in [20] of $\mathcal{I}_{m,n;\,e_1,\ldots,e_n}$ by formulas repeated as (4.17)–(4.29) was employed to establish in [20] that $\mathcal{I}_{m,n;\,e_1,\ldots,e_n}$ is a semi-invariant of the first kind for $\mathcal{C}_{m,n}$; the definition in [20] of $\mathcal{J}_{m,n;\,e_1,\ldots,e_n}$ by formulas repeated as (4.31)–(4.43) was used in [20] to prove that $\mathcal{J}_{m,n;\,e_1,\ldots,e_n}$ is a semi-invariant of the second kind for $\mathcal{C}_{m,n}$ with index $(e_1, e_2, \ldots, e_n)$; and the verification in [20, page 314, Theorem 27.7] of the identity $\mathcal{I}_{m,n;\,e_1,\ldots,e_n} \equiv \mathcal{J}_{m,n;\,e_1,\ldots,e_n}$ thereby showed that $\mathcal{I}_{m,n;\,e_1,\ldots,e_n}$ is a relative invariant for $\mathcal{C}_{m,n}$ having index $(e_1, e_2, \ldots, e_n)$.

Our definition of $\boldsymbol{R} \equiv \boldsymbol{C}_{p,q,r}(\boldsymbol{P}, \boldsymbol{Q})$ by (4.44)–(4.50) merely expresses $\boldsymbol{R}$ as a differential-polynomial combination over $\mathbb{Q}$ of semi-invariants of the first kind. That definition and the argument of Chapter 9 reveal the way Theorem 4.10 on page 36 was discovered and first established. The goal of this chapter is to provide Theorem 4.10 with a proof similar to the argument in [20] for $\mathcal{I}_{m,n;\,e_1,\ldots,e_n}$ as summarized in the preceding paragraph. To accomplish that, we shall: define $\boldsymbol{S}$ in $\mathcal{R}_{m,n}$ as a polynomial combination over $\mathbb{Q}$ of semi-invariants of the second kind for $\mathcal{C}_{m,n}$; show that $\boldsymbol{S}$ is a semi-invariant of the second kind; and establish $\boldsymbol{R} \equiv \boldsymbol{S}$.

11.1. Context employed and definition of S

Throughout, we assume that $r \geq 2$, $m \geq 2$, $n \geq 1$, $(m, n) \neq (2, 1)$, $\boldsymbol{P}$ and $\boldsymbol{Q}$ are relative invariants in $\mathcal{R}_{m,n}$ for $\mathcal{C}_{m,n}$ of respective weights p and q, $\boldsymbol{a}_{m,n}$ is defined in $\mathcal{R}_{m,n}$ by (4.45), and $\boldsymbol{R}$ is the differential polynomial $\boldsymbol{C}_{p,q,r}(\boldsymbol{P}, \boldsymbol{Q})$ defined in $\mathcal{R}_{m,n}$ with respect to (4.44)–(4.50) of page 36 by

$$(11.1) \qquad \boldsymbol{R} \equiv \sum_{s=0}^{r} \sum_{t=0}^{r-s} \boldsymbol{A}_{p,q,r,s,t}\, \boldsymbol{P}^{(t)}\, \boldsymbol{Q}^{(r-s-t)}.$$

Proposition 5.3 shows that $\boldsymbol{a}_{m,n}$ is a semi-invariant of the first kind for $\mathcal{C}_{m,n}$. Thus, as a consequence of Proposition 5.1, each of the nonconstant expression defined by (4.44)–(4.50) is a semi-invariant of the first kind. Moreover, $\boldsymbol{R}$ in (11.1) can not be a nonzero constant. Consequently, *either $\boldsymbol{R} \equiv 0$ or $\boldsymbol{R}$ is a semi-invariant of the first kind for $\mathcal{C}_{m,n}$*.

In view of Corollary 5.7, a semi-invariant $\boldsymbol{b}_{m,n}$ of the second kind for $\mathcal{C}_{m,n}$ having weight 2 is defined in $\mathcal{R}_{m,n}$ by

$$(11.2) \qquad \boldsymbol{b}_{m,n} \equiv \frac{1}{\binom{m+1}{3}} \left[\begin{array}{c} \boldsymbol{w}_{\underbrace{0,\ldots,0,2}_{n-1}} - \dfrac{m-2}{3} \boldsymbol{w}^{(1)}_{\underbrace{0,\ldots,0,1}_{n-1}} \\ - \dfrac{(3m-1)(m-2)}{6m(m-1)} (\boldsymbol{w}_{\underbrace{0,\ldots,0,1}_{n-1}})^2 \end{array} \right].$$

11. SYMMETRY WITH RESPECT TO SEMI-INVARIANTS

Thus, with reference to $z = f(\zeta)$ in (4.6) and Notation 1.0 of page 30, it yields

(11.3) $$b^{**}_{m,n}(\zeta) \equiv \left(f'(\zeta)\right)^2 b_{m,n}\bigl(f(\zeta)\bigr), \quad \text{on } \Omega^{**}.$$

We note that $\boldsymbol{b}_{m,n}$ is related to $\boldsymbol{a}_{m,n}$ of (4.45) through

(11.4) $$\boldsymbol{b}_{m,n} \equiv \boldsymbol{a}_{m,n} + \boldsymbol{d}^{(1)}_{m,n} + (\boldsymbol{d}_{m,n})^2,$$

where the definition of $\boldsymbol{d}_{m,n}$ in $\mathcal{R}_{m,n}$ by (4.19) of page 32 or (4.33) of page 34 or (5.15) of page 44 is repeated here as

(11.5) $$\boldsymbol{d}_{m,n} \equiv \frac{1}{m(m-1)} \, \boldsymbol{w}_{\underbrace{0,\ldots,0}_{n-1},1}.$$

With respect to the fixed integers m, n for $\mathcal{C}_{m,n}$, we define $\boldsymbol{S}$ in terms of

(11.6) $$\boldsymbol{H}_{h,i,j} \equiv 0, \quad \text{for } i \leq -1 \text{ and any } h, j,$$

(11.7) $$\boldsymbol{H}_{h,0,j} \equiv 1, \quad \text{for any } h, j,$$

(11.8) $$\boldsymbol{H}_{h,i+1,j} \equiv \sum_{k=i}^{j-1} \left[\begin{array}{l} \boldsymbol{H}^{(1)}_{h,i,k} + 2i\,\boldsymbol{d}_{m,n}\,\boldsymbol{H}_{h,i,k} \\ -\,k(2h+k-1)\,\boldsymbol{b}_{m,n}\,\boldsymbol{H}_{h,i-1,k-1} \end{array} \right],$$
$$\text{for } i \geq 0 \text{ and any } h, j,$$

the rational numbers $\mathfrak{C}_{p,q,r,\mu}$ repeated from (4.49) of page 36 in

(11.9) $$\mathfrak{C}_{p,q,r,\mu} \equiv (-1)^\mu \frac{\dbinom{r}{\mu}\dbinom{2q+r-1}{\mu}}{\dbinom{2p+\mu-1}{\mu}}, \quad \text{for } 0 \leq \mu \leq r,$$

as well as

(11.10) $$\boldsymbol{G}_{p,q,r,s,t} \equiv \sum_{k=0}^{s} \mathfrak{C}_{p,q,r,t+k}\,\boldsymbol{H}_{p,k,t+k}\,\boldsymbol{H}_{q,s-k,r-t-k},$$
$$\text{for } 0 \leq s \leq r \text{ and } 0 \leq t \leq r-s,$$

(11.11) $$\boldsymbol{\mathcal{P}}_0 \equiv \boldsymbol{P}, \quad \boldsymbol{\mathcal{P}}_k \equiv \boldsymbol{\mathcal{P}}^{(1)}_{k-1} + 2(p+k-1)\boldsymbol{d}_{m,n}\boldsymbol{\mathcal{P}}_{k-1}, \quad \text{for } k \geq 1,$$

and

(11.12) $$\boldsymbol{\mathcal{Q}}_0 \equiv \boldsymbol{Q}, \quad \boldsymbol{\mathcal{Q}}_k \equiv \boldsymbol{\mathcal{Q}}^{(1)}_{k-1} + 2(q+k-1)\boldsymbol{d}_{m,n}\boldsymbol{\mathcal{Q}}_{k-1}, \quad \text{for } k \geq 1,$$

by

(11.13) $$\boldsymbol{S} \equiv \sum_{s=0}^{r} \sum_{t=0}^{r-s} \boldsymbol{G}_{p,q,r,s,t}\,\boldsymbol{\mathcal{P}}_t\,\boldsymbol{\mathcal{Q}}_{r-s-t}.$$

11.2. When S is a semi-invariant of the second kind

PROPOSITION 11.1. *For S defined in $\mathcal{R}_{m,n}$ by means of (11.13), either $S \equiv 0$ or S is a semi-invariant of the second kind for $\mathcal{C}_{m,n}$ of weight $p+q+r$.*

PROOF. For any h and j, we note that (11.6)–(11.8) yield $\boldsymbol{H}_{h,1,j} \equiv 0$ and

$$\boldsymbol{H}_{h,2,j} \equiv -\boldsymbol{b}_{m,n} \sum_{k=1}^{j-1} \left[2\binom{k}{2} + 2h\binom{k}{1} \right].$$

Consequently, for $i_0 = 2$, it is true that:

(1) for any h and j, either $\boldsymbol{H}_{h,i_0,j} \equiv 0$ or $\boldsymbol{H}_{h,i_0,j}$ is a semi-invariant of the second kind of weight i_0, and
(2) for any h and j, either $\boldsymbol{H}_{h,i_0-1,j} \equiv 0$ or $\boldsymbol{H}_{h,i_0-1,j}$ is a semi-invariant of the second kind of weight $i_0 - 1$.

Let i_0 be an integer satisfying $i_0 \geq 2$ for which (1) and (2) are valid. Then, we use Proposition 5.5 to deduce that, for any h and k, either the polynomial $\boldsymbol{H}_{h,i_0,k}^{(1)} + 2i_0 \, \boldsymbol{d}_{m,n} \, \boldsymbol{H}_{h,i_0,k}$ in $\mathcal{R}_{m,n}$ is zero or it is a semi-invariant of the second kind having weight $i_0 + 1$. Moreover, for any h and k, either the polynomial $-k(2h + k - 1) \, \boldsymbol{b}_{m,n} \, \boldsymbol{H}_{h,i_0-1,k-1}$ in $\mathcal{R}_{m,n}$ is zero or it is a semi-invariant of the second kind having weight $i_0 + 1$. Thus, in view of (11.8), we see that, for any h and j, either $\boldsymbol{H}_{h,i_0+1,j}$ is zero or it is a semi-invariant of the second kind whose weight is $i_0 + 1$. This shows that, for $i \geq 1$ as well as any h and j, either $\boldsymbol{H}_{h,i,j}$ is zero or it is a semi-invariant of the second kind of weight i.

We apply (11.10), (11.7), and the preceding properties of $\boldsymbol{H}_{h,i,j}$ to deduce that: for $s = 0$ and any t, $\boldsymbol{G}_{p,q,r,0,t}$ is a rational number; and, for $s \geq 1$ and any t, either $\boldsymbol{G}_{p,q,r,s,t}$ is zero or $\boldsymbol{G}_{p,q,r,s,t}$ is a semi-invariant of the second kind having weight s.

We use the property of $\boldsymbol{P}$ as a semi-invariant of the second kind having weight p along with (11.11), Proposition 5.5, and mathematical induction on k to deduce that $\boldsymbol{\mathcal{P}}_k$ is a semi-invariant of the second kind having weight $p + k$, for $k \geq 0$.

Similarly, we employ (11.12) to verify that $\boldsymbol{\mathcal{Q}}_k$ is a semi-invariant of the second kind having weight $q + k$, for $k \geq 0$.

By combining the deductions of the three preceding paragraphs with (11.13), we conclude that either $S \equiv 0$ or S is a semi-invariant of the second kind having weight $p + q + r$. This completes the proof. $\square$

11.3. An expansion for differential polynomials like $\boldsymbol{\mathcal{P}}_k$ in (11.11)

For any integers h, i, and j, we define polynomials $\boldsymbol{K}_{h,i,j}$ in $\mathcal{R}_{m,n}$ with respect to $\boldsymbol{d}_{m,n}$ in (11.5) through

(11.14) $\quad\quad\quad \boldsymbol{K}_{h,i,j} \equiv 0, \quad$ for $i \leq -1$ and any h, j,

(11.15) $\quad\quad\quad \boldsymbol{K}_{h,0,j} \equiv 1, \quad$ for any h, j,

and

(11.16) $\quad\quad\quad \boldsymbol{K}_{h,i+1,j} \equiv \sum_{k=0}^{j} \left(\boldsymbol{K}_{h,i,k}^{(1)} + 2(h+i+k) \, \boldsymbol{d}_{m,n} \, \boldsymbol{K}_{h,i,k} \right),$

$\quad\quad\quad\quad\quad\quad\quad\quad\quad\quad$ for $i \geq 0$ and any h, j.

For later reference, we note that (11.16) yields

$$(11.17) \quad \boldsymbol{K}_{h,\mu+1,\nu} - \boldsymbol{K}_{h,\mu+1,\nu-1} \equiv \boldsymbol{K}^{(1)}_{h,\mu,\nu} + 2(h+\mu+\nu)\,d_{m,n}\,\boldsymbol{K}_{h,\mu,\nu},$$

for $\mu \geq 0$, $\nu \geq 0$ and any h.

PROPOSITION 11.2. *Suppose that $\boldsymbol{T}$ in $\mathcal{R}_{m,n}$ is a semi-invariant of the second kind of weight t for $\mathcal{C}_{m,n}$ and define $\boldsymbol{\mathcal{T}}_j$ in $\mathcal{R}_{m,n}$ by*

$$(11.18) \quad \boldsymbol{\mathcal{T}}_0 \equiv \boldsymbol{T} \quad \text{and} \quad \boldsymbol{\mathcal{T}}_j \equiv \boldsymbol{\mathcal{T}}^{(1)}_{j-1} + 2(t+j-1)\,d_{m,n}\,\boldsymbol{\mathcal{T}}_{j-1}, \quad \text{for } j \geq 1.$$

Then, $\boldsymbol{\mathcal{T}}_j$ is a semi-invariant of the second kind of weight $t+j$ for $\mathcal{C}_{m,n}$ and

$$(11.19) \quad \boldsymbol{\mathcal{T}}_j \equiv \sum_{i=0}^{j} \boldsymbol{K}_{t,i,j-i}\,\boldsymbol{T}^{(j-i)}, \quad \text{for } j \geq 0.$$

PROOF. For $j \geq 0$, repeated application of (11.18) and Proposition 5.5 shows that $\boldsymbol{\mathcal{T}}_j$ is a semi-invariant of the second kind for $\mathcal{C}_{m,n}$ having weight $t+j$.

We use (11.15) to see that (11.19) is true for $j=0$. Let j denote a nonnegative integer for which (11.19) is valid. Then, we employ (11.18) and (11.19) to see that

$$(11.20) \quad \boldsymbol{\mathcal{T}}_{j+1} \equiv \boldsymbol{\mathcal{T}}^{(1)}_j + 2(t+j)\,d_{m,n}\,\boldsymbol{\mathcal{T}}_j$$

$$\equiv \sum_{i=0}^{j} \boldsymbol{K}_{t,i,j-i}\,\boldsymbol{T}^{(j+1-i)} + \sum_{i=0}^{j} \boldsymbol{K}^{(1)}_{t,i,j-i}\,\boldsymbol{T}^{(j-i)}$$

$$+ \sum_{i=0}^{j} 2(t+j)\,d_{m,n}\,\boldsymbol{K}_{t,i,j-i}\,\boldsymbol{T}^{(j-i)}$$

$$\equiv \sum_{i=0}^{j} \boldsymbol{K}_{t,i,j-i}\,\boldsymbol{T}^{(j+1-i)} + \sum_{i=1}^{j+1} \boldsymbol{K}^{(1)}_{t,i-1,j+1-i}\,\boldsymbol{T}^{(j+1-i)}$$

$$+ \sum_{i=1}^{j+1} 2(t+j)\,d_{m,n}\,\boldsymbol{K}_{t,i-1,j+1-i}\,\boldsymbol{T}^{(j+1-i)}$$

$$\equiv \sum_{i=1}^{j} \left(\boldsymbol{K}_{t,i,j-i} + \boldsymbol{K}^{(1)}_{t,i-1,j+1-i} + 2(t+j)\,d_{m,n}\,\boldsymbol{K}_{t,i-1,j+1-i}\right)\boldsymbol{T}^{(j+1-i)}$$

$$+ \left(\boldsymbol{K}^{(1)}_{t,j,0} + 2(t+j)\,d_{m,n}\,\boldsymbol{K}_{t,j,0}\right)\boldsymbol{T}^{(0)} + \boldsymbol{K}_{t,0,j}\,\boldsymbol{T}^{(j+1)}.$$

In (11.17), we replace h with t, μ with $i-1$, and ν with $j+1-i$ to obtain

$$(11.21) \quad \boldsymbol{K}_{t,i,j-i} + \boldsymbol{K}^{(1)}_{t,i-1,j+1-i} + 2(t+j)\,d_{m,n}\,\boldsymbol{K}_{t,i-1,j+1-i} \equiv \boldsymbol{K}_{t,i,j+1-i},$$

for $1 \leq i \leq j+1$.

Replacing h with t, μ with j, and ν with 0 in (11.17), we use (11.16) to deduce

$$(11.22) \quad \boldsymbol{K}^{(1)}_{t,j,0} + 2(t+j)d_{m,n}\,\boldsymbol{K}_{t,j,0} \equiv \boldsymbol{K}_{t,j+1,0}, \quad \text{for } j \geq 0.$$

By rewriting (11.20) with the aid of (11.21), (11.22), and (11.15), we find that

$$\mathcal{T}_{j+1} \equiv \sum_{i=1}^{j} K_{t,i,j+1-i}\, T^{(j+1-i)} + K_{t,j+1,0}\, T^{(0)} + K_{t,0,j+1}\, T^{(j+1)}$$

$$\equiv \sum_{i=0}^{j+1} K_{t,i,j+1-i}\, T^{(j+1-i)}.$$

Thus, (11.19) is valid with j replaced by $j+1$. This completes the proof. $\square$

11.4. Formula for S that involves $F_{p,q,r,s,t}$ of (11.23)

Continuing the context for S in (11.13), we define $F_{p,q,r,s,t}$ in $\mathcal{R}_{m,n}$ with respect to (11.14)–(11.16) and (11.10) through

$$(11.23) \qquad F_{p,q,r,s,t} \equiv \sum_{\mu=0}^{s} \sum_{\nu=0}^{s-\mu} K_{p,\mu,t}\, K_{q,\nu,r-s-t}\, G_{p,q,r,s-\mu-\nu,t+\mu},$$
$$\text{for } 0 \le s \le r \text{ and } 0 \le t \le r-s.$$

THEOREM 11.3. *The differential polynomial S of (11.13) is given by*

$$(11.24) \qquad S \equiv \sum_{s=0}^{r} \sum_{t=0}^{r-s} F_{p,q,r,s,t}\, P^{(t)}\, Q^{(r-s-t)}.$$

PROOF. We apply Proposition 11.2 to $\mathcal{P}_t$ and $\mathcal{Q}_{r-s-t}$ to see that (11.13) yields

$$(11.25)\ S \equiv \sum_{s=0}^{r}\sum_{t=0}^{r-s} G_{p,q,r,s,t}\,\mathcal{P}_t\,\mathcal{Q}_{r-s-t}$$
$$\equiv \sum_{s=0}^{r}\sum_{t=0}^{r-s} G_{p,q,r,s,t}\sum_{\mu=0}^{t} K_{p,\mu,t-\mu}\, P^{(t-\mu)} \sum_{\nu=0}^{r-s-t} K_{q,\nu,r-s-t-\nu}\, Q^{(r-s-t-\nu)}$$
$$\equiv \sum_{s=0}^{r}\sum_{t=0}^{r-s}\sum_{\mu=0}^{t}\sum_{\nu=0}^{r-s-t} Z_{p,q,r,s,t,\mu,\nu},$$

where

$$(11.26) \qquad Z_{p,q,r,s,t,\mu,\nu} \equiv G_{p,q,r,s,t}\, K_{p,\mu,t-\mu}\, K_{q,\nu,r-s-t-\nu}\, P^{(t-\mu)}\, Q^{(r-s-t-\nu)}.$$

We use (11.25) to verify that

$$S \equiv \sum_{s=0}^{r}\sum_{t=0}^{r-s}\sum_{\mu=0}^{t}\sum_{\nu=0}^{r-s-t} Z_{p,q,r,s,t,\mu,\nu}$$
$$\equiv \sum_{s=0}^{r}\sum_{\mu=0}^{r-s}\sum_{t=\mu}^{r-s}\sum_{\nu=0}^{r-s-t} Z_{p,q,r,s,t,\mu,\nu}$$
$$\equiv \sum_{\mu=0}^{r}\sum_{s=0}^{r-\mu}\sum_{t=\mu}^{r-s}\sum_{\nu=0}^{r-s-t} Z_{p,q,r,s,t,\mu,\nu}$$
$$\equiv \sum_{\mu=0}^{r}\sum_{s=0}^{r-\mu}\sum_{\nu=0}^{r-s-\mu}\sum_{t=\mu}^{r-s-\nu} Z_{p,q,r,s,t,\mu,\nu}$$
$$\equiv \sum_{\mu=0}^{r}\sum_{\nu=0}^{r-\mu}\sum_{s=0}^{r-\mu-\nu}\sum_{t=\mu}^{r-s-\nu} Z_{p,q,r,s,t,\mu,\nu}$$

and

(11.27)
$$S \equiv \sum_{\mu=0}^{r} \sum_{\nu=0}^{r-\mu} \sum_{s=0}^{r-\mu-\nu} \sum_{t=0}^{r-s-\mu-\nu} Z_{p,q,r,s,t+\mu,\mu,\nu}$$
$$\equiv \sum_{\mu=0}^{r} \sum_{\nu=0}^{r-\mu} \sum_{t=0}^{r-\mu-\nu} \sum_{s=0}^{r-\mu-\nu-t} Z_{p,q,r,s,t+\mu,\mu,\nu}$$
$$\equiv \sum_{\mu=0}^{r} \sum_{\nu=0}^{r-\mu} \sum_{t=0}^{r-\mu-\nu} \sum_{s=\mu+\nu}^{r-t} Z_{p,q,r,s-\mu-\nu,t+\mu,\mu,\nu}$$
$$\equiv \sum_{\mu=0}^{r} \sum_{\nu=0}^{r-\mu} \sum_{s=\mu+\nu}^{r} \sum_{t=0}^{r-s} Z_{p,q,r,s-\mu-\nu,t+\mu,\mu,\nu}$$
$$\equiv \sum_{\mu=0}^{r} \sum_{s=\mu}^{r} \sum_{\nu=0}^{s-\mu} \sum_{t=0}^{r-s} Z_{p,q,r,s-\mu-\nu,t+\mu,\mu,\nu}$$
$$\equiv \sum_{s=0}^{r} \sum_{\mu=0}^{s} \sum_{\nu=0}^{s-\mu} \sum_{t=0}^{r-s} Z_{p,q,r,s-\mu-\nu,t+\mu,\mu,\nu}$$
$$\equiv \sum_{s=0}^{r} \sum_{\mu=0}^{s} \sum_{t=0}^{r-s} \sum_{\nu=0}^{s-\mu} Z_{p,q,r,s-\mu-\nu,t+\mu,\mu,\nu}$$
$$\equiv \sum_{s=0}^{r} \sum_{t=0}^{r-s} \sum_{\mu=0}^{s} \sum_{\nu=0}^{s-\mu} Z_{p,q,r,s-\mu-\nu,t+\mu,\mu,\nu}.$$

We employ (11.26) and (11.23) to rewrite (11.27) as
$$S \equiv \sum_{s=0}^{r} \sum_{t=0}^{r-s} \left[\sum_{\mu=0}^{s} \sum_{\nu=0}^{s-\mu} K_{p,\mu,t} K_{q,\nu,r-s-t} G_{p,q,r,s-\mu-\nu,t+\mu} \right] P^{(t)} Q^{(r-s-t)}$$
$$\equiv \sum_{s=0}^{r} \sum_{t=0}^{r-s} F_{p,q,r,s,t} P^{(t)} Q^{(r-s-t)}.$$

Thus, (11.24) is valid. This completes the proof. □

11.5. Reformulation in (11.29) for $F_{p,q,r,s,t}$ of (11.23)

We define $L_{h,i,j}$ in $\mathcal{R}_{m,n}$ with respect to $K_{h,i,j}$ in (11.14)–(11.16) and $H_{h,i,j}$ in (11.6)–(11.8) by means of

(11.28) $$L_{h,i,j} \equiv \sum_{k=0}^{i} K_{h,k,j} H_{h,i-k,i+j}, \quad \text{for any } h, i, j.$$

PROPOSITION 11.4. *The differential polynomials $F_{p,q,r,s,t}$ of (11.23) are expressed in terms of (11.9) and (11.28) by*

(11.29) $$F_{p,q,r,s,t} \equiv \sum_{\lambda=0}^{s} \mathfrak{C}_{p,q,r,t+\lambda} L_{p,\lambda,t} L_{q,s-\lambda,r-s-t},$$

for $0 \leq s \leq r$ and $0 \leq t \leq r - s$.

PROOF. After introducing the abbreviations

(11.30) $$U_{p,t,\mu,k} \equiv K_{p,\mu,t} H_{p,k,t+\mu+k}$$

and

(11.31) $$V_{q,r,s,t,\mu,\nu,k} \equiv K_{q,\nu,r-s-t} H_{q,s-\mu-\nu-k,r-t-\mu-k},$$

we apply them with (11.23) and (11.10) to verify that

(11.32) $$F_{p,q,r,s,t} \equiv \sum_{\mu=0}^{s} \sum_{\nu=0}^{s-\mu} K_{p,\mu,t} K_{q,\nu,r-s-t} G_{p,q,r,s-\mu-\nu,t+\mu}$$
$$\equiv \sum_{\mu=0}^{s} \sum_{\nu=0}^{s-\mu} \sum_{k=0}^{s-\mu-\nu} \mathfrak{C}_{p,q,r,t+\mu+k} U_{p,t,\mu,k} V_{q,r,s,t,\mu,\nu,k}$$
$$\equiv \sum_{\mu=0}^{s} \sum_{k=0}^{s-\mu} \sum_{\nu=0}^{s-\mu-k} \mathfrak{C}_{p,q,r,t+\mu+k} U_{p,t,\mu,k} V_{q,r,s,t,\mu,\nu,k}$$
$$\equiv \sum_{\lambda=0}^{s} \sum_{\mu=0}^{\lambda} \sum_{\nu=0}^{s-\lambda} \mathfrak{C}_{p,q,r,t+\lambda} U_{p,t,\mu,\lambda-\mu} V_{q,r,s,t,\mu,\nu,\lambda-\mu}$$
$$\equiv \sum_{\lambda=0}^{s} \mathfrak{C}_{p,q,r,t+\lambda} \sum_{\mu=0}^{\lambda} U_{p,t,\mu,\lambda-\mu} \sum_{\nu=0}^{s-\lambda} V_{q,r,s,t,\mu,\nu,\lambda-\mu}.$$

Application of (11.31) with (11.28) and (11.30) with (11.28) yields

(11.33) $$\sum_{\nu=0}^{s-\lambda} V_{q,r,s,t,\mu,\nu,\lambda-\mu} \equiv \sum_{\nu=0}^{s-\lambda} K_{q,\nu,r-s-t} H_{q,s-\lambda-\nu,r-t-\lambda} \equiv L_{q,s-\lambda,r-s-t}$$

and

(11.34) $$\sum_{\mu=0}^{\lambda} U_{p,t,\mu,\lambda-\mu} \equiv \sum_{\mu=0}^{\lambda} K_{p,\mu,t} H_{p,\lambda-\mu,t+\lambda} \equiv L_{p,\lambda,t}.$$

We use (11.32), (11.33), and (11.34) to obtain (11.29) and complete the proof. □

11.6. Initial reformulation for $L_{h,i,j}$

PROPOSITION 11.5. *The differential polynomials $L_{h,i,j}$ of (11.28) satisfy*

(11.35) $$L_{h,i+1,j} - L_{h,i+1,j-1} \equiv L_{h,i,j}^{(1)} + 2(h+i+j) d_{m,n} L_{h,i,j}$$
$$- (i+j)(2h+i+j-1) b_{m,n} L_{h,i-1,j},$$
for $i \geq 0$, $j \geq 0$, and any h.

PROOF. For $k \geq 0$, $j \geq 0$, and any h, we rewrite (11.17) as

(11.36) $$K_{h,k,j}^{(1)} \equiv K_{h,k+1,j} - K_{h,k+1,j-1} - 2(h+k+j) d_{m,n} K_{h,k,j}.$$

For $0 \leq i \leq j$, we note that (11.8) of page 108 yields

$$H_{h,i+1,j+1} - H_{h,i+1,j} \equiv H_{h,i,j}^{(1)} + 2i d_{m,n} H_{h,i,j} - j(2h+j-1) b_{m,n} H_{h,i-1,j-1}.$$

Consequently, for $0 \leq k \leq i$ and $j \geq 0$, we have $i+j \geq i \geq i-k \geq 0$ and

(11.37) $\quad \boldsymbol{H}^{(1)}_{h,i-k,i+j} \equiv \boldsymbol{H}_{h,i+1-k,i+1+j} - \boldsymbol{H}_{h,i+1-k,i+j} - 2(i-k)\,\boldsymbol{d}_{m,n}\,\boldsymbol{H}_{h,i-k,i+j}$
$\qquad\qquad\qquad + (i+j)(2h+i+j-1)\,\boldsymbol{b}_{m,n}\,\boldsymbol{H}_{h,i-1-k,i-1+j}.$

We apply (11.28), (11.36), and (11.37) to deduce

(11.38) $\quad \boldsymbol{L}^{(1)}_{h,i,j} \equiv \sum_{k=0}^{i} \boldsymbol{K}^{(1)}_{h,k,j}\,\boldsymbol{H}_{h,i-k,i+j} + \sum_{k=0}^{i} \boldsymbol{K}_{h,k,j}\,\boldsymbol{H}^{(1)}_{h,i-k,i+j}$

$\qquad\qquad \equiv \mathcal{L}_1 + \mathcal{L}_2 + \mathcal{L}_3 + \mathcal{L}_4 + \mathcal{L}_5 + \mathcal{L}_6 + \mathcal{L}_7,$

where

$$\mathcal{L}_1 \equiv \sum_{k=0}^{i} \boldsymbol{K}_{h,k+1,j}\,\boldsymbol{H}_{h,i-k,i+j},$$

$$\mathcal{L}_2 \equiv \sum_{k=0}^{i} -\boldsymbol{K}_{h,k+1,j-1}\,\boldsymbol{H}_{h,i-k,i+j},$$

$$\mathcal{L}_3 \equiv \sum_{k=0}^{i} -2(h+k+j)\,\boldsymbol{d}_{m,n}\,\boldsymbol{K}_{h,k,j}\,\boldsymbol{H}_{h,i-k,i+j},$$

$$\mathcal{L}_4 \equiv \sum_{k=0}^{i} \boldsymbol{K}_{h,k,j}\,\boldsymbol{H}_{h,i+1-k,i+1+j},$$

$$\mathcal{L}_5 \equiv \sum_{k=0}^{i} -\boldsymbol{K}_{h,k,j}\,\boldsymbol{H}_{h,i+1-k,i+j},$$

$$\mathcal{L}_6 \equiv \sum_{k=0}^{i} -2(i-k)\,\boldsymbol{d}_{m,n}\,\boldsymbol{K}_{h,k,j}\,\boldsymbol{H}_{h,i-k,i+j},$$

and

$$\mathcal{L}_7 \equiv \sum_{k=0}^{i} (i+j)(2h+i+j-1)\,\boldsymbol{b}_{m,n}\,\boldsymbol{K}_{h,k,j}\,\boldsymbol{H}_{h,i-1-k,i-1+j}.$$

By using these expressions with (11.7), (11.15), (11.28), and (11.6), we find that

$$\mathcal{L}_1 + \mathcal{L}_5 \equiv \sum_{k=1}^{i+1} \boldsymbol{K}_{h,k,j}\,\boldsymbol{H}_{h,i+1-k,i+j} - \sum_{k=0}^{i} \boldsymbol{K}_{h,k,j}\,\boldsymbol{H}_{h,i+1-k,i+j}$$

$$\equiv \boldsymbol{K}_{h,i+1,j} - \boldsymbol{H}_{h,i+1,i+j},$$

$$\mathcal{L}_2 \equiv \sum_{k=1}^{i+1} -K_{h,k,j-1}\, H_{h,i+1-k,i+j}$$

$$\equiv \left[\sum_{k=0}^{i+1} -K_{h,k,j-1}\, H_{h,i+1-k,(i+1)+(j-1)}\right] + H_{h,i+1,i+j}$$

$$\equiv -L_{h,i+1,j-1} + H_{h,i+1,i+j},$$

$$\mathcal{L}_4 \equiv \left[\sum_{k=0}^{i+1} K_{h,k,j}\, H_{h,i+1-k,i+1+j}\right] - K_{h,i+1,j}$$

$$\equiv L_{h,i+1,j} - K_{h,i+1,j},$$

$$\mathcal{L}_3 + \mathcal{L}_6 \equiv \sum_{k=0}^{i} -2(h+i+j)\, d_{m,n}\, K_{h,k,j}\, H_{h,i-k,i+j}$$

$$\equiv -2(h+i+j)\, d_{m,n}\, L_{h,i,j},$$

and, with $H_{h,-1,i-1+j} \equiv 0$,

$$\mathcal{L}_7 \equiv \sum_{k=0}^{i-1} (i+j)(2h+i+j-1)\, b_{m,n}\, K_{h,k,j}\, H_{h,i-1-k,i-1+j}$$

$$\equiv (i+j)(2h+i+j-1)\, b_{m,n}\, L_{h,i-1,j}.$$

Hence, for (11.38), we have

$$L_{h,i,j}^{(1)} \equiv L_{h,i+1,j} - L_{h,i+1,j-1} - 2(h+i+j)\, d_{m,n}\, L_{h,i,j}$$
$$+ (i+j)(2h+i+j-1)\, b_{m,n}\, L_{h,i-1,j}.$$

We rewrite this to obtain (11.35) and complete the proof. □

LEMMA 11.6. *The differential polynomials $L_{h,i,j}$ of (11.28) satisfy*

(11.39) $\qquad L_{h,i+1,j} \equiv 0, \quad \text{for } i \geq 0,\ j \leq -1,\ \text{and any } h.$

PROOF. Let i denote a nonnegative integer. We apply (11.28) to obtain

(11.40) $\qquad L_{h,i+1,j} \equiv \sum_{k=0}^{i+1} K_{h,k,j}\, H_{h,i+1-k,i+1+j}, \quad \text{for any } h \text{ and } j.$

If $k \geq 1$ and $j \leq -1$, then (11.16) yields $K_{h,k,j} \equiv 0$. If $k = 0$ and $j \leq -1$, then we use $(i+1+j)-1 = i+j < i$ with (11.8) to verify that $H_{h,i+1,i+1+j} \equiv 0$. Combining these results with (11.40), we obtain $L_{h,i+1,j} \equiv 0$, when $j \leq -1$. This completes the proof. □

PROPOSITION 11.7. *The differential polynomials $\boldsymbol{L}_{h,i,j}$ of (11.28) satisfy*

(11.41) $$\boldsymbol{L}_{h,i,j} \equiv 0, \quad \text{for } i \leq -1 \text{ and any } h, j,$$

(11.42) $$\boldsymbol{L}_{h,0,j} \equiv 1, \quad \text{for any } h, j,$$

and

(11.43) $$\boldsymbol{L}_{h,i+1,j} \equiv \sum_{k=0}^{j} \left[\begin{array}{l} \boldsymbol{L}_{h,i,k}^{(1)} + 2(h+i+k)\,\boldsymbol{d}_{m,n}\,\boldsymbol{L}_{h,i,k} \\ -(i+k)(2h+i+k-1)\,\boldsymbol{b}_{m,n}\,\boldsymbol{L}_{h,i-1,k} \end{array} \right],$$
$$\text{for } i \geq 0 \text{ and any } h, j.$$

PROOF. In view of (11.28), we see that (11.41) is valid. We employ (11.28), (11.15), and (11.7) to deduce (11.42). For $i \geq 0$ and $j \leq -1$, Lemma 11.6 shows that (11.43) reduces to $0 \equiv 0$. Suppose that $i \geq 0$ and $j \geq 0$. Then, (11.39) yields

(11.44) $$\boldsymbol{L}_{h,i+1,j} \equiv \boldsymbol{L}_{h,i+1,j} - \boldsymbol{L}_{h,i+1,-1} \equiv \sum_{k=0}^{j} \left(\boldsymbol{L}_{h,i+1,k} - \boldsymbol{L}_{h,i+1,k-1} \right).$$

By setting $j = k$ in (11.35), we use the resulting expression to rewrite (11.44). This shows that (11.43) is valid as stated and completes the proof. $\square$

11.7. Final reformulation for $\boldsymbol{L}_{h,i,j}$

With reference to $\boldsymbol{d}_{m,n}$ of (11.5) on page 108, we define $\boldsymbol{e}_{m,n}$ in $\mathcal{R}_{m,n}$ by

(11.45) $$\boldsymbol{e}_{m,n} \equiv \boldsymbol{d}_{m,n}^{(1)} + (\boldsymbol{d}_{m,n})^2$$

in order to have $\boldsymbol{b}_{m,n} = \boldsymbol{a}_{m,n} + \boldsymbol{e}_{m,n}$ as a simplified writing for the identity (11.4). We define $\boldsymbol{M}_{h,i,j,\nu}$ in $\mathcal{R}_{m,n}$ with respect to $\boldsymbol{a}_{m,n}$ and $\boldsymbol{e}_{m,n}$ by

(11.46) $$\boldsymbol{M}_{h,i,j,\nu} \equiv 0, \quad \text{for any } h, i, j, \nu \text{ satisfying } i < 0 \text{ or } \nu < 0 \text{ or } \nu > i,$$

(11.47) $$\boldsymbol{M}_{h,0,j,0} \equiv 1, \quad \text{for any } h, j,$$

and

(11.48) $$\boldsymbol{M}_{h,i+1,j,\nu} \equiv \sum_{k=0}^{j} \left[\begin{array}{l} \boldsymbol{M}_{h,i,k,\nu}^{(1)} + (\nu+1)\,\boldsymbol{e}_{m,n}\,\boldsymbol{M}_{h,i,k,\nu+1} \\ + (2h+2i+2k+1-\nu)\,\boldsymbol{M}_{h,i,k,\nu-1} \\ -(i+k)(2h+i+k-1)(\boldsymbol{a}_{m,n}+\boldsymbol{e}_{m,n})\,\boldsymbol{M}_{h,i-1,k,\nu} \end{array} \right],$$

for any h, i, j, ν satisfying $i \geq 0$ and $0 \leq \nu \leq i+1$.

PROPOSITION 11.8. *The $\boldsymbol{L}_{h,i,j}$ defined in $\mathcal{R}_{m,n}$ through (11.28) are given by*

(11.49) $$\boldsymbol{L}_{h,i,j} \equiv \sum_{\nu=0}^{i} \boldsymbol{M}_{h,i,j,\nu} (\boldsymbol{d}_{m,n})^{\nu}, \quad \text{for any } h, i, j.$$

PROOF. In view of (11.41), (11.42), and (11.47), we see that (11.49) is valid for $i \leq 0$ and any h, j. Let i_0 denote a fixed nonnegative integer such that (11.49) is true for $i \leq i_0$ and any h, j. Then, for any integer k, (11.49), (11.45), and (11.46) yield

$$(11.50) \quad \boldsymbol{L}^{(1)}_{h,i_0,k} \equiv \sum_{\nu=0}^{i_0} \boldsymbol{M}^{(1)}_{h,i_0,k,\nu}(\boldsymbol{d}_{m,n})^\nu + \sum_{\nu=0}^{i_0} \boldsymbol{M}_{h,i_0,k,\nu}\,\nu\,(\boldsymbol{d}_{m,n})^{\nu-1}\,\boldsymbol{d}^{(1)}_{m,n}$$

$$\equiv \sum_{\nu=0}^{i_0} \boldsymbol{M}^{(1)}_{h,i_0,k,\nu}(\boldsymbol{d}_{m,n})^\nu + \sum_{\nu=0}^{i_0} \nu\,\boldsymbol{M}_{h,i_0,k,\nu}\bigl[\boldsymbol{e}_{m,n} - (\boldsymbol{d}_{m,n})^2\bigr](\boldsymbol{d}_{m,n})^{\nu-1}$$

$$\equiv \sum_{\nu=0}^{i_0+1} \left[\begin{array}{l} \boldsymbol{M}^{(1)}_{h,i_0,k,\nu} + (\nu+1)\,\boldsymbol{e}_{m,n}\,\boldsymbol{M}_{h,i_0,k,\nu+1} \\ -\,(\nu-1)\boldsymbol{M}_{h,i_0,k,\nu-1} \end{array} \right] (\boldsymbol{d}_{m,n})^\nu.$$

We note that (11.49) and (11.46) give

$$(11.51) \quad 2(h+i_0+k)\,\boldsymbol{d}_{m,n}\,\boldsymbol{L}_{h,i_0,k} \equiv \sum_{\nu=0}^{i_0} 2(h+i_0+k)\,\boldsymbol{M}_{h,i_0,k,\nu}\,(\boldsymbol{d}_{m,n})^{\nu+1}$$

$$\equiv \sum_{\nu=0}^{i_0+1} 2(h+i_0+k)\,\boldsymbol{M}_{h,i_0,k,\nu-1}\,(\boldsymbol{d}_{m,n})^\nu.$$

Using (11.49), (11.46), and $\boldsymbol{b}_{m,n} \equiv \boldsymbol{a}_{m,n} + \boldsymbol{e}_{m,n}$ from (11.4) via (11.45), we obtain

$$(11.52) \quad -(i_0+k)(2h+i_0+k-1)\,\boldsymbol{b}_{m,n}\,\boldsymbol{L}_{h,i_0-1,k}$$

$$\equiv \sum_{\nu=0}^{i_0+1} -(i_0+k)(2h+i_0+k-1)(\boldsymbol{a}_{m,n}+\boldsymbol{e}_{m,n})\,\boldsymbol{M}_{h,i_0-1,k,\nu}\,(\boldsymbol{d}_{m,n})^\nu.$$

Consequently, (11.50), (11.51), and (11.52) yield

$$\boldsymbol{L}^{(1)}_{h,i_0,k} + 2(h+i_0+k)\,\boldsymbol{d}_{m,n}\,\boldsymbol{L}_{h,i_0,k}$$
$$-\,(i_0+k)(2h+i_0+k-1)\,\boldsymbol{b}_{m,n}\,\boldsymbol{L}_{h,i_0-1,k}$$

$$\equiv \sum_{\nu=0}^{i_0+1} \left[\begin{array}{l} \boldsymbol{M}^{(1)}_{h,i_0,k,\nu} + (\nu+1)\,\boldsymbol{e}_{m,n}\,\boldsymbol{M}_{h,i_0,k,\nu+1} \\ +\,(2h+2i_0+2k+1-\nu)\,\boldsymbol{M}_{h,i_0,k,\nu-1} \\ -\,(i_0+k)(2h+i_0+k-1)(\boldsymbol{a}_{m,n}+\boldsymbol{e}_{m,n})\,\boldsymbol{M}_{h,i_0-1,k,\nu} \end{array} \right] (\boldsymbol{d}_{m,n})^\nu.$$

We employ (11.43) as well as the preceding formula and (11.48) to deduce

$$\boldsymbol{L}_{h,i_0+1,j} \equiv \sum_{k=0}^{j} \left[\begin{array}{l} \boldsymbol{L}^{(1)}_{h,i_0,k} + 2(h+i_0+k)\boldsymbol{d}_{m,n}\,\boldsymbol{L}_{h,i_0,k} \\ -(i_0+k)(2h+i_0+k-1)\boldsymbol{b}_{m,n}\,\boldsymbol{L}_{h,i_0-1,k} \end{array} \right]$$

$$\equiv \sum_{\nu=0}^{i_0+1} \sum_{k=0}^{j} \left[\begin{array}{l} \boldsymbol{M}^{(1)}_{h,i_0,k,\nu} + (\nu+1)\boldsymbol{e}_{m,n}\,\boldsymbol{M}_{h,i_0,k,\nu+1} \\ + (2h+2i_0+2k+1-\nu)\boldsymbol{M}_{h,i_0,k,\nu-1} \\ -(i_0+k)(2h+i_0+k-1)(\boldsymbol{a}_{m,n}+\boldsymbol{e}_{m,n})\boldsymbol{M}_{h,i_0-1,k,\nu} \end{array} \right] (\boldsymbol{d}_{m,n})^{\nu}$$

$$\equiv \sum_{\nu=0}^{i_0+1} \boldsymbol{M}_{h,i_0+1,j,\nu}\,(\boldsymbol{d}_{m,n})^{\nu}.$$

Thus, by induction on i, (11.49) is valid for any h, i, j. This completes the proof. $\square$

We recall that the notation $\mathfrak{P}_\nu$ introduced for (9.53) on page 89 yields: $\mathfrak{P}_\nu = \nu!$, when $\nu \geq 1$; and, $\mathfrak{P}_\nu = 1$, when $\nu \leq 0$.

PROPOSITION 11.9. *The coefficients $\boldsymbol{M}_{h,i,j,\nu}$ for (11.49) and the $\boldsymbol{B}_{h,i,j}$ defined in $\mathcal{R}_{m,n}$ by (4.46)–(4.48) of page 36 are related by*

$$(11.53) \qquad \boldsymbol{M}_{h,i,j,\nu} \equiv \mathfrak{P}_\nu \binom{i+j}{\nu}\binom{2h+i+j-1}{\nu}\boldsymbol{B}_{h,i-\nu,i+j-\nu},$$

for any h, i, j, ν satisfying $j \geq 0$.

PROOF. If $i < 0$ or $\nu < 0$ or $\nu > i$, then (11.46), the binomial coefficients in (11.53), and (4.46) of page 36 yield $0 \equiv 0$ for (11.53). If $i = 0$ and $\nu = 0$, then (11.47) and (4.47) give $1 \equiv 1$ for (11.53). It remains for us to establish (11.53) when h, i, j, ν satisfy $i \geq 1$, $j \geq 0$, and $0 \leq \nu \leq i$.

Let i_0 be a nonnegative integer such that (11.53) is valid for any h, i, j, ν subject to $i \leq i_0$ and $j \geq 0$. Henceforth, let h, j, ν denote fixed integers that satisfy $j \geq 0$ and $0 \leq \nu \leq i_0 + 1$. Using (11.48) and (11.53) with $i = i_0$, we obtain

$$(11.54) \qquad \boldsymbol{M}_{h,i_0+1,j,\nu} \equiv \mathcal{M}_1 + \mathcal{M}_2 + \mathcal{M}_3 + \mathcal{M}_4,$$

where

$$\mathcal{M}_1 \equiv \sum_{k=0}^{j} \boldsymbol{M}^{(1)}_{h,i_0,k,\nu}$$
$$\equiv \sum_{k=0}^{j} \mathfrak{P}_\nu \binom{i_0+k}{\nu}\binom{2h+i_0+k-1}{\nu} \boldsymbol{B}^{(1)}_{h,i_0-\nu,i_0+k-\nu},$$

$$\boldsymbol{\mathcal{M}}_2 \equiv \sum_{k=0}^{j} (\nu+1)\, \boldsymbol{e}_{m,n}\, \boldsymbol{M}_{h,i_0,k,\nu+1}$$

$$\equiv \sum_{k=0}^{j} (\nu+1)\, \boldsymbol{e}_{m,n}\, \mathfrak{P}_{\nu+1} \binom{i_0+k}{\nu+1}\binom{2h+i_0+k-1}{\nu+1} \boldsymbol{B}_{h,i_0-\nu-1,i_0+k-\nu-1}$$

$$\equiv \sum_{k=0}^{j} \mathfrak{P}_\nu \binom{i_0+k}{\nu}\binom{2h+i_0+k-1}{\nu} \begin{bmatrix} (i_0+k-\nu)\, \boldsymbol{e}_{m,n} \\ \times (2h+i_0+k-\nu-1) \\ \times \boldsymbol{B}_{h,i_0-\nu-1,i_0+k-\nu-1} \end{bmatrix},$$

$$\boldsymbol{\mathcal{M}}_3 \equiv \sum_{k=0}^{j} (2h+2i_0+2k+1-\nu)\, \boldsymbol{M}_{h,i_0,k,\nu-1}$$

$$\equiv \sum_{k=0}^{j} \mathfrak{P}_{\nu-1} \begin{bmatrix} (2h+2i_0+2k+1-\nu) \binom{i_0+k}{\nu-1}\binom{2h+i_0+k-1}{\nu-1} \\ \times \boldsymbol{B}_{h,i_0+1-\nu,i_0+k+1-\nu} \end{bmatrix},$$

and

$$\boldsymbol{\mathcal{M}}_4 \equiv \sum_{k=0}^{j} -(i_0+k)(2h+i_0+k-1)(\boldsymbol{a}_{m,n}+\boldsymbol{e}_{m,n})\, \boldsymbol{M}_{h,i_0-1,k,\nu}$$

$$\equiv \sum_{k=0}^{j} -\mathfrak{P}_\nu \binom{i_0-1+k}{\nu}\binom{2h+i_0+k-2}{\nu} \begin{bmatrix} (i_0+k)(\boldsymbol{a}_{m,n}+\boldsymbol{e}_{m,n}) \\ \times (2h+i_0+k-1) \\ \times \boldsymbol{B}_{h,i_0-1-\nu,i_0+k-\nu-1} \end{bmatrix}$$

$$\equiv \sum_{k=0}^{j} -\mathfrak{P}_\nu \binom{i_0+k}{\nu}\binom{2h+i_0+k-1}{\nu} \begin{bmatrix} (i_0+k-\nu)(\boldsymbol{a}_{,n}+\boldsymbol{e}_{m,n}) \\ \times (2h+i_0+k-\nu-1) \\ \times \boldsymbol{B}_{h,i_0-\nu-1,i_0+k-\nu-1} \end{bmatrix}.$$

We note that

$$\boldsymbol{\mathcal{M}}_2 + \boldsymbol{\mathcal{M}}_4 \equiv \sum_{k=0}^{j} \mathfrak{P}_\nu \binom{i_0+k}{\nu}\binom{2h+i_0+k-1}{\nu} \begin{bmatrix} -(i_0+k-\nu)\, \boldsymbol{a}_{m,n} \\ \times (2h+i_0+k-\nu-1) \\ \times \boldsymbol{B}_{h,i_0-\nu-1,i_0+k-\nu-1} \end{bmatrix}.$$

We employ (9.29) of page 84 to verify that: for $k \geq 0$,

$$\boldsymbol{B}^{(1)}_{h,i_0-\nu,i_0+k-\nu} - (i_0+k-\nu)(2h+i_0+k-\nu-1)\, \boldsymbol{a}_{m,n}\, \boldsymbol{B}_{h,i_0-\nu-1,i_0+k-\nu-1}$$
$$\equiv \boldsymbol{B}_{h,i_0+1-\nu,i_0+k+1-\nu} - \boldsymbol{B}_{h,i_0+1-\nu,i_0+k-\nu}.$$

Hence, we have

$$\boldsymbol{\mathcal{M}}_1 + (\boldsymbol{\mathcal{M}}_2 + \boldsymbol{\mathcal{M}}_4) \equiv \sum_{k=0}^{j} \mathfrak{P}_\nu \binom{i_0+k}{\nu}\binom{2h+i_0+k-1}{\nu} \begin{bmatrix} \boldsymbol{B}_{h,i_0+1-\nu,i_0+k+1-\nu} \\ -\boldsymbol{B}_{h,i_0+1-\nu,i_0+k-\nu} \end{bmatrix}.$$

In view of (11.54), this gives

(11.55) $\boldsymbol{M}_{h,i_0+1,j,\nu} \equiv (\mathcal{M}_1 + (\mathcal{M}_2 + \mathcal{M}_4)) + \mathcal{M}_3$

$$\equiv \sum_{k=0}^{j} \mathfrak{P}_\nu \binom{i_0+k}{\nu}\binom{2h+i_0+k-1}{\nu}\begin{bmatrix}\boldsymbol{B}_{h,i_0+1-\nu,i_0+k+1-\nu}\\-\boldsymbol{B}_{h,i_0+1-\nu,i_0+k-\nu}\end{bmatrix}$$

$$+\sum_{k=0}^{j}\mathfrak{P}_{\nu-1}\begin{bmatrix}(2h+2i_0+2k+1-\nu)\binom{i_0+k}{\nu-1}\\ \times\binom{2h+i_0+k-1}{\nu-1}\boldsymbol{B}_{h,i_0+1-\nu,i_0+k+1-\nu}\end{bmatrix}.$$

If $\nu = 0$, then we use $\boldsymbol{B}_{h,i_0+1,i_0} \equiv 0$ from (4.48) to see that (11.55) yields

(11.56) $\boldsymbol{M}_{h,i_0+1,j,0} \equiv \sum_{k=0}^{j}(\boldsymbol{B}_{h,i_0+1,i_0+k+1} - \boldsymbol{B}_{h,i_0+1,i_0+k}) \equiv \boldsymbol{B}_{h,i_0+1,i_0+1+j}.$

Suppose that $1 \leq \nu \leq i_0+1$. Then, we obtain

$$\frac{(2h+2i_0+2k+1-\nu)}{\nu}\binom{i_0+k}{\nu-1}\binom{2h+i_0+k-1}{\nu-1}$$
$$+\binom{i_0+k}{\nu}\binom{2h+i_0+k-1}{\nu} \equiv \binom{i_0+k+1}{\nu}\binom{2h+i_0+k}{\nu},$$

by setting $\kappa = i_0+k$ as well as $\omega = 2h$ in the identity of Lemma 9.11 on page 88 and we use it with $\mathfrak{P}_{\nu-1} = \mathfrak{P}_\nu/\nu$ to rewrite (11.55) as

(11.57) $\boldsymbol{M}_{h,i_0+1,j,\nu} \equiv \mathfrak{P}_\nu \sum_{k=0}^{j}\begin{bmatrix}\binom{i_0+k+1}{\nu}\binom{2h+i_0+k}{\nu}\boldsymbol{B}_{h,i_0+1-\nu,i_0+k+1-\nu}\\-\binom{i_0+k}{\nu}\binom{2h+i_0+k-1}{\nu}\boldsymbol{B}_{h,i_0+1-\nu,i_0+k-\nu}\end{bmatrix}$

$$\equiv \mathfrak{P}_\nu\begin{bmatrix}\binom{i_0+j+1}{\nu}\binom{2h+i_0+j}{\nu}\boldsymbol{B}_{h,i_0+1-\nu,i_0+1+j-\nu}\\-\binom{i_0}{\nu}\binom{2h+i_0-1}{\nu}\boldsymbol{B}_{h,i_0+1-\nu,i_0-\nu}\end{bmatrix}.$$

For the situation $1 \leq \nu \leq i_0$, (4.48) of page 36 yields $\boldsymbol{B}_{h,i_0+1-\nu,i_0-\nu} \equiv 0$. While, if $\nu = i_0+1$, then $\binom{i_0}{\nu} = 0$. Thus, (11.57) gives

(11.58) $\boldsymbol{M}_{h,i_0+1,j,\nu} \equiv \mathfrak{P}_\nu \binom{i_0+1+j}{\nu}\binom{2h+i_0+1+j-1}{\nu}\boldsymbol{B}_{h,i_0+1-\nu,i_0+1+j-\nu},$

when $1 \leq \nu \leq i_0+1$.

In view of (11.56) and (11.58), (11.53) is valid for $i = i_0+1$ and any integers h, j, ν satisfying $0 \leq \nu \leq i_0+1$ and $j \geq 0$. This induction on i completes the proof. □

11.8. Simplification for $F_{p,q,r,s,t}$ that yields $S \equiv R$

THEOREM 11.10. *The $F_{p,q,r,s,t}$ of (11.23) on page 111 and the $A_{p,q,r,s,t}$ of (4.50) on page 36 satisfy*

(11.59) $$F_{p,q,r,s,t} \equiv A_{p,q,r,s,t}, \quad \text{for } 0 \leq s \leq r, \text{ and } 0 \leq t \leq r-s.$$

PROOF. Using (11.29) of page 112 and (11.49) of page 116, we obtain

(11.60) $$F_{p,q,r,s,t} \equiv \sum_{\nu=0}^{s} \mathfrak{C}_{p,q,r,t+\nu} \, L_{p,\nu,t} \, L_{q,s-\nu,r-s-t}$$

$$\equiv \sum_{\nu=0}^{s} \mathfrak{C}_{p,q,r,t+\nu} \sum_{\mu=0}^{\nu} M_{p,\nu,t,\mu} \, (d_{m,n})^{\mu} \sum_{\lambda=0}^{s-\nu} M_{q,s-\nu,r-s-t,\lambda} \, (d_{m,n})^{\lambda}$$

$$\equiv \sum_{\nu=0}^{s} \sum_{\mu=0}^{\nu} \sum_{\lambda=0}^{s-\nu} \mathfrak{C}_{p,q,r,t+\nu} \, M_{p,\nu,t,\mu} \, M_{q,s-\nu,r-s-t,\lambda} \, (d_{m,n})^{\lambda+\mu}$$

$$\equiv \sum_{\mu=0}^{s} \sum_{\nu=\mu}^{s} \sum_{\lambda=0}^{s-\nu} \mathfrak{C}_{p,q,r,t+\nu} \, M_{p,\nu,t,\mu} \, M_{q,s-\nu,r-s-t,\lambda} \, (d_{m,n})^{\lambda+\mu}$$

$$\equiv \sum_{\mu=0}^{s} \sum_{\lambda=0}^{s-\mu} \sum_{\nu=\mu}^{s-\lambda} \mathfrak{C}_{p,q,r,t+\nu} \, M_{p,\nu,t,\mu} \, M_{q,s-\nu,r-s-t,\lambda} \, (d_{m,n})^{\lambda+\mu}$$

$$\equiv \sum_{k=0}^{s} \sum_{\mu=0}^{k} \sum_{\nu=\mu}^{s+\mu-k} \mathfrak{C}_{p,q,r,t+\nu} \, M_{p,\nu,t,\mu} \, M_{q,s-\nu,r-s-t,k-\mu} \, (d_{m,n})^{k}$$

$$\equiv \sum_{k=0}^{s} \sum_{\mu=0}^{k} \sum_{\nu=0}^{s-k} \mathfrak{C}_{p,q,r,t+\mu+\nu} M_{p,\mu+\nu,t,\mu} M_{q,s-\mu-\nu,r-s-t,k-\mu} (d_{m,n})^{k}$$

$$\equiv \sum_{k=0}^{s} N_{p,q,r,s,t,k} \, (d_{m,n})^{k},$$

where

(11.61) $$N_{p,q,r,s,t,k} \equiv \sum_{\nu=0}^{s-k} \sum_{\mu=0}^{k} \mathfrak{C}_{p,q,r,t+\mu+\nu} \, M_{p,\mu+\nu,t,\mu} \, M_{q,s-\mu-\nu,r-s-t,k-\mu}.$$

Since (11.53) of page 118 yields

$$M_{p,\mu+\nu,t,\mu} \equiv \mu! \binom{t+\mu+\nu}{\mu} \binom{2p+t+\mu+\nu-1}{\mu} B_{p,\nu,t+\nu}$$

and

$$M_{q,s-\mu-\nu,r-s-t,k-\mu} \equiv \left[\begin{array}{c} (k-\mu)! \binom{r-t-\mu-\nu}{k-\mu} \\ \times \binom{2q+r-t-\mu-\nu-1}{k-\mu} \end{array} \right] B_{q,s-k-\nu,r-t-k-\nu},$$

we introduce the rational numbers $\Theta_{p,q,r,t,k,\nu}$ defined by

$$(11.62) \quad \Theta_{p,q,r,t,k,\nu} \equiv \sum_{\mu=0}^{k} \mu!(k-\mu)! \left[\begin{array}{c} \mathfrak{C}_{p,q,r,t+\mu+\nu} \binom{t+\mu+\nu}{\mu} \\ \times \binom{2p+t+\mu+\nu-1}{\mu} \binom{r-t-\mu-\nu}{k-\mu} \\ \times \binom{2q+r-t-\mu-\nu-1}{k-\mu} \end{array} \right]$$

in order to rewrite (11.61) as

$$(11.63) \quad N_{p,q,r,s,t,k} \equiv \sum_{\nu=0}^{s-k} \Theta_{p,q,r,t,k,\nu} \, B_{p,\nu,t+\nu} \, B_{q,s-k-\nu,r-t-k-\nu}.$$

For $p, q, r \geq 2$ and $t, \mu, \nu \geq 0$, we have

$$\mathfrak{C}_{p,q,r,t+\mu+\nu} \equiv (-1)^{t+\mu+\nu} \frac{\binom{r}{t+\mu+\nu}\binom{2q+r-1}{t+\mu+\nu}}{\binom{2p+t+\mu+\nu-1}{t+\mu+\nu}}$$

from (4.49) of page 36 and we check, as we did for (9.62) of page 93, that

$$\binom{r}{t+\mu+\nu}\binom{r-t-\mu-\nu}{k-\mu} \equiv \binom{r}{t+k+\nu}\binom{t+k+\nu}{k-\mu},$$

$$\binom{2q+r-1}{t+\mu+\nu}\binom{2q+r-t-\mu-\nu-1}{k-\mu} \equiv \binom{2q+r-1}{t+k+\nu}\binom{t+k+\nu}{k-\mu},$$

and

$$\frac{\binom{2p+t+\mu+\nu-1}{\mu}}{\binom{2p+t+\mu+\nu-1}{t+\mu+\nu}} \equiv \frac{\binom{t+\mu+\nu}{\mu}}{\binom{2p+t+\nu-1}{t+\nu}}$$

are valid identities. Since the expression $\kappa_{p,q,r,t,k,\nu}$ defined by

$$\kappa_{p,q,r,t,k,\nu} \equiv (-1)^{t+\nu} k! \, \frac{\binom{r}{t+k+\nu}\binom{2q+r-1}{t+k+\nu}}{\binom{2p+t+\nu-1}{t+\nu}}$$

is independent of μ, we verify that (11.62) yields

$$(11.64) \quad \Theta_{p,q,r,t,k,\nu} \equiv \kappa_{p,q,r,t,k,\nu} \sum_{\mu=0}^{k} \frac{(-1)^{\mu}}{\binom{k}{\mu}} \left[\binom{t+k+\nu}{k-\mu}\binom{t+\mu+\nu}{\mu} \right]^2.$$

In view of the identity

$$\binom{t+k+\nu}{k-\mu}\binom{t+\mu+\nu}{\mu} \equiv \binom{t+k+\nu}{k}\binom{k}{\mu},$$

we rewrite (11.64) as

$$\Theta_{p,q,r,t,k,\nu} \equiv \left[\binom{t+k+\nu}{k}\right]^2 \kappa_{p,q,r,t,k,\nu} \sum_{\mu=0}^{k}(-1)^\mu \binom{k}{\mu}. \qquad (11.65)$$

We have

$$\sum_{\mu=0}^{k}(-1)^\mu \binom{k}{\mu} \equiv (1+(-1))^k \equiv 0, \quad \text{for } 1 \leq k \leq s.$$

Thus, (11.65) yields

$$\Theta_{p,q,r,t,k,\nu} \equiv 0, \quad \text{for } 1 \leq k \leq s. \qquad (11.66)$$

We use (11.63) and (11.66) to deduce

$$\boldsymbol{N}_{p,q,r,s,t,k} \equiv 0, \quad \text{for } 1 \leq k \leq s. \qquad (11.67)$$

Moreover, for $k=0$, (11.62) gives

$$\Theta_{p,q,r,t,0,\nu} \equiv \mathfrak{C}_{p,q,r,t+\nu}. \qquad (11.68)$$

As a consequence of (11.60), (11.67), (11.63), (11.68), and (4.50), we obtain

$$\boldsymbol{F}_{p,q,r,s,t} \equiv \sum_{k=0}^{s} \boldsymbol{N}_{p,q,r,s,t,k} \left(\boldsymbol{d}_{m,n}\right)^k$$

$$\equiv \boldsymbol{N}_{p,q,r,s,t,0}$$

$$\equiv \sum_{\nu=0}^{s} \Theta_{p,q,r,t,0,\nu} \boldsymbol{B}_{p,\nu,t+\nu} \boldsymbol{B}_{q,s-\nu,r-t-\nu}$$

$$\equiv \sum_{\nu=0}^{s} \mathfrak{C}_{p,q,r,t+\nu} \boldsymbol{B}_{p,\nu,t+\nu} \boldsymbol{B}_{q,s-\nu,r-t-\nu}$$

$$\equiv \boldsymbol{A}_{p,q,r,s,t}.$$

This yields (11.59) and completes the proof. $\square$

THEOREM 11.11. *The differential polynomials $\boldsymbol{R}$ of (11.1) and $\boldsymbol{S}$ of (11.13) are related by $\boldsymbol{R} \equiv \boldsymbol{S}$. Moreover, either $\boldsymbol{R} \equiv 0$ or $\boldsymbol{R}$ is a relative invariant in $\mathcal{R}_{m,n}$ for $\mathcal{C}_{m,n}$ of weight $p+q+r$.*

PROOF. We employ (11.24), (11.59), and (11.1) to verify that

$$\boldsymbol{S} \equiv \sum_{s=0}^{r}\sum_{t=0}^{r-s} \boldsymbol{F}_{p,q,r,s,t}\, \boldsymbol{P}^{(t)}\, \boldsymbol{Q}^{(r-s-t)} \equiv \sum_{s=0}^{r}\sum_{t=0}^{r-s} \boldsymbol{A}_{p,q,r,s,t}\, \boldsymbol{P}^{(t)}\, \boldsymbol{Q}^{(r-s-t)} \equiv \boldsymbol{R}.$$

In view of $\boldsymbol{S} \equiv \boldsymbol{R}$, the observations of Section 11.1, and Proposition 11.1, we see that: if $\boldsymbol{R} \not\equiv 0$, then $\boldsymbol{R}$ is a semi-invariant of the first kind, $\boldsymbol{S} \not\equiv 0$, $\boldsymbol{S}$ is a semi-invariant of the second kind of weight $p+q+r$, and $\boldsymbol{R}$ is therefore a relative invariant in $\mathcal{R}_{m,n}$ for $\mathcal{C}_{m,n}$ of weight $p+q+r$. This completes the proof. $\square$

OBSERVATION 11.12. For the situation $m, r \geq 2$, merely assume that $\boldsymbol{P}$ and $\boldsymbol{Q}$ are isobaric polynomials in $\mathcal{R}_{m,n}$ of respective positive weights p and q. Then, for $\boldsymbol{R}$ defined in $\mathcal{R}_{m,n}$ by (11.1) as well as (4.45)–(4.50) and for $\boldsymbol{S}$ defined in $\mathcal{R}_{m,n}$ by (11.13) as well as (11.2)–(11.12), our arguments show that $\boldsymbol{R} \equiv \boldsymbol{S}$. Moreover, if they are nonzero, then $\boldsymbol{R}$ and $\boldsymbol{S}$ are isobaric polynomials of weight $p+q+r$.

In particular, suppose that $\boldsymbol{P}$ and $\boldsymbol{Q}$ are isobaric semi-invariants of the first kind in $\mathcal{R}_{m,n}$ for $\mathcal{C}_{m,n}$ of respective weights p and q such that $\boldsymbol{S} \not\equiv 0$. Then, $\boldsymbol{S}$ is an isobaric semi-invariant of the first kind in $\mathcal{R}_{m,n}$ for $\mathcal{C}_{m,n}$ of weight $p+q+r$.

As another condition, suppose that $\boldsymbol{P}$ and $\boldsymbol{Q}$ are semi-invariants of the second kind in $\mathcal{R}_{m,n}$ for $\mathcal{C}_{m,n}$ having respective weights p and q such that $\boldsymbol{R} \not\equiv 0$. Then, $\boldsymbol{R}$ is a semi-invariant of the second kind in $\mathcal{R}_{m,n}$ for $\mathcal{C}_{m,n}$ of weight $p+q+r$.

OBSERVATION 11.13. For $r = 1$ and the context of Chapter 8, we note that (4.51) of page 36 expresses $\boldsymbol{C}_{p,q,1}(\boldsymbol{P}, \boldsymbol{Q})$ as the difference of two semi-invariants of the first kind. Using (4.51) as well as (5.15) and Proposition 5.5 of page 44, we see that the identity

$$(11.69) \quad \boldsymbol{C}_{p,q,1}(\boldsymbol{P}, \boldsymbol{Q}) \equiv \boldsymbol{P}\boldsymbol{Q}^{(1)} - \frac{q}{p}\boldsymbol{P}^{(1)}\boldsymbol{Q}$$
$$\equiv \boldsymbol{P}\left(\boldsymbol{Q}^{(1)} + 2q\,\boldsymbol{d}_{m,n}\boldsymbol{Q}\right) - \frac{q}{p}\left(\boldsymbol{P}^{(1)} + 2p\,\boldsymbol{d}_{m,n}\boldsymbol{P}\right)\boldsymbol{Q}$$

expresses $\boldsymbol{C}_{p,q,1}(\boldsymbol{P}, \boldsymbol{Q})$ as the difference of two semi-invariants of the second kind. In particular, (11.69) for $r = 1$ is analogous to $\boldsymbol{S}$ for $m, r \geq 2$.

Part 4

Relative Invariants of a Given Weight

CHAPTER 12

Representations Involving $C_{p,q,r}(P, Q)$

In Chapter 7, we employed $C_{p,q,r}(P, Q)$ to provide useful representations for the historically significant relative invariants D_2, E_6, and E_7 in $\mathcal{R}_{2,2}$ for $\mathcal{C}_{2,2}$. In Part 4, we give additional evidence for the belief that representations analogous to (7.20), (7.18), and (7.19) for D_2, E_6, and E_7 exist for the relative invariants in the $\mathcal{R}_{m,n}$ for each $\mathcal{C}_{m,n}$ having $m \geq 2$ and $(m, n) \neq (2, 1)$.

We recall from page 38 that the subset $\mathcal{V}_{m,n;s}$ of $\mathcal{R}_{m,n}$, consisting of 0 and all of the relative invariants in $\mathcal{R}_{m,n}$ for $\mathcal{C}_{m,n}$ having weight s, forms a vector space over $\mathbb{Q}$ whose dimension $d_{m,n}(s)$ is finite. For $n = 1$ and $m \geq 3$ or $n = 2$ and $m \geq 2$ with m, s of modest size, the programs presented in Sections 13.1 and 14.1 compute the dimension $d_{m,n}(s)$ of $\mathcal{V}_{m,n;s}$ and, when $d_{m,n}(s)$ is positive, they produce machine representations for a basis of $\mathcal{V}_{m,n;s}$. The remainders of Chapters 13 and 14 show how machine-generated bases can be replaced with ones conveniently represented by $C_{p,q,r}(P, Q)$ in terms of basic relative invariants. Summaries of selected results are given: for $(m, n) = (3, 1)$ in Section 4.8; for $(m, n) = (4, 1)$ in Section 12.1; for $(m, n) = (5, 1)$ in Section 12.2; and for $(m, n) = (2, 2)$ in Section 12.3.

12.1. The relative invariants in $\mathcal{R}_{4,1}$ for $\mathcal{C}_{4,1}$ of weight $s \leq 12$

The basic relative invariants in $\mathcal{R}_{4,1}$ for $\mathcal{C}_{4,1}$ are obtained by applying the technique of Section 6.1 to Theorem 4.6 with $m = 4$ and $n = 1$. They are

$$(12.1) \quad \mathcal{I}_{4,1;3} \equiv w_3 - \tfrac{1}{2} w_1 w_2 + \tfrac{1}{8}(w_1)^3 - w_2^{(1)} + \tfrac{3}{4} w_1 w_1^{(1)} + \tfrac{1}{2} w_1^{(2)}$$

and

$$(12.2) \quad \begin{aligned}\mathcal{I}_{4,1;4} \equiv{} & w_4 - \tfrac{1}{4} w_1 w_3 - \tfrac{1}{2} w_3^{(1)} - \tfrac{9}{100}(w_2)^2 + \tfrac{1}{5} w_2^{(2)} + \tfrac{13}{100}(w_1)^2 w_2 \\ & + \tfrac{27}{100} w_1^{(1)} w_2 + \tfrac{1}{4} w_1 w_2^{(1)} - \tfrac{39}{1600}(w_1)^4 - \tfrac{39}{200}(w_1)^2 w_1^{(1)} \\ & - \tfrac{33}{200}(w_1^{(1)})^2 - \tfrac{3}{20} w_1 w_1^{(2)} - \tfrac{1}{20} w_1^{(3)}.\end{aligned}$$

Here, as a convenience, we write $\mathcal{I}_3$ for $\mathcal{I}_{4,1;3}$ and $\mathcal{I}_4$ for $\mathcal{I}_{4,1;4}$. Moreover, (4.18) or (4.45) yield $a_{4,1} \equiv \tfrac{1}{10}[w_2 - \tfrac{3}{2} w_1^{(1)} - \tfrac{3}{8}(w_1)^2]$. The technique of Section 13.5 yields $d_{4,1}(2) = d_{4,1}(5) = 0$ and shows that the relative invariants in $\mathcal{R}_{4,1}$ for $\mathcal{C}_{4,1}$ having weight $s \leq 12$ are uniquely given by

$$(12.3) \quad \sum_{i=1}^{d_{4,1}(s)} K_{s,i} M_{s,i}, \quad \text{with } K_{s,1}, K_{s,2}, \ldots, K_{s,d_{4,1}(s)} \text{ in } \mathbb{Q} \text{ not all zero, where:}$$

the weight $s = 3$ has $d_{4,1}(3) = 1$ and $M_{3,1} \equiv \mathcal{I}_3$;

the weight $s = 4$ has $d_{4,1}(4) = 1$ and $M_{4,1} \equiv \mathcal{I}_4$;

the weight $s = 6$ has $d_{4,1}(6) = 1$ and $M_{6,1} \equiv (\mathcal{I}_3)^2$;

the weight $s = 7$ has $d_{4,1}(7) = 1$ and $M_{7,1} \equiv \mathcal{I}_3 \mathcal{I}_4$;

the weight $s = 8$ has $d_{4,1}(8) = 3$ and

$$M_{8,1} \equiv (\mathcal{I}_4)^2, \qquad M_{8,2} \equiv C_{3,3,2}(\mathcal{I}_3, \mathcal{I}_3),$$
$$M_{8,3} \equiv C_{3,4,1}(\mathcal{I}_3, \mathcal{I}_4);$$

the weight $s = 9$ has $d_{4,1}(9) = 2$ and

$$M_{9,1} \equiv (\mathcal{I}_3)^3, \qquad M_{9,2} \equiv C_{3,4,2}(\mathcal{I}_3, \mathcal{I}_4);$$

the weight $s = 10$ has $d_{4,1}(10) = 4$ and

$$M_{10,1} \equiv (\mathcal{I}_3)^2 \mathcal{I}_4, \qquad M_{10,2} \equiv C_{3,3,4}(\mathcal{I}_3, \mathcal{I}_3),$$
$$M_{10,3} \equiv C_{3,4,3}(\mathcal{I}_3, \mathcal{I}_4), \qquad M_{10,4} \equiv C_{4,4,2}(\mathcal{I}_4, \mathcal{I}_4);$$

the weight $s = 11$ has $d_{4,1}(11) = 4$ and

$$M_{11,1} \equiv \mathcal{I}_3(\mathcal{I}_4)^2, \qquad M_{11,2} \equiv C_{3,4,4}(\mathcal{I}_3, \mathcal{I}_4),$$
$$M_{11,3} \equiv C_{3,6,2}(\mathcal{I}_3, (\mathcal{I}_3)^2), \qquad M_{11,4} \equiv C_{3,7,1}(\mathcal{I}_3, (\mathcal{I}_3 \mathcal{I}_4));$$

the weight $s = 12$ has $d_{4,1}(12) = 9$ and

$$M_{12,1} \equiv (\mathcal{I}_3)^4, \qquad M_{12,2} \equiv (\mathcal{I}_4)^3,$$
$$M_{12,3} \equiv C_{3,3,6}(\mathcal{I}_3, \mathcal{I}_3), \qquad M_{12,4} \equiv C_{3,4,5}(\mathcal{I}_3, \mathcal{I}_4),$$
$$M_{12,5} \equiv C_{4,4,4}(\mathcal{I}_4, \mathcal{I}_4), \qquad M_{12,6} \equiv C_{3,6,3}(\mathcal{I}_3, (\mathcal{I}_3)^2),$$
$$M_{12,7} \equiv C_{3,7,2}(\mathcal{I}_3, (\mathcal{I}_3 \mathcal{I}_4)), \qquad M_{12,8} \equiv C_{4,6,2}(\mathcal{I}_4, (\mathcal{I}_3)^2),$$
$$M_{12,9} \equiv C_{3,8,1}(\mathcal{I}_3, (\mathcal{I}_4)^2).$$

12.2. The relative invariants in $\mathcal{R}_{5,1}$ for $\mathcal{C}_{5,1}$ of weight $s \leq 12$

We apply the technique of Section 6.1 or Section 6.2 to see that the basic relative invariants in $\mathcal{R}_{5,1}$ for $\mathcal{C}_{5,1}$ are given by

(12.4) $\mathcal{I}_{5,1;3} \equiv w_3 - \frac{3}{5} w_1 w_2 + \frac{4}{25}(w_1)^3 - \frac{3}{2} w_2^{(1)} + \frac{6}{5} w_1 w_1^{(1)} + w_1^{(2)},$

(12.5) $\mathcal{I}_{5,1;4} \equiv w_4 - \frac{2}{5} w_1 w_3 - \frac{4}{25}(w_2)^2 + \frac{31}{125}(w_1)^2 w_2 - \frac{31}{625}(w_1)^4 - w_3^{(1)}$
$\qquad + \frac{3}{5} w_2^{(2)} + \frac{16}{25} w_1^{(1)} w_2 + \frac{3}{5} w_1 w_2^{(1)} - \frac{62}{125}(w_1)^2 w_1^{(1)} - \frac{12}{25} w_1 w_1^{(2)}$
$\qquad - \frac{13}{25}(w_1^{(1)})^2 - \frac{1}{5} w_1^{(3)},$

and

(12.6) $\mathcal{I}_{5,1;5} \equiv w_5 - \frac{1}{5} w_1 w_4 - \frac{4}{35} w_2 w_3 + \frac{3}{35}(w_1)^2 w_3 + \frac{12}{175} w_1 (w_2)^2$
$\qquad - \frac{47}{875}(w_1)^3 w_2 + \frac{188}{21875}(w_1)^5 - \frac{1}{2} w_4^{(1)} + \frac{3}{14} w_3^{(2)} + \frac{1}{5} w_1 w_3^{(1)}$
$\qquad + \frac{8}{35} w_1^{(1)} w_3 - \frac{1}{14} w_2^{(3)} - \frac{9}{70} w_1 w_2^{(2)} + \frac{6}{35} w_2 w_2^{(1)} - \frac{9}{70}(w_1)^2 w_2^{(1)}$
$\qquad - \frac{3}{10} w_1^{(1)} w_2^{(1)} - \frac{1}{7} w_1^{(2)} w_2 - \frac{48}{175} w_1 w_1^{(1)} w_2 + \frac{1}{70} w_1^{(4)} + \frac{2}{35} w_1 w_1^{(3)}$
$\qquad + \frac{1}{5} w_1^{(1)} w_1^{(2)} + \frac{19}{175}(w_1)^2 w_1^{(2)} + \frac{6}{25} w_1 (w_1^{(1)})^2 + \frac{94}{875}(w_1)^3 w_1^{(1)}.$

Here, we write $\mathcal{I}_3$ for $\mathcal{I}_{5,1;3}$, $\mathcal{I}_4$ for $\mathcal{I}_{5,1;4}$, and $\mathcal{I}_5$ for $\mathcal{I}_{5,1;5}$. Using the technique of Section 13.6 with $a_{5,1} \equiv \frac{1}{20}[w_2 - 2w_1^{(1)} - \frac{2}{5}(w_1)^2]$ from (4.18) or (4.45), we see

12.2. THE RELATIVE INVARIANTS IN $\mathcal{R}_{5,1}$ FOR $\mathcal{C}_{5,1}$ OF WEIGHT $s \leq 12$

that the relative invariants of weight $s \leq 12$ in $\mathcal{R}_{5,1}$ for $\mathcal{C}_{5,1}$ are uniquely given by

$$(12.7) \quad \sum_{i=1}^{d_{5,1}(s)} K_{s,i} \, \boldsymbol{N}_{s,i}, \quad \text{with } K_{s,1}, K_{s,2}, \ldots, K_{s,d_{5,1}(s)} \text{ in } \mathbb{Q} \text{ not all zero, where:}$$

the weight $s = 3$ has $d_{5,1}(3) = 1$ and $\boldsymbol{N}_{3,1} \equiv \boldsymbol{\mathcal{I}}_3$;

the weight $s = 4$ has $d_{5,1}(4) = 1$ and $\boldsymbol{N}_{4,1} \equiv \boldsymbol{\mathcal{I}}_4$;

the weight $s = 5$ has $d_{5,1}(5) = 1$ and $\boldsymbol{N}_{5,1} \equiv \boldsymbol{\mathcal{I}}_5$;

the weight $s = 6$ has $d_{5,1}(6) = 1$ and $\boldsymbol{N}_{6,1} \equiv (\boldsymbol{\mathcal{I}}_3)^2$;

the weight $s = 7$ has $d_{5,1}(7) = 1$ and $\boldsymbol{N}_{7,1} \equiv \boldsymbol{\mathcal{I}}_3 \boldsymbol{\mathcal{I}}_4$;

the weight $s = 8$ has $d_{5,1}(8) = 4$ and

$\boldsymbol{N}_{8,1} \equiv \boldsymbol{\mathcal{I}}_3 \boldsymbol{\mathcal{I}}_5$, $\qquad\qquad \boldsymbol{N}_{8,2} \equiv (\boldsymbol{\mathcal{I}}_4)^2$,

$\boldsymbol{N}_{8,3} \equiv \boldsymbol{C}_{3,3,2}(\boldsymbol{\mathcal{I}}_3, \boldsymbol{\mathcal{I}}_3)$, $\qquad \boldsymbol{N}_{8,4} \equiv \boldsymbol{C}_{3,4,1}(\boldsymbol{\mathcal{I}}_3, \boldsymbol{\mathcal{I}}_4)$;

the weight $s = 9$ has $d_{5,1}(9) = 4$ and

$\boldsymbol{N}_{9,1} \equiv (\boldsymbol{\mathcal{I}}_3)^3$, $\qquad\qquad \boldsymbol{N}_{9,2} \equiv \boldsymbol{\mathcal{I}}_4 \boldsymbol{\mathcal{I}}_5$,

$\boldsymbol{N}_{9,3} \equiv \boldsymbol{C}_{3,4,2}(\boldsymbol{\mathcal{I}}_3, \boldsymbol{\mathcal{I}}_4)$, $\qquad \boldsymbol{N}_{9,4} \equiv \boldsymbol{C}_{3,5,1}(\boldsymbol{\mathcal{I}}_3, \boldsymbol{\mathcal{I}}_5)$;

the weight $s = 10$ has $d_{5,1}(10) = 7$ and

$\boldsymbol{N}_{10,1} \equiv (\boldsymbol{\mathcal{I}}_3)^2 \boldsymbol{\mathcal{I}}_4$, $\qquad\qquad \boldsymbol{N}_{10,2} \equiv (\boldsymbol{\mathcal{I}}_5)^2$,

$\boldsymbol{N}_{10,3} \equiv \boldsymbol{C}_{3,3,4}(\boldsymbol{\mathcal{I}}_3, \boldsymbol{\mathcal{I}}_3)$, $\qquad \boldsymbol{N}_{10,4} \equiv \boldsymbol{C}_{3,4,3}(\boldsymbol{\mathcal{I}}_3, \boldsymbol{\mathcal{I}}_4)$,

$\boldsymbol{N}_{10,5} \equiv \boldsymbol{C}_{3,5,2}(\boldsymbol{\mathcal{I}}_3, \boldsymbol{\mathcal{I}}_5)$, $\qquad \boldsymbol{N}_{10,6} \equiv \boldsymbol{C}_{4,4,2}(\boldsymbol{\mathcal{I}}_4, \boldsymbol{\mathcal{I}}_4)$,

$\boldsymbol{N}_{10,7} \equiv \boldsymbol{C}_{4,5,1}(\boldsymbol{\mathcal{I}}_4, \boldsymbol{\mathcal{I}}_5)$;

the weight $s = 11$ has $d_{5,1}(11) = 7$ and

$\boldsymbol{N}_{11,1} \equiv (\boldsymbol{\mathcal{I}}_3)^2 \boldsymbol{\mathcal{I}}_5$, $\qquad\qquad \boldsymbol{N}_{11,2} \equiv \boldsymbol{\mathcal{I}}_3 (\boldsymbol{\mathcal{I}}_4)^2$,

$\boldsymbol{N}_{11,3} \equiv \boldsymbol{C}_{3,4,4}(\boldsymbol{\mathcal{I}}_3, \boldsymbol{\mathcal{I}}_4)$, $\qquad \boldsymbol{N}_{11,4} \equiv \boldsymbol{C}_{3,5,3}(\boldsymbol{\mathcal{I}}_3, \boldsymbol{\mathcal{I}}_5)$,

$\boldsymbol{N}_{11,5} \equiv \boldsymbol{C}_{3,6,2}(\boldsymbol{\mathcal{I}}_3, (\boldsymbol{\mathcal{I}}_3)^2)$ $\qquad \boldsymbol{N}_{11,6} \equiv \boldsymbol{C}_{4,5,2}(\boldsymbol{\mathcal{I}}_4, \boldsymbol{\mathcal{I}}_5)$,

$\boldsymbol{N}_{11,7} \equiv \boldsymbol{C}_{3,7,1}(\boldsymbol{\mathcal{I}}_3, (\boldsymbol{\mathcal{I}}_3 \boldsymbol{\mathcal{I}}_4))$;

the weight $s = 12$ has $d_{5,1}(12) = 14$ and

$\boldsymbol{N}_{12,1} \equiv (\boldsymbol{\mathcal{I}}_3)^4$, $\qquad\qquad \boldsymbol{N}_{12,2} \equiv (\boldsymbol{\mathcal{I}}_4)^3$,

$\boldsymbol{N}_{12,3} \equiv \boldsymbol{\mathcal{I}}_3 \boldsymbol{\mathcal{I}}_4 \boldsymbol{\mathcal{I}}_5$, $\qquad\qquad \boldsymbol{N}_{12,4} \equiv \boldsymbol{C}_{3,3,6}(\boldsymbol{\mathcal{I}}_3, \boldsymbol{\mathcal{I}}_3)$,

$\boldsymbol{N}_{12,5} \equiv \boldsymbol{C}_{3,4,5}(\boldsymbol{\mathcal{I}}_3, \boldsymbol{\mathcal{I}}_4)$, $\qquad \boldsymbol{N}_{12,6} \equiv \boldsymbol{C}_{3,5,4}(\boldsymbol{\mathcal{I}}_3, \boldsymbol{\mathcal{I}}_5)$,

$\boldsymbol{N}_{12,7} \equiv \boldsymbol{C}_{4,4,4}(\boldsymbol{\mathcal{I}}_4, \boldsymbol{\mathcal{I}}_4)$, $\qquad \boldsymbol{N}_{12,8} \equiv \boldsymbol{C}_{3,6,3}(\boldsymbol{\mathcal{I}}_3, (\boldsymbol{\mathcal{I}}_3)^2)$,

$\boldsymbol{N}_{12,9} \equiv \boldsymbol{C}_{4,5,3}(\boldsymbol{\mathcal{I}}_4, \boldsymbol{\mathcal{I}}_5)$, $\qquad \boldsymbol{N}_{12,10} \equiv \boldsymbol{C}_{3,7,2}(\boldsymbol{\mathcal{I}}_3, (\boldsymbol{\mathcal{I}}_3 \boldsymbol{\mathcal{I}}_4))$,

$\boldsymbol{N}_{12,11} \equiv \boldsymbol{C}_{4,6,2}(\boldsymbol{\mathcal{I}}_4, (\boldsymbol{\mathcal{I}}_3)^2)$, $\qquad \boldsymbol{N}_{12,12} \equiv \boldsymbol{C}_{5,5,2}(\boldsymbol{\mathcal{I}}_5, \boldsymbol{\mathcal{I}}_5)$,

$\boldsymbol{N}_{12,13} \equiv \boldsymbol{C}_{3,8,1}(\boldsymbol{\mathcal{I}}_3, (\boldsymbol{\mathcal{I}}_3 \boldsymbol{\mathcal{I}}_5))$, $\qquad \boldsymbol{N}_{12,14} \equiv \boldsymbol{C}_{3,8,1}(\boldsymbol{\mathcal{I}}_3, (\boldsymbol{\mathcal{I}}_4)^2)$.

12.3. The relative invariants in $\mathcal{R}_{2,2}$ for $\mathcal{C}_{2,2}$ of weight $s \leq 12$

The equations of $\mathcal{C}_{2,2}$ have the form given by (7.1) on page 65. The three basic relative invariants for $\mathcal{C}_{2,2}$ are $\boldsymbol{\mathcal{I}}_{2,2;\,1,1}$ of weight 2 in (7.14) of page 67 as well as $\boldsymbol{\mathcal{I}}_{2,2;\,1,2}$ of weight 3 in (7.15) and $\boldsymbol{\mathcal{I}}_{2,2;\,2,2}$ of weight 4 in (7.16). Here, we write $\boldsymbol{\mathcal{I}}_2$ for $\boldsymbol{\mathcal{I}}_{2,2;\,1,1}$, $\boldsymbol{\mathcal{I}}_3$ for $\boldsymbol{\mathcal{I}}_{2,2;\,1,2}$, and $\boldsymbol{\mathcal{I}}_4$ for $\boldsymbol{\mathcal{I}}_{2,2;\,2,2}$. Moreover, (4.18) or (4.45) yield $a_{2,2} \equiv w_{0,2} - \frac{1}{2} w_{0,1}^{(1)} - \frac{1}{4} (w_{0,1})^2$. Using the technique of Section 14.3, we discover that the relative invariants of weight $s \leq 12$ for $\mathcal{C}_{2,2}$ are uniquely specified through

$$(12.8) \quad \sum_{i=1}^{d_{2,2}(s)} K_{s,i} \boldsymbol{F}_{s,i}, \quad \text{with } K_{s,1}, K_{s,2}, \ldots, K_{s,d_{2,2}(s)} \text{ in } \mathbb{Q} \text{ not all zero, where:}$$

the weight $s = 2$ has $d_{2,2}(2) = 1$ and $\boldsymbol{F}_{2,1} \equiv \boldsymbol{\mathcal{I}}_2$;

the weight $s = 3$ has $d_{2,2}(3) = 1$ and $\boldsymbol{F}_{3,1} \equiv \boldsymbol{\mathcal{I}}_3$;

the weight $s = 4$ has $d_{2,2}(4) = 2$ and

$$\boldsymbol{F}_{4,1} \equiv (\boldsymbol{\mathcal{I}}_2)^2, \qquad \boldsymbol{F}_{4,2} \equiv \boldsymbol{\mathcal{I}}_4;$$

the weight $s = 5$ has $d_{2,2}(5) = 1$ and

$$\boldsymbol{F}_{5,1} \equiv \boldsymbol{\mathcal{I}}_2 \boldsymbol{\mathcal{I}}_3;$$

the weight $s = 6$ has $d_{2,2}(6) = 5$ and

$$\boldsymbol{F}_{6,1} \equiv (\boldsymbol{\mathcal{I}}_2)^3, \qquad \boldsymbol{F}_{6,2} \equiv (\boldsymbol{\mathcal{I}}_3)^2,$$
$$\boldsymbol{F}_{6,3} \equiv \boldsymbol{\mathcal{I}}_2 \boldsymbol{\mathcal{I}}_4, \qquad \boldsymbol{F}_{6,4} \equiv C_{2,3,1}(\boldsymbol{\mathcal{I}}_2, \boldsymbol{\mathcal{I}}_3),$$
$$\boldsymbol{F}_{6,5} \equiv C_{2,2,2}(\boldsymbol{\mathcal{I}}_2, \boldsymbol{\mathcal{I}}_2);$$

the weight $s = 7$ has $d_{2,2}(7) = 4$ and

$$\boldsymbol{F}_{7,1} \equiv (\boldsymbol{\mathcal{I}}_2)^2 \boldsymbol{\mathcal{I}}_3, \qquad \boldsymbol{F}_{7,2} \equiv \boldsymbol{\mathcal{I}}_3 \boldsymbol{\mathcal{I}}_4,$$
$$\boldsymbol{F}_{7,3} \equiv C_{2,4,1}(\boldsymbol{\mathcal{I}}_2, \boldsymbol{\mathcal{I}}_4), \qquad \boldsymbol{F}_{7,4} \equiv C_{2,3,2}(\boldsymbol{\mathcal{I}}_2, \boldsymbol{\mathcal{I}}_3);$$

the weight $s = 8$ has $d_{2,2}(8) = 11$ and

$$\boldsymbol{F}_{8,1} \equiv (\boldsymbol{\mathcal{I}}_2)^4, \qquad \boldsymbol{F}_{8,2} \equiv \boldsymbol{\mathcal{I}}_2 (\boldsymbol{\mathcal{I}}_3)^2,$$
$$\boldsymbol{F}_{8,3} \equiv (\boldsymbol{\mathcal{I}}_2)^2 \boldsymbol{\mathcal{I}}_4, \qquad \boldsymbol{F}_{8,4} \equiv (\boldsymbol{\mathcal{I}}_4)^2,$$
$$\boldsymbol{F}_{8,5} \equiv \boldsymbol{\mathcal{I}}_2 C_{2,3,1}(\boldsymbol{\mathcal{I}}_2, \boldsymbol{\mathcal{I}}_3), \qquad \boldsymbol{F}_{8,6} \equiv \boldsymbol{\mathcal{I}}_2 C_{2,2,2}(\boldsymbol{\mathcal{I}}_2, \boldsymbol{\mathcal{I}}_2),$$
$$\boldsymbol{F}_{8,7} \equiv C_{3,4,1}(\boldsymbol{\mathcal{I}}_3, \boldsymbol{\mathcal{I}}_4), \qquad \boldsymbol{F}_{8,8} \equiv C_{3,3,2}(\boldsymbol{\mathcal{I}}_3, \boldsymbol{\mathcal{I}}_3),$$
$$\boldsymbol{F}_{8,9} \equiv C_{2,4,2}(\boldsymbol{\mathcal{I}}_2, \boldsymbol{\mathcal{I}}_4), \qquad \boldsymbol{F}_{8,10} \equiv C_{2,3,3}(\boldsymbol{\mathcal{I}}_2, \boldsymbol{\mathcal{I}}_3),$$
$$\boldsymbol{F}_{8,11} \equiv C_{2,2,4}(\boldsymbol{\mathcal{I}}_2, \boldsymbol{\mathcal{I}}_2);$$

the weight $s = 9$ has $d_{2,2}(9) = 11$ and

$$F_{9,1} \equiv (\mathcal{I}_2)^3 \mathcal{I}_3, \qquad F_{9,2} \equiv (\mathcal{I}_3)^3,$$
$$F_{9,3} \equiv \mathcal{I}_2 \mathcal{I}_3 \mathcal{I}_4, \qquad F_{9,4} \equiv \mathcal{I}_3 C_{2,3,1}(\mathcal{I}_2, \mathcal{I}_3),$$
$$F_{9,5} \equiv \mathcal{I}_3 C_{2,2,2}(\mathcal{I}_2, \mathcal{I}_2), \qquad F_{9,6} \equiv \mathcal{I}_2 C_{2,4,1}(\mathcal{I}_2, \mathcal{I}_4),$$
$$F_{9,7} \equiv \mathcal{I}_2 C_{2,3,2}(\mathcal{I}_2, \mathcal{I}_3), \qquad F_{9,8} \equiv C_{3,4,2}(\mathcal{I}_3, \mathcal{I}_4),$$
$$F_{9,9} \equiv C_{2,4,3}(\mathcal{I}_2), (\mathcal{I}_2)^2), \qquad F_{9,10} \equiv C_{2,4,3}(\mathcal{I}_2, \mathcal{I}_4),$$
$$F_{9,11} \equiv C_{2,3,4}(\mathcal{I}_2, \mathcal{I}_3);$$

the weight $s = 10$ has $d_{2,2}(10) = 22$ and

$$F_{10,1} \equiv (\mathcal{I}_2)^5, \qquad F_{10,2} \equiv (\mathcal{I}_2)^2 (\mathcal{I}_3)^2,$$
$$F_{10,3} \equiv (\mathcal{I}_2)^3 \mathcal{I}_4, \qquad F_{10,4} \equiv \mathcal{I}_2 (\mathcal{I}_4)^2,$$
$$F_{10,5} \equiv (\mathcal{I}_3)^2 \mathcal{I}_4, \qquad F_{10,6} \equiv (\mathcal{I}_2)^2 C_{2,3,1}(\mathcal{I}_2, \mathcal{I}_3),$$
$$F_{10,7} \equiv (\mathcal{I}_2)^2 C_{2,2,2}(\mathcal{I}_2, \mathcal{I}_2), \qquad F_{10,8} \equiv \mathcal{I}_2 C_{3,4,1}(\mathcal{I}_3, \mathcal{I}_4),$$
$$F_{10,9} \equiv \mathcal{I}_2 C_{3,3,2}(\mathcal{I}_3, \mathcal{I}_3), \qquad F_{10,10} \equiv \mathcal{I}_2 C_{2,4,2}(\mathcal{I}_2, \mathcal{I}_4),$$
$$F_{10,11} \equiv \mathcal{I}_2 C_{2,3,3}(\mathcal{I}_2, \mathcal{I}_3), \qquad F_{10,12} \equiv \mathcal{I}_2 C_{2,2,4}(\mathcal{I}_2, \mathcal{I}_2),$$
$$F_{10,13} \equiv \mathcal{I}_3 C_{2,4,1}(\mathcal{I}_2, \mathcal{I}_4), \qquad F_{10,14} \equiv \mathcal{I}_3 C_{2,3,2}(\mathcal{I}_2, \mathcal{I}_3),$$
$$F_{10,15} \equiv \mathcal{I}_4 C_{2,2,2}(\mathcal{I}_2, \mathcal{I}_2), \qquad F_{10,16} \equiv C_{2,5,3}(\mathcal{I}_2, (\mathcal{I}_2 \mathcal{I}_3)),$$
$$F_{10,17} \equiv C_{4,4,2}(\mathcal{I}_4, \mathcal{I}_4), \qquad F_{10,18} \equiv C_{3,4,3}(\mathcal{I}_3, \mathcal{I}_4),$$
$$F_{10,19} \equiv C_{2,4,4}(\mathcal{I}_2, \mathcal{I}_4), \qquad F_{10,20} \equiv C_{3,3,4}(\mathcal{I}_3, \mathcal{I}_3),$$
$$F_{10,21} \equiv C_{2,3,5}(\mathcal{I}_2, \mathcal{I}_3), \qquad F_{10,22} \equiv C_{2,2,6}(\mathcal{I}_2, \mathcal{I}_2);$$

the weight $s = 11$ has $d_{2,2}(11) = 26$ and

$$F_{11,1} \equiv (\mathcal{I}_2)^4 \mathcal{I}_3, \qquad F_{11,2} \equiv \mathcal{I}_2 (\mathcal{I}_3)^3,$$
$$F_{11,3} \equiv (\mathcal{I}_2)^2 \mathcal{I}_3 \mathcal{I}_4, \qquad F_{11,4} \equiv \mathcal{I}_3 (\mathcal{I}_4)^2,$$
$$F_{11,5} \equiv (\mathcal{I}_2 \mathcal{I}_3) C_{2,3,1}(\mathcal{I}_2, \mathcal{I}_3), \qquad F_{11,6} \equiv (\mathcal{I}_2 \mathcal{I}_3) C_{2,2,2}(\mathcal{I}_2, \mathcal{I}_2),$$
$$F_{11,7} \equiv (\mathcal{I}_2)^2 C_{2,4,1}(\mathcal{I}_2, \mathcal{I}_4), \qquad F_{11,8} \equiv (\mathcal{I}_2)^2 C_{2,3,2}(\mathcal{I}_2, \mathcal{I}_3),$$
$$F_{11,9} \equiv \mathcal{I}_2 C_{3,4,2}(\mathcal{I}_3, \mathcal{I}_4), \qquad F_{11,10} \equiv \mathcal{I}_2 C_{2,4,3}(\mathcal{I}_2, (\mathcal{I}_2)^2),$$
$$F_{11,11} \equiv \mathcal{I}_2 C_{2,4,3}(\mathcal{I}_2, \mathcal{I}_4), \qquad F_{11,12} \equiv \mathcal{I}_2 C_{2,3,4}(\mathcal{I}_2, \mathcal{I}_3),$$
$$F_{11,13} \equiv \mathcal{I}_3 C_{3,4,1}(\mathcal{I}_3, \mathcal{I}_4), \qquad F_{11,14} \equiv \mathcal{I}_3 C_{3,3,2}(\mathcal{I}_3, \mathcal{I}_3),$$
$$F_{11,15} \equiv \mathcal{I}_3 C_{2,4,2}(\mathcal{I}_2, \mathcal{I}_4), \qquad F_{11,16} \equiv \mathcal{I}_3 C_{2,3,3}(\mathcal{I}_2, \mathcal{I}_3),$$
$$F_{11,17} \equiv \mathcal{I}_3 C_{2,2,4}(\mathcal{I}_2, \mathcal{I}_2), \qquad F_{11,18} \equiv \mathcal{I}_4 C_{2,4,1}(\mathcal{I}_2, \mathcal{I}_4),$$

$$F_{11,19} \equiv \mathcal{I}_4\, C_{2,3,2}(\mathcal{I}_2, \mathcal{I}_3), \qquad F_{11,20} \equiv C_{2,3,6}(\mathcal{I}_2, \mathcal{I}_3),$$
$$F_{11,21} \equiv C_{2,4,5}(\mathcal{I}_2, \mathcal{I}_4), \qquad F_{11,22} \equiv C_{2,4,5}(\mathcal{I}_2, (\mathcal{I}_2)^2),$$
$$F_{11,23} \equiv C_{3,4,4}(\mathcal{I}_3, \mathcal{I}_4), \qquad F_{11,24} \equiv C_{3,4,4}(\mathcal{I}_3, (\mathcal{I}_2)^2),$$
$$F_{11,25} \equiv C_{3,5,3}(\mathcal{I}_3, (\mathcal{I}_2\mathcal{I}_3)), \qquad F_{11,26} \equiv C_{4,4,3}(\mathcal{I}_4, (\mathcal{I}_2)^2);$$

the weight $s = 12$ has $d_{2,2}(12) = 49$ and

$$F_{12,1} \equiv (\mathcal{I}_2)^6, \qquad F_{12,2} \equiv (\mathcal{I}_2)^4\,\mathcal{I}_4,$$
$$F_{12,3} \equiv (\mathcal{I}_2)^3\,(\mathcal{I}_3)^2, \qquad F_{12,4} \equiv (\mathcal{I}_2)^2\,(\mathcal{I}_4)^2,$$
$$F_{12,5} \equiv \mathcal{I}_2\,(\mathcal{I}_3)^2\,\mathcal{I}_4, \qquad F_{12,6} \equiv (\mathcal{I}_3)^4,$$
$$F_{12,7} \equiv (\mathcal{I}_4)^3, \qquad F_{12,8} \equiv \mathcal{I}_2\,C_{2,2,6}(\mathcal{I}_2, \mathcal{I}_2),$$
$$F_{12,9} \equiv \mathcal{I}_2\,C_{2,3,5}(\mathcal{I}_2, \mathcal{I}_3), \qquad F_{12,10} \equiv \mathcal{I}_2\,C_{2,4,4}(\mathcal{I}_2, \mathcal{I}_4),$$
$$F_{12,11} \equiv \mathcal{I}_2\,C_{3,3,4}(\mathcal{I}_3, \mathcal{I}_3), \qquad F_{12,12} \equiv \mathcal{I}_2\,C_{3,4,3}(\mathcal{I}_3, \mathcal{I}_4),$$
$$F_{12,13} \equiv \mathcal{I}_2\,C_{4,4,2}(\mathcal{I}_4, \mathcal{I}_4), \qquad F_{12,14} \equiv \mathcal{I}_2\,C_{2,5,3}(\mathcal{I}_2, (\mathcal{I}_2\mathcal{I}_3)),$$
$$F_{12,15} \equiv (\mathcal{I}_2)^2\,C_{2,2,4}(\mathcal{I}_2, \mathcal{I}_2), \qquad F_{12,16} \equiv (\mathcal{I}_2)^2\,C_{2,3,3}(\mathcal{I}_2, \mathcal{I}_3),$$
$$F_{12,17} \equiv (\mathcal{I}_2)^2\,C_{2,4,2}(\mathcal{I}_2, \mathcal{I}_4), \qquad F_{12,18} \equiv (\mathcal{I}_2)^2\,C_{3,3,2}(\mathcal{I}_3, \mathcal{I}_3),$$
$$F_{12,19} \equiv (\mathcal{I}_2)^2\,C_{3,4,1}(\mathcal{I}_3, \mathcal{I}_4), \qquad F_{12,20} \equiv \mathcal{I}_2\,\mathcal{I}_3\,C_{2,3,2}(\mathcal{I}_2, \mathcal{I}_3),$$
$$F_{12,21} \equiv \mathcal{I}_2\,\mathcal{I}_3\,C_{2,4,1}(\mathcal{I}_2, \mathcal{I}_4), \qquad F_{12,22} \equiv (\mathcal{I}_2)^3\,C_{2,2,2}(\mathcal{I}_2, \mathcal{I}_2),$$
$$F_{12,23} \equiv (\mathcal{I}_2)^3\,C_{2,3,1}(\mathcal{I}_2, \mathcal{I}_3), \qquad F_{12,24} \equiv \mathcal{I}_2\,\mathcal{I}_4\,C_{2,2,2}(\mathcal{I}_2, \mathcal{I}_2),$$
$$F_{12,25} \equiv \mathcal{I}_3\,C_{2,3,4}(\mathcal{I}_2, \mathcal{I}_3), \qquad F_{12,26} \equiv \mathcal{I}_3\,C_{2,4,3}(\mathcal{I}_2, (\mathcal{I}_2)^2),$$
$$F_{12,27} \equiv \mathcal{I}_3\,C_{2,4,3}(\mathcal{I}_2, \mathcal{I}_4), \qquad F_{12,28} \equiv \mathcal{I}_3\,C_{3,4,2}(\mathcal{I}_3, \mathcal{I}_4),$$
$$F_{12,29} \equiv (\mathcal{I}_3)^2\,C_{2,2,2}(\mathcal{I}_2, \mathcal{I}_2), \qquad F_{12,30} \equiv (\mathcal{I}_3)^2\,C_{2,3,1}(\mathcal{I}_2, \mathcal{I}_3),$$
$$F_{12,31} \equiv \mathcal{I}_4\,C_{2,2,4}(\mathcal{I}_2, \mathcal{I}_2), \qquad F_{12,32} \equiv \mathcal{I}_4\,C_{2,3,3}(\mathcal{I}_2, \mathcal{I}_3),$$
$$F_{12,33} \equiv \mathcal{I}_4\,C_{2,4,2}(\mathcal{I}_2, \mathcal{I}_4), \qquad F_{12,34} \equiv \mathcal{I}_4\,C_{3,3,2}(\mathcal{I}_3, \mathcal{I}_3),$$
$$F_{12,35} \equiv \mathcal{I}_4\,C_{3,4,1}(\mathcal{I}_3, \mathcal{I}_4), \qquad F_{12,36} \equiv C_{2,2,8}(\mathcal{I}_2, \mathcal{I}_2),$$
$$F_{12,37} \equiv C_{2,3,7}(\mathcal{I}_2, \mathcal{I}_3), \qquad F_{12,38} \equiv C_{2,4,6}(\mathcal{I}_2, (\mathcal{I}_2)^2),$$
$$F_{12,39} \equiv C_{2,4,6}(\mathcal{I}_2, \mathcal{I}_4), \qquad F_{12,40} \equiv C_{3,3,6}(\mathcal{I}_3, \mathcal{I}_3),$$
$$F_{12,41} \equiv C_{2,5,5}(\mathcal{I}_2, (\mathcal{I}_2\mathcal{I}_3)), \qquad F_{12,42} \equiv C_{3,4,5}(\mathcal{I}_3, (\mathcal{I}_2)^2),$$
$$F_{12,43} \equiv C_{3,4,5}(\mathcal{I}_3, \mathcal{I}_4), \qquad F_{12,44} \equiv C_{2,6,4}(\mathcal{I}_2, (\mathcal{I}_2)^3),$$
$$F_{12,45} \equiv C_{2,6,4}(\mathcal{I}_2, (\mathcal{I}_3)^2), \qquad F_{12,46} \equiv C_{4,4,4}(\mathcal{I}_4, (\mathcal{I}_2)^2),$$
$$F_{12,47} \equiv C_{4,4,4}(\mathcal{I}_4, \mathcal{I}_4), \qquad F_{12,48} \equiv C_{2,7,3}(\mathcal{I}_2, (\mathcal{I}_3\mathcal{I}_4)),$$
$$F_{12,49} \equiv C_{3,6,3}(\mathcal{I}_3, (\mathcal{I}_3)^2).$$

In particular, Theorem 4.10 has enabled us to systematically present and extend the isolated results in [**20**, Section 14.3] about (12.8) when $2 \leq s \leq 9$.

CHAPTER 13

Computer Algebra for $\mathcal{V}_{3,1;s}$, $\mathcal{V}_{4,1;s}$, $\mathcal{V}_{5,1;s}$, ...

The vector space $\mathcal{V}_{m,n;s}$ was introduced in Section 4.7 on page 38. The program of Section 13.1, enables the dimension of $\mathcal{V}_{m,1;s}$ to be computed for integers m, s of modest size and, when $\mathcal{V}_{m,1;s} \neq \{0\}$, it yields a machine representation for a basis of $\mathcal{V}_{m,1;s}$. We use it in Sections 13.2 and 13.4–13.6 to verify the properties of the relative invariants $\boldsymbol{L}_{s,i}$, $\boldsymbol{M}_{s,i}$, and $\boldsymbol{N}_{s,i}$ presented in Sections 4.8, 12.1, and 12.2.

13.1. The relative invariants of weight s in $\mathcal{R}_{m,1}$ for $\mathcal{C}_{m,1}$

When integers m, s satisfy $m, s \geq 3$, the program presented in this section specifies linearly independent relative invariants over $\mathbb{Q}$ of weight s in $\mathcal{R}_{m,1}$ for $\mathcal{C}_{m,1}$ such that their nonzero linear combinations over $\mathbb{Q}$ uniquely yield each relative invariant of weight s for $\mathcal{C}_{m,1}$. However, it requires that a specific integral value m be assigned to m in the first input statement. We have selected m = 3 on page 134 as the first input statement; and the program as written therefore applies to $\mathcal{R}_{3,1}$ for $\mathcal{C}_{3,1}$. By merely changing the first input statement to m = 4 or ... and making no other alterations, the program then applies to $\mathcal{R}_{4,1}$ for $\mathcal{C}_{4,1}$ or ....

We select a version of *Mathematica* from [55, 56, 57, 58, 59] as the system of computer algebra and represent the derivation ′ for $\mathcal{R}_{m,1}$ as differentiation with respect to a fictitious independent variable denoted by z. This enables us to represent the variables $w_{i_1}^{(k)}$ of (4.9) for $\mathcal{R}_{m,1}$ by D[w[i1][z],{z,k}] as i1 ranges from 1 to m and k ranges over the nonnegative integers.

13.1.1. The monic monomials of weight s.
For a fixed integer $s \geq 1$, let a nonnegative integer $e_{s,i_1,k}$ be given for each pair (i_1, k) of integers that satisfy $1 \leq i_1 \leq m$ and $0 \leq k \leq s - i_1$. Then, the expression

$$(13.1) \qquad M = \prod_{\substack{1 \leq i_1 \leq m \\ 0 \leq k \leq s-i_1}} \left[w_{i_1}^{(k)}\right]^{e_{s,i_1,k}}$$

is a monic monomial of weight s in $\mathcal{R}_{m,1}$ if and only if the condition

$$(13.2) \qquad \sum_{\substack{1 \leq i_1 \leq m \\ 0 \leq k \leq s-i_1}} (i_1 + k)\, e_{s,i_1,k} = s$$

is satisfied. Moreover, each monic monomial of weight s in $\mathcal{R}_{m,1}$ is expressible in the form (13.1) for some nonnegative integers $e_{s,i_1,k}$ that satisfy (13.2).

The relative invariants of weight s are special isobaric polynomials of weight s. Moreover, any isobaric polynomial of weight s is a linear combination over $\mathbb{Q}$ of the monic monomials of weight s in the variables (4.9) having $n = 1$. Our first task is to find and enumerate all such monomials. The trial-and-error assignment of integers to the $e_{s,i_1,k}$ must be efficiently restricted to reduce computation times. After starting a new session, the first three of the following four *Mathematica* commands

```
m = 3;

u[s_,p1_,r_] := Floor[ (1/(p1+r))*(s
  -Sum[(p1+k)*e[s,p1,k], {k,r+1,s-p1}]
  -Sum[(i1+k)*e[s,i1,k], {i1,p1+1,m}, {k,0,s-i1}])]

nM[s_] := Module[ {counter, seq}, counter = 0;
  seq = Apply[ Sequence, Reverse[ Flatten[
  Table[{e[s,i1,k],0,u[s,i1,k]},
  {i1,1,m},{k,0,s-i1}],1]]];
  Do[If[(Sum[(i1+k)*e[s,i1,k],{i1,1,m},
  {k,0,s-i1}])==s, (counter = counter + 1;
  monM[s,counter]
  = Product[D[w[i1][z],{z,k}]^e[s,i1,k],
  {i1,1,m},{k,0,s-i1}] ),
  counter = counter], Evaluate[seq]]; counter ]

printM[s_] := Do[Print["monM[",s,",",j,"] = ",
  monM[s,j]], {j,1,nM[s]}]
```

accomplish that purpose. For instance, after the preceding four input statements are evaluated, the additional evaluation of `printM[4]` computes, numbers, and displays the monic monomials of weight 4.

13.1.2. Condition for relative invariants. An isobaric polynomial F of weight s in $\mathcal{R}_{m,1}$ is representable by the evaluation of

```
wPolynomial[s_] := Sum[x[s,j]*monM[s,j],{j,1,nM[s]}]
```

for some rational numbers `x[s,i]`. Due to (4.13) and (4.14) of page 31, the condition

(13.3) $\quad \rho(z)^s \left[F^*(z) - F(z) \right] \equiv 0 \quad \text{and} \quad \left(f'(\zeta) \right)^s \left[F^{**}(\zeta) - (f'(\zeta))^s F(f(\zeta)) \right] \equiv 0$

is satisfied for any suitable $\rho(z)$ and $f(\zeta)$ if and only if F is a relative invariant. Representations for $F(z)$, $F^*(z)$, $F^{**}(\zeta)$, and the expressions in (13.3) based on (3.4), (3.5), (3.21), (3.22), (3.23), and (3.24) are obtained through evaluations of

```
sub[s_] := Flatten[Table[
  D[w[i1][z],{z,k}]->D[c[i1][z],
  {z,k}],{i1,1,m},{k,0,s-i1}]]

cPoly[s_] := wPolynomial[s] /. sub[s]

c[0][z_] = 1;

cOneStar[i1_][z_] := Sum[ Binomial[m-j1,i1-j1]*
  (D[rho[z],{z,i1-j1}]/rho[z])*c[j1][z],{j1,0,i1}]
```

13.1. THE RELATIVE INVARIANTS OF WEIGHT s IN $\mathcal{R}_{m,1}$ FOR $\mathcal{C}_{m,1}$

```
sub1[s_] := Flatten[ Table[
    D[w[i1][z],{z,k}]->D[cOneStar[i1][z],{z,k}],
    {i1,1,m},{k,0,s-i1}] ]

cOneStarPoly[s_] := wPolynomial[s] /. sub1[s]

alpha[0,j_][zet_] = 1;

alpha[i_,j_][zet_] := Sum[
    D[alpha[i-1,k][zet],zet]
    -(i-1+k)(f''[zet]/f'[zet])alpha[i-1,k][zet],
    {k,1,j}] /; i>=1

cTwoStar[i1_][zet_] := Sum[
    f'[zet]^mu*alpha[i1-mu,m-i1][zet]*
    c[mu][f[zet]],{mu,0,i1}]

sub2[s_] := Flatten[ Table[
    D[w[i1][z],{z,k}]->D[cTwoStar[i1][zet],{zet,k}],
    {i1,1,m},{k,0,s-i1}] ]

cTwoStarPoly[s_] := wPolynomial[s] /. sub2[s]

firstCondition[s_] :=
  firstCondition[s] = Expand[
  rho[z]^s*(cOneStarPoly[s] - cPoly[s])]

secondCondition[s_] :=
  secondCondition[s] = Expand[
  f'[zet]^s*( cTwoStarPoly[s]
  -f'[zet]^s*(cPoly[s] /. z->f[zet]) )]
```

by the selected version of *Mathematica* from [**55, 56, 57, 58, 59**]. The condition (13.3) requires that the coefficients x[s,1], ..., x[s,nM[s]] of wPolynomial[s] be such that firstCondition[s] and secondCondition[s] are identically zero.

13.1.3. Details for the first condition. We use (3.4)–(3.5) of page 19 to see that $\rho(z)^s\bigl[F^*(z) - F(z)\bigr]$ is representable as a linear combination over $\mathbb{Q}$ of the various expressions

$$(13.4) \qquad \left[\prod_{0\le t\le s} \left[\rho^{(t)}(z)\right]^{h_{s,t}}\right] \times \prod_{\substack{1\le i_1\le m \\ 0\le k\le s-i_1}} \left[c_{i_1}^{(k)}(z)\right]^{e_{s,i_1,k}}$$

that correspond to nonnegative integers $e_{s,i_1,k}$, $h_{s,t}$ that satisfy

$$(13.5) \qquad \sum_{\substack{1\le i_1\le m \\ 0\le k\le s-i_1}} (i_1+k)\, e_{s,i_1,k} + \sum_{t=1}^{s} t\, h_{s,t} = s \quad \text{and} \quad \sum_{t=0}^{s} h_{s,t} = s.$$

After the evaluation of the preceding eighteen input statements that begin with the assignment for m as the first input on page 134, the expressions corresponding to (13.4)–(13.5) needed for firstCondition[s] are obtained from the evaluation of

```
v[s_,p1_] := Floor[(1/p1)*(s
  -Sum[(i1+k)*e[s,i1,k],{i1,1,m},{k,0,s-i1}]
  -Sum[t*h[s,t],{t,p1+1,s}])]

nA[s_] := Module[ {counter,seq1,seq2}, counter = 0;
  seq1 = Apply[ Sequence, Reverse[ Flatten[
  Table[ {e[s,i1,k], 0, u[s,i1,k]},
  {i1,1,m},{k,0,s-i1}], 1] ] ];
  seq2 = Apply[ Sequence, Reverse[
  Table[ {h[s,t],0,v[s,t]},{t,1,s}]]];
  Do[ If[(Sum[(i1+k)*e[s,i1,k],{i1,1,m},{k,0,s-i1}]
  +Sum[t*h[s,t],{t,1,s}]) == s,
  ( counter = counter + 1;
  monA[s,counter] =
  Product[ D[c[i1][z],{z,k}]^e[s,i1,k],
  {i1,1,m},{k,0,s-i1}]*
  rho[z]^(s - Sum[h[s,t],{t,1,s}])*
  Product[D[rho[z],{z,t}]^h[s,t],{t,1,s}]),
  counter = counter ],
  Evaluate[seq1], Evaluate[seq2] ]; counter ]

printA[s_] := Do[ Print["monA[",s,", ",j,"] = ",
  monA[s,j]],{j,1,nA[s]}]
```

by the same version from [55, 56, 57, 58, 59]. For example, when the three preceding input statements are evaluated after the ones on page 134, the evaluation of printA[2] displays computer representations for the expressions of the form (13.4) that satisfy (13.5) with $s = 2$.

13.1.4. Details for the second condition. We use (3.21)–(3.24) of page 24 to see that the expression

$$\left(f'(\zeta)\right)^s \left[F^{**}(\zeta) - (f'(\zeta))^s F(f(\zeta))\right]$$

is representable as a linear combination over $\mathbb{Q}$ of the various terms

$$(13.6) \qquad \left[\prod_{0 \leq t \leq s} \left[f^{(t+1)}(\zeta)\right]^{h_{s,t}}\right] \times \prod_{\substack{1 \leq i_1 \leq m \\ 0 \leq k \leq s-i_1}} \left[c_{i_1}^{(k)}(f(\zeta))\right]^{e_{s,i_1,k}}$$

that correspond to nonnegative integers $e_{s,i_1,k}$, $h_{s,t}$ satisfying

$$(13.7) \qquad \sum_{\substack{1 \leq i_1 \leq m \\ 0 \leq k \leq s-i_1}} (i_1+k)\, e_{s,i_1,k} + \sum_{t=1}^{s} t\, h_{s,t} = s \quad \text{and} \quad \sum_{t=0}^{s}(t+1)\, h_{s,t} = 2s.$$

When the input statements of the three preceding subsections are evaluated, the expressions corresponding to (13.6)–(13.7) needed for secondCondition[s] can be displayed after the evaluation of the following two input commands

```
nB[s_] := Module[ {counter, seq1, seq2}, counter = 0;
  seq1 = Apply[ Sequence, Reverse[ Flatten[
  Table[ {e[s,i1,k], 0, u[s,i1,k]},
  {i1,1,m},{k,0,s-i1}], 1] ] ];
  seq2 = Apply[ Sequence, Reverse[
  Table[ {h[s,t],0,v[s,t]},{t,1,s}]]];
  Do[ If[(Sum[(i1+k)*e[s,i1,k],{i1,1,m},{k,0,s-i1}]
  + Sum[ t*h[s,t],{t,1,s}]) == s,
  (counter = counter + 1;
  monB[s,counter] = Product[
  (D[c[i1][z],{z,k}] /. z->f[zet] )^e[s,i1,k],
  {i1,1,m},{k,0,s-i1}]*
  f'[zet]^(2s - Sum[(t+1)*h[s,t],{t,1,s}])*
  Product[D[f[zet],{zet,t+1}]^h[s,t],{t,1,s}]),
  counter = counter ],
  Evaluate[seq1], Evaluate[seq2] ]; counter ]

printB[s_] := Do[ Print["monB[",s,", ",j,"] = ",
  monB[s,j]],{j,1,nB[s]}]
```

by the same version from [**55, 56, 57, 58, 59**]. For instance, after the preceding evaluations, the evaluation of printB[2] displays computer representations for the expressions of the form (13.6) that satisfy (13.7) when $s = 2$.

13.1.5. The remaining input commands. Next, for i = 1, ..., nA[s] and j = 1, ..., nB[s], we shall define a[s,i] as the coefficient of monA[s,i] in firstCondition[s] and shall define b[s,j] as the coefficient of monB[s,j] in secondCondition[s]. Then, each a[s,i] and b[s,j] will be given as a linear combinations over $\mathbb{Q}$ of the coefficients designated by x[s,1], ..., x[s,nM[s]] for wPolynomial[s]. A system of homogeneous linear algebraic equations over $\mathbb{Q}$ in x[s,1], ..., x[s,nM[s]] is then obtained by equating to zero each of the a[s,i] and b[s,j]. Any nonzero solution over $\mathbb{Q}$ of this system specifies a relative invariant. The evaluations of

```
firstCoefficients[s_] := Do[ a[s,i] =
  Coefficient[firstCondition[s],monA[s,i]],{i,1,nA[s]}]

firstZeroCheck[s_] := Expand[ firstCondition[s]
  -Sum[a[s,i]*monA[s,i],{i,1,nA[s]}]]

secondCoefficients[s_] := Do[ b[s,j] =
  Coefficient[secondCondition[s],monB[s,j]],{j,1,nB[s]}]

secondZeroCheck[s_] := Expand[ secondCondition[s]
  -Sum[ b[s,j]*monB[s,j], {j,1,nB[s]}] ]
```

```
eqs[s_] := Join[ Table[ a[s,i] == 0, {i,1,nA[s]}],
  Table[ b[s,j] == 0, {j,1,nB[s]}] ]

rules[s_] := If[TrueQ[$VersionNumber < 7.5],
  Flatten[Solve[eqs[s],Table[x[s,j],{j,1,nM[s]}]]],
  Flatten[Solve[eqs[s],Table[x[s,nM[s]+1-j],{j,1,nM[s]}]]]]

invariants[s_] :=
  Module[ {expansion, counter}, counter = 0; Print["    "];
  Print["   Computation of invariants[",s,"]"];
  Print["Number of Monic Monomials = ", nM[s]];
  Print["Number of firstCondition Terms = ",nA[s]];
  Print["Number of secondCondition Terms = ",nB[s]];
  firstCondition[s];  firstCoefficients[s];
  Print["firstZeroCheck[",s,"] = ",
  firstZeroCheck[s]];
  secondCondition[s]; secondCoefficients[s];
  Print["secondZeroCheck[",s,"] = ",
  secondZeroCheck[s]]; expansion = Expand[
  (Sum[x[s,j]*monM[s,j],{j,1,nM[s]}]) /. rules[s]];
  Do[ If[Length[Coefficient[expansion,x[s,j]]] != 0,
  (counter = counter + 1;
  inv[s, counter] = Coefficient[expansion, x[s,j]];
  Print["inv[", s, ", ", counter, "] = ",
  inv[s,counter]]), counter = counter],
  {j,1,nM[s]} ];  d[s] = counter;
  Print["The vector space for weight ", s,
  " has dimension d[",s,"] = ", d[s], "."] ]
```

by the same version from [55, 56, 57, 58, 59] enable us to obtain machine representations for all of the relative invariants in $\mathcal{R}_{m,1}$ for $\mathcal{C}_{m,1}$ of a given weight s. Before providing details in the following paragraph, we note that the unusual form of the next to the last of the preceding input commands is explained in Section 13.7.

When s *is a positive integer, the evaluation of* invariants[s] *yields an integer labeled* d[s] *that is equal to the dimension of the vector space* $\mathcal{V}_{m,1;s}$ *introduced in Section 4.7. Moreover, when* d[s] *is positive, that evaluation also yields machine representations labeled*

(13.8) inv[s,1], inv[s,2], ..., inv[s,d[s]]

for d[s] *linearly independent relative invariants over* $\mathbb{Q}$ *of weight s for* $\mathcal{C}_{m,1}$. Thus, when d[s] is positive, we have a machine representation for a basis of $\mathcal{V}_{m,1;s}$.

With the selection m = 3 as the first input command on page 134, the preceding summary applies to the context of relative invariants in $\mathcal{V}_{3,1;s}$ for $\mathcal{C}_{3,1}$. However, when m = 3 is replaced by m = 4 or m = 5 or ... in the first input statement of page 134 and no other alterations are made to the other twenty-nine input commands, the italicized summary for the evaluated program then applies respectively to the context of relative invariants in $\mathcal{V}_{4,1;s}$ for $\mathcal{C}_{4,1}$ or in $\mathcal{V}_{5,1;s}$ for $\mathcal{C}_{5,1}$ or ....

13.2. Simple verifications for Section 4.8 about $\mathcal{V}_{3,1;\,s}$

Theorem 4.10 of page 36 shows that each $\boldsymbol{L}_{s,i}$ explicitly defined on page 39 is a relative invariant in $\mathcal{R}_{3,1}$ for $\mathcal{C}_{3,1}$ of weight s and therefore belongs to $\mathcal{V}_{3,1;\,s}$. By using (4.61) to see that each of $\boldsymbol{L}_{12,1}$, $\boldsymbol{L}_{12,2}$, and $\boldsymbol{L}_{12,3}$ contains a term not present in the other two, we verify that $\{\boldsymbol{L}_{12,1}, \boldsymbol{L}_{12,2}, \boldsymbol{L}_{12,3}\}$ is linearly independent over $\mathbb{Q}$. Hence, for $s = 3, 6, 8, 9, 10, 11, 12, 13$, the subset $\{\boldsymbol{L}_{s,1}, \ldots, \boldsymbol{L}_{d_{3,1}}(s)\}$ of $\mathcal{V}_{3,1;\,s}$ is linearly independent over $\mathbb{Q}$.

With a selected version of *Mathematica* from [55, 56, 57, 58, 59], we evaluate the thirty input commands of Section 13.1 beginning with m = 3 on page 134. Then, the evaluation of Do[invariants[s],{s,{2,4,5,7}}] yields d[2] = 0, d[4] = 0, d[5] = 0, and d[7] = 0. Hence, there are no relative invariants in $\mathcal{R}_{3,1}$ for $\mathcal{C}_{3,1}$ of weight 2, 4, 5, or 7. Next, the evaluation of

Do[invariants[s], {s,{3,6,8,9,10,11,12,13}}]

yields d[12] = 3 and d[s] = 1, for s = 3, 6, 8, 9, 10, 11, 13. Consequently, for $2 \leq s \leq 13$, the integer $d_{3,1}(s)$ defined in Section 4.8 is the dimension of $\mathcal{V}_{3,1;\,s}$. Thus, for $2 \leq s \leq 13$ and $d_{3,1}(s) \geq 1$, $\{\boldsymbol{L}_{s,1}, \ldots, \boldsymbol{L}_{s,d_{3,1}(s)}\}$ is a basis of $\mathcal{V}_{3,1;\,s}$.

Alternatively, the assertions of Section 4.8 can be verified by showing that each machine generated relative invariant of a basis for $\mathcal{V}_{3,1;\,s}$ is expressible as a linear combination of machine representations for each of $\boldsymbol{L}_{s,1}, \ldots, \boldsymbol{L}_{s,d_{3,1}(s)}$. This more informative approach in Section 13.4 will use the program of the next section to obtain machine representations for the $\boldsymbol{L}_{s,i}$ of Section 4.8

13.3. Representation of $\boldsymbol{C}_{p,q,r}(\boldsymbol{P}, \boldsymbol{Q})$ in $\mathcal{R}_{m,1}$ with respect to (4.9)

For $m \geq 3$, $r \geq 0$, and given machine representations P and Q of relative invariants $\boldsymbol{P}$ and $\boldsymbol{Q}$ in $\mathcal{R}_{m,1}$ for $\mathcal{C}_{m,1}$, we provide a program based on Section 4.5 to obtain a machine representation Com[r,P,Q] of $\boldsymbol{C}_{p,q,r}(\boldsymbol{P}, \boldsymbol{Q})$.

After the evaluations of Section 13.1 have been made for a particular initial assignment of m, we successive evaluate each of

```
weight[iso_] := Module[{h,seq,t1,t2,t3,t4,t5,t6,t7},
  t1 = Expand[iso];
  t2 = If[Head[t1]===Plus, w[0][z]*t1[[1]],
  Expand[w[0][z]*t1]];
  t3 = Apply[List, t2];
  seq = {w[i_][z] -> h[i,0,1],
  Power[w[i_][z],k_] -> h[i,0,k],
  Derivative[j_][w[i_]][z] -> h[i,j,1],
  Power[Derivative[j_][w[i_]][z],k_]->h[i,j,k]};
  t4 = t3 /. seq;
  t5 = Cases[t4,h[i_,j_,k_] ];
  t6 = t5 /. h[i_,j_,k_]->(i+j)k;
  t7 = Apply[Plus, t6]]

aFirstKind[m,1][z_] := (1/Binomial[m+1,3])( w[2][z]
  -((m-1)/2)D[w[1][z],z]-((m-1)/(2m))w[1][z]^2 )

B[h_,i_,j_][z_] := 0  /;  i <= -1
```

```
B[h_,0,j_][z_] := 1

B[h_,i_,j_][z_] := Sum[ D[B[h,i-1,k][z], z]
    -k(2h+k-1)aFirstKind[m,1][z]*B[h,i-2,k-1][z],
    {k,i-1,j-1}] /; i >= 1

Ce[p_,q_,r_,mu_] := (-1)^mu*Binomial[r,mu]*
    Binomial[2q+r-1,mu]/Binomial[2p+mu-1,mu]

A[p_,q_,r_,s_,t_][z_] :=
    Sum[Ce[p,q,r,t+k]*B[p,k,t+k][z]*
    B[q,s-k,r-t-k][z], {k,0,s} ]

Com[r_,P_,Q_] := Module[ {p, q},
    p = weight[P]; q = weight[Q];
    Sum[ A[p,q,r,s,t][z]*D[P,{z,t}]*
    D[Q,{z,r-s-t}],{s,0,r},{t,0,r-s}] ]
```

with the same version from [55, 56, 57, 58, 59]. This enables machine representations to be easily obtained for the $L_{s,i}$ of page 38, the $M_{s,i}$ of pages 127–128, and the $N_{s,i}$ of pages 129–130. In particular, the weights of P and Q for Com[r,P,Q] are deduced from the machine representations P and Q for P and Q.

13.4. Alternative verifications for Section 4.8 about $\mathcal{V}_{3,1;\,s}$

After selecting one of [55, 56, 57, 58, 59], we use it to evaluate the thirty input commands of Section 13.1 beginning with m = 3 on page 134 and then have it evaluate the eight input commands of Section 13.3. Next, the evaluation of

```
invariants[3]
```

yields d[3] = 1 as well as a machine representation labeled inv[3,1] for the basic relative invariant $\mathcal{I}_{3,1;\,3}$ displayed in (4.61). Consequently, in view of Section 13.3, the evaluations for each of

```
L[3,1]  := inv[3,1];

L[6,1]  := L[3,1]^2;

L[8,1]  := Com[2,L[3,1],L[3,1]];

L[9,1]  := L[3,1]^3;

L[10,1] := Com[4,L[3,1],L[3,1]];

L[11,1] := Com[2,L[3,1],L[3,1]^2];

L[12,1] := L[3,1]^4;
```

```
L[12,2] := Com[6,L[3,1],L[3,1]];

L[12,3] := Com[3,L[3,1],L[3,1]^2];

L[13,1] := Com[4,L[3,1],L[3,1]^2];
```

provide machine representations for the corresponding $L_{s,i}$ defined on page 39. With d[3] = 1, the relative invariant $L_{3,1} \equiv \mathcal{I}_{3,1;3}$ forms a basis for $\mathcal{V}_{3,1;3}$.

An evaluation of invariants[6] yields d[6] = 1 and a machine generated relative invariant labeled inv[6,1] of weight 6 that is equal to the evaluation of L[6,1]. Thus, $\{L_{6,1}\}$ is a basis for $\mathcal{V}_{3,1;6}$.

An evaluation of invariants[8] yields d[8] = 1 and a machine generated relative invariant labeled inv[8,1] of weight 8 that is equal to the evaluation of (1/2)L[8,1]. Thus, $\{L_{8,1}\}$ is a basis for $\mathcal{V}_{3,1;8}$.

An evaluation of invariants[9] yields d[9] = 1 and a machine generated relative invariant labeled inv[9,1] of weight 9 that is equal to the evaluation of L[9,1]. Thus, $\{L_{9,1}\}$ is a basis for $\mathcal{V}_{3,1;9}$.

An evaluation of invariants[10] yields d[10] = 1 and a machine generated relative invariant labeled inv[10,1] of weight 10 that is equal to the evaluation of (1/2)L[10,1]. Thus, $\{L_{10,1}\}$ is a basis for $\mathcal{V}_{3,1;10}$.

An evaluation of invariants[11] yields d[11] = 1 and a machine generated relative invariant labeled inv[11,1] of weight 11 that is equal to the evaluation of (7/40)L[11,1]. Thus, $\{L_{11,1}\}$ is a basis for $\mathcal{V}_{3,1;11}$.

An evaluation of invariants[12] yields d[12] = 3 as well as three linearly independent machine generated relative invariants labeled inv[12,1], inv[12,2], and inv[12,3] of weight 12 that are respectively equal to the evaluations of L[12,1], (-2/9)L[12,3], and (1/2)L[12,2]. Hence, $\{L_{12,1}, L_{12,2}, L_{12,3}\}$ is a basis for $\mathcal{V}_{3,1;12}$.

An evaluation of invariants[13] yields d[13] = 1 and a machine generated relative invariant labeled inv[13,1] of weight 13 that is equal to the evaluation of (6/77)L[13,1]. Thus, $\{L_{13,1}\}$ is a basis for $\mathcal{V}_{3,1;11}$.

13.5. Verifications for Section 12.1 about $\mathcal{V}_{4,1;s}$

For the $M_{s,i}$ of pages 127–128 defined when $1 \leq s \leq 12$ and $1 \leq i \leq d_{4,1}(s)$, we apply Theorem 4.10 and (4.51)–(4.52) of page 36 to verify that each $M_{s,i}$ is a relative invariant in $\mathcal{R}_{4,1}$ for $\mathcal{C}_{4,1}$ of weight s and therefore belongs to $\mathcal{V}_{4,1;s}$.

Using a version of *Mathematica* from [55, 56, 57, 58, 59], we begin by evaluating m = 4 in place of m = 3 as the first input command on page 134 and then we evaluate the remaining twenty-nine input items of Section 13.1 as well as the eight input items of Section 13.3. Then, an evaluation of invariants[3] yields d[3] = 1 and a machine representation labeled inv[3,1] for the basic relative invariant $\mathcal{I}_{4,1;3}$ displayed in (12.1). Next, we observe that an evaluation of invariants[4] yields d[4] = 1 and a machine representation labeled inv[4,1] for the basic relative invariant $\mathcal{I}_{4,1;4}$ displayed in (12.2). Consequently, machine representations for the relative invariants $M_{s,i}$ are given by the evaluations of

```
M[3,1] := inv[3,1];

M[4,1] := inv[4,1];
```

as well as the evaluations of

```
M[6,1] := (M[3,1])^2;

M[7,1] := M[3,1]*M[4,1];

M[8,1] := M[4,1]^2;

M[8,2] := Com[2,M[3,1],M[3,1]];

M[8,3] := Com[1,M[3,1],M[4,1]];
```

and similar expressions for M[9,1], ..., M[12,0] as required by the definitions for the various $M_{s,i}$ on page 128. The evaluation of

```
Do[ invariants[s],{s,5,12}]
```

yields d[5] = 0, d[6] = 1, d[7] = 1, d[8] = 3, d[9] = 2, d[10] = 4, d[11] = 4, and d[12] = 9. Thus, for $2 \leq s \leq 12$, the dimension of $\mathcal{V}_{4,1;s}$ is given by $d_{4,1}(s)$ as defined on pages 127–128. For $6 \leq s \leq 12$ and s = s, that evaluation also yields machine representations labeled inv[s,1], ..., inv[s,d[s]] for relative invariants that specify a basis of $\mathcal{V}_{4,1;s}$. The verifications for Section 12.1 are completed by showing that each one of those basis vectors is equal to the evaluation of some explicit linear combination over $\mathbb{Q}$ of M[s,1], ..., M[s,d[s]]. For that, we use a natural modification of the procedure illustrated on page 153.

13.6. Verifications for Section 12.2 about $\mathcal{V}_{5,1;s}$

For the $N_{s,i}$ of page 129 defined when $1 \leq s \leq 12$ and $1 \leq i \leq d_{5,1}(s)$, we apply Theorem 4.10 and (4.51)–(4.52) of page 36 to verify that each $N_{s,i}$ is a relative invariant in $\mathcal{R}_{5,1}$ for $\mathcal{C}_{5,1}$ of weight s and therefore belongs to $\mathcal{V}_{5,1;s}$.

Using a version of *Mathematica* from [55, 56, 57, 58, 59], we begin by evaluating m = 5 in place of m = 3 on page 134 as the first input command and then we evaluate the remaining twenty-nine input items of Section 13.1 as well as the eight input items of Section 13.3. After that, an evaluation of invariants[3] yields d[3] = 1 and a machine representation labeled inv[3,1] for the basic relative invariant $\mathcal{I}_{5,1;3}$ in (12.4). An evaluation of invariants[4] yields d[4] = 1 and a machine representation labeled inv[4,1] for the basic relative invariant $\mathcal{I}_{5,1;4}$ in (12.5). An evaluation of invariants[5] yields d[5] = 1 and a machine representation labeled inv[5,1] for the basic relative invariant $\mathcal{I}_{5,1;5}$ in (12.6). Thus, each relative invariant $N_{s,i}$ has a corresponding machine representation given by the evaluation of n[s,i] where

```
n[3,1] := inv[3,1];          n[4,1] := inv[4,1];

n[5,1] := inv[5,1];          n[6,1] := n[3,1]^2;

n[7,1] := n[3,1]*n[4,1];     n[8,1] := n[3,1]*n[5,1];

n[8,2] := n[4,1]^2;          n[8,3] := Com[2,n[3,1],n[3,1]];
```

and there are similar expressions for n[9,1], ..., n[12,14] as required by the definitions for the various $N_{s,i}$ on page 129. The evaluation of

    Do[invariants[s],{s,6,12}]

yields d[6] = 1, d[7] = 1, d[8] = 4, d[9] = 4, d[10] = 7, d[11] = 7, and d[12] = 14. Thus, for $3 \leq s \leq 12$, the dimension of $\mathcal{V}_{5,1;s}$ is given by $d_{5,1}(s)$ as defined on page 129. For $6 \leq s \leq 12$ and s = s, that evaluation also yields machine representations labeled inv[s,1], ..., inv[s,d[s]] for relative invariants that specify a basis of $\mathcal{V}_{5,1;s}$. The verifications for Section 12.2 are completed by showing that each one of those basis vectors is equal to the evaluation of some explicit linear combination over $\mathbb{Q}$ of n[s,1], ..., n[s,d[s]]. For that, we use a natural modification of the procedure illustrated on page 153.

13.7. Observations about versions of *Mathematica*

The unusual form of the second input command on page 138 and the sixth input command on page 150 is required so that the output does not depend on the particular version of *Mathematica* from [55, 56, 57, 58, 59]. Namely, when d denotes a specific positive integer and eqs represents a list of homogeneous linear equations in variables V[1], ..., V[d] having rational coefficient, the output that Version 7.0 yields as the evaluation of Solve[eqs, Table[V[i],{i,1,d}] is obtained with later versions as the evaluation of Solve[eqs, Table[V[d+1-i],{i,1,d}].

In the *Mathematica* notebooks for [19, 20], there is just one occurrence of the Solve command. When its input statement given in [20, page 176, lines 14–15] as

    rules[s_] := Flatten[
            Solve[eqs[s], Table[x[s,i], {i,1,nM[s]}]]]

is replaced by

    rules[s_] := If[TrueQ[$VersionNumber < 7.5],
      Flatten[Solve[eqs[s],Table[x[s,j],{j,1,nM[s]}]]],
      Flatten[Solve[eqs[s],Table[x[s,nM[s]+1-j],{j,1,nM[s]}]]]]

and no other alterations are made, we have found that Versions 3.0, 7.0.1, 8.0.1, 9.0.1, 10.1, and 11.2 then yield the same correct evaluation for each notebook of [19] and [20].

CHAPTER 14

Computer Algebra for $\mathcal{V}_{2,2;s}$, $\mathcal{V}_{3,2;s}$, $\mathcal{V}_{4,2;s}$, ...

The machine instructions in this chapter enable us to establish the results of Section 12.3 about relative invariants in $\mathcal{R}_{2,2}$ for $\mathcal{C}_{2,2}$. They also yield analogous representations of relative invariants in $\mathcal{R}_{3,2}$ for $\mathcal{C}_{3,2}$, in $\mathcal{R}_{4,2}$ for $\mathcal{C}_{4,2}$, ...

Simple modifications can be made to obtain representations of the relative invariants in $\mathcal{R}_{m,n}$ for $\mathcal{C}_{m,n}$ when $m \geq 2$ with $n = 3$ or $n = 4$ or ....

14.1. The relative invariants of weight s in $\mathcal{R}_{m,2}$ for $\mathcal{C}_{m,2}$

When integers m, s satisfy $m, s \geq 2$, the program presented in this section specifies linearly independent relative invariants over $\mathbb{Q}$ of weight s in $\mathcal{R}_{m,2}$ for $\mathcal{C}_{m,2}$ such that their nonzero linear combinations over $\mathbb{Q}$ uniquely yield each relative invariant of weight s in $\mathcal{R}_{m,2}$ for $\mathcal{C}_{m,2}$. However, it requires that a specific integral value m be assigned to m in the first input statement. We have selected m = 2 on page 146 as the first input statement; and the program as written therefore applies to $\mathcal{R}_{2,2}$ for $\mathcal{C}_{2,2}$. By merely changing the first input statement to m = 3 or m = 4 or ... and making no other alterations, the program then applies to $\mathcal{R}_{3,2}$ for $\mathcal{C}_{3,2}$ or $\mathcal{R}_{4,2}$ for $\mathcal{C}_{4,2}$ or ....

We select a version of *Mathematica* from [55, 56, 57, 58, 59] as the system of computer algebra and represent the derivation ' for $\mathcal{R}_{m,2}$ as differentiation with respect to a fictitious independent variable denoted by z. This enables us to represent the variables $w_{i_1,i_2}^{(k)}$ of (4.9) for $\mathcal{R}_{m,2}$ by D[w[i1,i2][z],{z,k}] as i1 ranges from 0 to m, i2 ranges from Max[1,i1] to m, and k ranges over the nonnegative integers.

14.1.1. The monic monomials of weight s.
For a fixed integer $s \geq 1$, let a nonnegative integer $e_{s,i_1,i_2,k}$ be given for each triple (i_1, i_2, k) of integers that satisfy $0 \leq i_1 \leq m$, $max\{1, i_1\} \leq i_2 \leq m$, and $0 \leq k \leq s - i_1 - i_2$. Then, the expression

$$(14.1) \qquad M = \prod_{\substack{0 \leq i_1 \leq m \\ max\{1,i_1\} \leq i_2 \leq m \\ 0 \leq k \leq s-i_1-i_2}} \left[w_{i_1,i_2}^{(k)} \right]^{e_{s,i_1,i_2,k}}$$

is a monic monomial of weight s in $\mathcal{R}_{m,2}$ if and only if the condition

$$(14.2) \qquad \sum_{\substack{0 \leq i_1 \leq m \\ max\{1,i_1\} \leq i_2 \leq m \\ 0 \leq k \leq s-i_1-i_2}} (i_1 + i_2 + k)\, e_{s,i_1,i_2,k} = s$$

is satisfied. Moreover, each monic monomial of weight s in $\mathcal{R}_{m,2}$ is expressible in the form (14.1) for some nonnegative integers $e_{s,i_1,i_2,k}$ that satisfy (14.2).

The relative invariants of weight s are special isobaric polynomials of weight s. Moreover, any isobaric polynomial of weight s is a linear combination over $\mathbb{Q}$ of the monic monomials of weight s in the variables (4.9) having $n = 2$. Our first task is to find and enumerate all such monomials. The trial-and-error assignment of integers to the $e_{s,i_1,i_2,k}$ must therefore be efficiently restricted to reduce computation times. After starting a new session, the first three of the following four input commands

```
m = 2;

u[s_,p1_,p2_,r_] := Floor[ (1/(p1+p2+r))*(s
   -Sum[(p1+p2+k)*e[s,p1,p2,k],{k,r+1,s-p1-p2}]
   -Sum[(p1+i2+k)*e[s,p1,i2,k],{i2,p2+1,m},
   {k,0,s-p1-i2}] - Sum[(i1+i2+k)*e[s,i1,i2,k],
   {i1,p1+1,m},{i2,i1,m},{k,0,s-i1-i2}])]

nM[s_] := Module[ {counter, seq}, counter = 0;
   seq = Apply[ Sequence, Reverse[ Flatten[
   Table[{e[s,i1,i2,k],0,u[s,i1,i2,k]},
   {i1,0,m},{i2,Max[1,i1],m},{k,0,s-i1-i2}],2]]];
   Do[If[(Sum[(i1+i2+k)*e[s,i1,i2,k],{i1,0,m},
   {i2,Max[1,i1],m},{k,0,s-i1-i2}])==s,
   (counter = counter + 1; monM[s,counter]
   = Product[D[w[i1,i2][z],{z,k}]^e[s,i1,i2,k],
   {i1,0,m},{i2,Max[1,i1],m},{k,0,s-i1-i2}] ),
   counter = counter], Evaluate[seq]]; counter ]

printM[s_] := Do[Print["monM[",s,",",j,"] = ",
   monM[s,j]], {j,1,nM[s]}]
```

accomplish that purpose. For instance, after the preceding four input statements are evaluated, the additional evaluation of `printM[4]` computes, numbers, and displays the monic monomials for $\mathcal{R}_{m,2}$ of weight 4.

14.1.2. Conditions for relative invariants. An isobaric polynomial F of weight s in $\mathcal{R}_{m,2}$ is representable by the evaluation of

```
wPolynomial[s_] := Sum[x[s,j]*monM[s,j],{j,1,nM[s]}]
```

for some rational numbers `x[s,i]` that are not all zero. Due to (4.13) and (4.14) of page 31, the condition

$$(14.3) \quad \rho(z)^s \left[F^*(z) - F(z) \right] \equiv 0 \quad \text{and} \quad (f'(\zeta))^s \left[F^{**}(\zeta) - (f'(\zeta))^s F(f(\zeta)) \right] \equiv 0$$

is satisfied for any suitable $\rho(z)$ and $f(\zeta)$ if and only if F is a relative invariant. Representations for $F(z)$, $F^*(z)$, $F^{**}(\zeta)$, and the expressions in (14.3) based on (3.4), (3.5), (3.21), (3.22), (3.23), and (3.24) are obtained through evaluations of the following fourteen input commands

```
sub[s_] := Flatten[Table[D[w[i1,i2][z],{z,k}]->D[c[i1,i2][z],
   {z,k}],{i1,0,m},{i2,Max[1,i1],m},{k,0,s-i1-i2}]]
```

14.1. THE RELATIVE INVARIANTS OF WEIGHT s IN $\mathcal{R}_{m,2}$ FOR $\mathcal{C}_{m,2}$

```
cPoly[s_] := wPolynomial[s] /. sub[s]

c[0,0][z_] = 1;

c[i1_,i2_][z_] := Apply[c,Sort[{i1,i2}]][z] /;
  {i1,i2} != Sort[{i1,i2}]

cOneStar[i1_,i2_][z_] := Sum[ Binomial[m-j1,i1-j1]*
  Binomial[m-j2,i2-j2]*(D[rho[z],{z,i1-j1}]/rho[z])*
  (D[rho[z],{z,i2-j2}]/rho[z])*
  c[j1,j2][z],{j1,0,i1},{j2,0,i2}]

sub1[s_] := Flatten[ Table[
  D[w[i1,i2][z],{z,k}]->D[cOneStar[i1,i2][z],{z,k}],
  {i1,0,m},{i2,Max[1,i1],m},{k,0,s-i1-i2}] ]

cOneStarPoly[s_] := wPolynomial[s] /. sub1[s]

alpha[0,j_][zet_] = 1;

alpha[i_,j_][zet_] := Sum[D[alpha[i-1,k][zet],zet]
  -(i-1+k)(f''[zet]/f'[zet])alpha[i-1,k][zet],
  {k,1,j}] /; i>=1

cTwoStar[i1_,i2_][zet_] := Sum[
  f'[zet]^(j1+j2)*alpha[i1-j1,m-i1][zet]*
  alpha[i2-j2,m-i2][zet]*c[j1,j2][f[zet]],
  {j1,0,i1},{j2,0,i2}]

sub2[s_] := Flatten[ Table[
  D[w[i1,i2][z],{z,k}]->
  D[cTwoStar[i1,i2][zet],{zet,k}],
  {i1,0,m},{i2,Max[1,i1],m},{k,0,s-i1-i2}] ]

cTwoStarPoly[s_] := wPolynomial[s] /. sub2[s]

firstCondition[s_] := firstCondition[s] = Expand[
  rho[z]^s*(cOneStarPoly[s] - cPoly[s])]

secondCondition[s_] := secondCondition[s] = Expand[
  f'[zet]^s*( cTwoStarPoly[s]
  -f'[zet]^s*(cPoly[s] /. z->f[zet]) )]
```

by the selected version of *Mathematica* from [55, 56, 57, 58, 59]. The condition (14.3) requires that the coefficients `x[s,1]`, ..., `x[s,nM[s]]` of `wPolynomial[s]` be such that `firstCondition[s]` and `secondCondition[s]` are identically zero.

14.1.3. Details for the first condition.
We use (3.4)–(3.5) of page 19 to see that the expression $\rho(z)^s\left[F^*(z)-F(z)\right]$ in (14.3) is representable as a linear combination over $\mathbb{Q}$ of the various expressions

$$(14.4) \qquad \left[\prod_{0\leq t\leq s}\left[\rho^{(t)}(z)\right]^{h_{s,t}}\right]\times\prod_{\substack{0\leq i_1\leq m \\ max\{1,i_1\}\leq i_2\leq m \\ 0\leq k\leq s-i_1-i_2}}\left[c_{i_1,i_2}^{(k)}(z)\right]^{e_{s,i_1,i_2,k}}$$

corresponding to the nonnegative integers $e_{s,i_1,i_2,k}$, $h_{s,t}$ that satisfy

$$(14.5) \qquad \sum_{\substack{0\leq i_1\leq m \\ max\{1,i_1\}\leq i_2\leq m \\ 0\leq k\leq s-i_1-i_2}}(i_1+i_2+k)\,e_{s,i_1,i_2,k}+\sum_{t=1}^{s}t\,h_{s,t}=s \quad\text{and}\quad \sum_{t=0}^{s}h_{s,t}=s.$$

After the evaluation of the preceding nineteen input statements beginning with the assignment for m on page 146, representations for the expressions of the form (14.4) that satisfy (14.5) are obtained for firstCondition[s] through the evaluation of the following three input commands

```
v[s_,p1_] := Floor[(1/p1)*(s
  -Sum[(i1+i2+k)*e[s,i1,i2,k],{i1,0,m},
  {i2,Max[1,i1],m},{k,0,s-i1-i2}]
  -Sum[t*h[s,t],{t,p1+1,s}])]

nA[s_] := Module[ {counter,seq1,seq2}, counter = 0;
  seq1 = Apply[ Sequence, Reverse[ Flatten[
  Table[ {e[s,i1,i2,k], 0, u[s,i1,i2,k]},
  {i1,0,m},{i2,Max[1,i1],m},{k,0,s-i1-i2}], 2] ] ];
  seq2 = Apply[ Sequence, Reverse[
  Table[ {h[s,t],0,v[s,t]},{t,1,s}]]];
  Do[ If[(Sum[(i1+i2+k)*e[s,i1,i2,k],{i1,0,m},
  {i2,Max[1,i1],m},{k,0,s-i1-i2}]
  +Sum[t*h[s,t],{t,1,s}]) == s,
  ( counter = counter + 1;
  monA[s,counter] =
  Product[ D[c[i1,i2][z],{z,k}]^e[s,i1,i2,k],
  {i1,0,m},{i2,Max[1,i1],m},{k,0,s-i1-i2}]*
  rho[z]^(s - Sum[h[s,t],{t,1,s}])*
  Product[D[rho[z],{z,t}]^h[s,t],{t,1,s}]),
  counter = counter ],
  Evaluate[seq1], Evaluate[seq2] ]; counter ]

printA[s_] := Do[ Print["monA[",s,", ",j,"] = ",
  monA[s,j]],{j,1,nA[s]}]
```

by the selected version. For example, after the preceding evaluations are made, the evaluation of printA[2] displays computer representations for each expression of the form (14.4) subject to (14.5) and $s=2$.

14.1.4. Details for the second condition.
We use (3.21)–(3.24) of page 24 to see that the expression

$$\left(f'(\zeta)\right)^s \left[F^{**}(\zeta) - (f'(\zeta)^s F(f(\zeta))\right]$$

in (14.3) is representable as a linear combination over $\mathbb{Q}$ of the various terms

(14.6)
$$\left[\prod_{0 \le t \le s} \left[f^{(t+1)}(\zeta)\right]^{h_{s,t}}\right] \times \prod_{\substack{0 \le i_1 \le m \\ max\{1,i_1\} \le i_2 \le m \\ 0 \le k \le s-i_1-i_2}} \left[c_{i_1,i_2}^{(k)}(f(\zeta))\right]^{e_{s,i_1,i_2,k}}$$

corresponding to the nonnegative integers $e_{s,i_1,i_2,k}$, $h_{s,t}$ that satisfy

(14.7)
$$\sum_{\substack{0 \le i_1 \le m \\ max\{1,i_1\} \le i_2 \le m \\ 0 \le k \le s-i_1-i_2}} (i_1 + i_2 + k)\, e_{s,i_1,i_2,k} + \sum_{t=1}^{s} t\, h_{s,t} = s$$

and

(14.8)
$$\sum_{t=0}^{s} (t+1) h_{s,t} = 2s.$$

After all of the input commands in the three preceding subsections are evaluated by the selected version from [55, 56, 57, 58, 59], representations for the expressions of the form (14.6) that satisfy (14.7) and (14.8) are provided for secondCondition[s] through the evaluation of the following two input commands

```
nB[s_] := Module[ {counter, seq1, seq2}, counter = 0;
  seq1 = Apply[ Sequence, Reverse[ Flatten[
  Table[ {e[s,i1,i2,k], 0, u[s,i1,i2,k]},
  {i1,0,m},{i2,Max[1,i1],m},{k,0,s-i1-i2}], 2] ] ];
  seq2 = Apply[ Sequence, Reverse[
  Table[ {h[s,t],0,v[s,t]},{t,1,s}]]];
  Do[ If[(Sum[(i1+i2+k)*e[s,i1,i2,k],{i1,0,m},
  {i2,Max[1,i1],m},{k,0,s-i1-i2}]
  + Sum[ t*h[s,t],{t,1,s}]) == s,
  (counter = counter + 1;
  monB[s,counter] = Product[
  (D[c[i1,i2][z],{z,k}] /. z->f[zet] )^e[s,i1,i2,k],
  {i1,0,m},{i2,Max[1,i1],m},{k,0,s-i1-i2}]*
  f'[zet]^(2s - Sum[(t+1)*h[s,t],{t,1,s}])*
  Product[D[f[zet],{zet,t+1}]^h[s,t],{t,1,s}]),
  counter = counter ],Evaluate[seq1],Evaluate[seq2]
  ]; counter ]

printB[s_] := Do[ Print["monB[",s,", ",j,"] = ",
  monB[s,j]],{j,1,nB[s]}]
```

by that same version. For example, after the preceding evaluations are made, the evaluation of printB[2] displays computer representations for each expression of the form (14.6) subject to (14.7), (14.8), and $s = 2$.

14.1.5. The remaining input commands. Next, for i = 1, ..., nA[s] and j = 1, ..., nB[s], we introduce a[s,i] as the coefficient of monA[s,i] in firstCondition[s] and we introduce b[s,j] as the coefficient of monB[s,j] in secondCondition[s]. Then, each a[s,i] and b[s,j] is a linear combination over $\mathbb{Q}$ of coefficients designated by x[s,1], ..., x[s,nM[s]] for wPolynomial[s]. A system of homogeneous linear algebraic equations in x[s,1], ..., x[s,nM[s]] over $\mathbb{Q}$ is obtained by equating to zero each of the a[s,i] and b[s,j]. Any nonzero solution over $\mathbb{Q}$ of this system specifies a relative invariant. The evaluations of

```
firstCoefficients[s_] := Do[ a[s,i] = Coefficient[
  firstCondition[s],monA[s,i]], {i,1,nA[s]}]

firstZeroCheck[s_] := Expand[ firstCondition[s]
  -Sum[a[s,i]*monA[s,i],{i,1,nA[s]}]]

secondCoefficients[s_] := Do[ b[s,j] = Coefficient[
  secondCondition[s],monB[s,j]], {j,1,nB[s]}]

secondZeroCheck[s_] := Expand[ secondCondition[s]
  -Sum[ b[s,j]*monB[s,j], {j,1,nB[s]}] ]

eqs[s_] := Join[ Table[ a[s,i] == 0, {i,1,nA[s]}],
  Table[ b[s,j] == 0, {j,1,nB[s]}] ]

rules[s_] := If[ TrueQ[$VersionNumber < 7.5],
  Flatten[Solve[eqs[s],Table[x[s,j],{j,1,nM[s]}]]],
  Flatten[Solve[eqs[s],Table[x[s,nM[s]+1-j],{j,1,nM[s]}]]]]

invariants[s_] :=
  Module[ {expansion, counter}, counter = 0;
  Print["       "];
  Print["   Computation of invariants[",s,"]"];
  Print["Number of Monic Monomials = ", nM[s]];
  Print["Number of firstCondition Terms = ",nA[s]];
  Print["Number of secondCondition Terms = ",nB[s]];
  firstCondition[s];  firstCoefficients[s];
  Print["firstZeroCheck[",s,"] = ", firstZeroCheck[s]];
  secondCondition[s]; secondCoefficients[s];
  Print["secondZeroCheck[",s,"] = ", secondZeroCheck[s]];
  expansion = Expand[
  (Sum[x[s,j]*monM[s,j],{j,1,nM[s]}]) /. rules[s]];
  Do[ If[Length[Coefficient[expansion,x[s,j]]] != 0,
  (counter = counter + 1;
  inv[s, counter] = Coefficient[expansion, x[s,j]];
  Print["inv[", s, ", ", counter, "] = ",
  inv[s,counter]]), counter = counter],{j,1,nM[s]}];
  d[s] = counter;
  Print["The vector space for weight ", s,
  " has dimension d[",s,"] = ", d[s], "."] ]
```

by the same version from [55, 56, 57, 58, 59] enable us to obtain machine representations for all the relative invariants in $\mathcal{R}_{m,2}$ for $\mathcal{C}_{m,2}$ of a given weight s. Before providing details in the following paragraph, we note that the unusual form of the next to the last of the preceding input commands is explained in Section 13.7.

When s *is a positive integer, the evaluation of* invariants[s] *yields an integer labeled* d[s] *that is equal to the dimension of the vector space* $\mathcal{V}_{m,2;\,s}$ *introduced in Section 4.7. Moreover, when* d[s] *is positive, that evaluation also yields machine representations labeled*

(14.9) $\qquad\qquad$ inv[s,1], inv[s,2], ..., inv[s,d[s]]

for d[s] *linearly independent relative invariants of weight* s *in* $\mathcal{V}_{m,2;\,s}$ *for* $\mathcal{C}_{m,2}$. Thus, when d[s] is positive, (14.9) provides representations for the elements in a basis of $\mathcal{V}_{m,2;\,s}$.

Due to m = 2 as the first input command of page 146, we see that the preceding summary applies to relative invariants in $\mathcal{V}_{2,2;\,s}$ for $\mathcal{C}_{2,2}$. However, when m = 2 is replaced by m = 3 or m = 4 or ... in that first input statement of page 146 and no alterations are made to any of the other thirty input commands of pages 146–150, the italicized summary for the evaluated program then applies respectively to the context of relative invariants in $\mathcal{V}_{3,2;\,s}$ for $\mathcal{C}_{3,2}$ or in $\mathcal{V}_{4,2;\,s}$ for $\mathcal{C}_{4,2}$ or ....

14.2. Representation of $C_{p,q,r}(P, Q)$ in $\mathcal{R}_{m,2}$ with respect to (4.9)

For $m \geq 2$, $r \geq 0$, and given machine representations P and Q of relative invariants P and Q in $\mathcal{R}_{m,2}$ for $\mathcal{C}_{m,2}$, we provide a program based on Section 4.5 to obtain a machine representation Com[r,P,Q] of $C_{p,q,r}(P, Q)$. The notation Com[r,P,Q] does not need more detail because the program computes the weights p of P and q of Q directly from the machine representations P for P and Q for Q.

After the evaluations of Section 14.1 have been made for a particular initial assignment of m, we evaluate each of

```
weight[iso_] := Module[
  {h, seq, t1, t2, t3, t4, t5, t6, t7},
  t1 = Expand[iso];
  t2 = If[Head[t1]===Plus, w[0,0][z]*t1[[1]],
  Expand[w[0,0][z]*t1]];
  t3 = Apply[ List, t2];
  seq = { w[i1_,i2_][z] -> h[i1,i2,0,1],
  Power[w[i1_,i2_][z],k_] -> h[i1,i2,0,k],
  Derivative[j_][w[i1_,i2_]][z] -> h[i1,i2,j,1],
  Power[Derivative[j_][w[i1_,i2_]][z],k_]
  ->h[i1,i2,j,k] }; t4 = t3 /. seq;
  t5 = Cases[ t4, h[i1_,i2_,j_,k_] ];
  t6 = t5 /. h[i1_,i2_,j_,k_] -> (i1+i2+j)k;
  t7 = Apply[ Plus, t6] ]

aFirstKind[m_,2][z_] := (1/Binomial[m+1,3])(w[0,2][z]
  -((m-1)/2)D[w[0,1][z],z] - ((m-1)/(2m))w[0,1][z]^2)
```

```
B[h_,i_,j_][z_] := 0   /; i <= -1

B[h_,0,j_][z_] := 1

B[h_,i_,j_][z_] := Sum[ D[B[h,i-1,k][z], z]
  -k(2h+k-1)aFirstKind[m,2][z]*B[h,i-2,k-1][z],
  {k,i-1,j-1}]  /; i >= 1

Ce[p_,q_,r_,mu_] := (-1)^mu*Binomial[r,mu]*
  Binomial[2q+r-1,mu]/Binomial[2p+mu-1,mu]

A[p_,q_,r_,s_,t_][z_] := Sum[ Ce[p,q,r,t+k]*
  B[p,k,t+k][z]*B[q,s-k,r-t-k][z], {k,0,s} ]

Com[r_,P_,Q_] := Module[ {p, q},
  p = weight[P]; q = weight[Q];
  Sum[ A[p,q,r,s,t][z]*D[P,{z,t}]*
  D[Q,{z,r-s-t}], {s,0,r}, {t,0,r-s}] ]
```

with the same version of *Mathematica* from [55, 56, 57, 58, 59].

For example, if m = 2 while P and Q are both machine representations of the relative invariant $\mathcal{I}_{2,2;\,1,1}$ of (7.14), then the evaluation of Com[8,P,Q] is a machine representation for $F_{12,36} \equiv C_{2,2,8}(\mathcal{I}_{2,2;\,1,1}, \mathcal{I}_{2,2;\,1,1})$ of page 132.

14.3. Verifications for Section 12.3 about $\mathcal{V}_{2,2;\,s}$

Our task is to check that: for each s satisfying $2 \leq s \leq 12$, the relative invariants $F_{s,1}, \ldots, F_{s,d_{2,2}(s)}$ in $\mathcal{R}_{2,2}$ for $\mathcal{C}_{2,2}$ specified on pages 130–132 form a basis of the vector space $\mathcal{V}_{2,2;\,s}$. We use the program of Section 14.1 with the assignments of 2 for m and one of 2, 3, ..., 12 for s. Then, the evaluation of invariants[s] yields an integer d[s] as the dimension of $\mathcal{V}_{2,2;\,s}$ and it provides representations labeled inv[s,1], ..., inv[s,d[s]] for a basis of $\mathcal{V}_{2,2;\,s}$. After observing that the number $d_{2,2}(s)$ of relative invariants $F_{s,1}, \ldots, F_{s,d_{2,2}(s)}$ in Section 12.3 is equal to d[s], we see that the verification can be completed by merely checking that computer representations for $F_{s,1}, \ldots, F_{s,d_{2,2}(s)}$ are linearly independent over $\mathbb{Q}$. However, we prefer the more thorough verification where each of inv[s,1], ..., inv[s,d[s]] is expressed as an explicit linear combination over $\mathbb{Q}$ of the machine representations for $F_{s,1}, \ldots, F_{s,d_{2,2}(s)}$.

We select one of the versions of *Mathematica* from [55, 56, 57, 58, 59] and use it to evaluate each the thirty-one input commands on pages 146–150 as well as each of the eight input commands in Section 14.2. Next, the evaluation of

```
Do[ invariants[s], {s,2,5}]
```

is done within seconds and its output shows that: d[2] equals 1 while inv[2,1] is a machine representation for the basic relative invariant $\mathcal{I}_{2,2;\,1,1}$ in (7.14); also, d[3] equals 1 and inv[3,1] is a machine representation for the basic relative invariant $\mathcal{I}_{2,2;\,1,2}$ in (7.15); moreover, d[4] equals 2, inv[4,1] is a machine representation for $(\mathcal{I}_{2,2;\,1,1})^2$, and inv[4,2] is a machine representation for the basic relative invariant $\mathcal{I}_{2,2;\,2,2}$ in (7.16); d[5] = 1 and inv[5,1] is a machine representation of

14.3. VERIFICATIONS FOR SECTION 12.3 ABOUT $\mathcal{V}_{2,2;s}$

the relative invariant $(\mathcal{I}_{2,2;1,1}\,\mathcal{I}_{2,2;1,2})$. We use the selected version of *Mathematica* to successively evaluate each of

```
f[2,1] = inv[2,1];              f[3,1] = inv[3,1];

f[4,1] = f[2,1]^2;              f[4,2] = inv[4,2];

f[5,1] = f[2,1]*f[3,1];

f[6,1] = f[2,1]^3;              f[6,2] = f[3,1]^2;

f[6,3] = f[2,1]*f[4,2];         f[6,4] = Com[1,f[2,1],f[3,1]];

f[6,5] = Com[2,f[2,1],f[2,1]];
```

and additional expressions f[7,1], f[7,2], ..., f[12,48], f[12,49] such that each relative invariant $\boldsymbol{F}_{s,i}$ defined in Section 12.3 has a corresponding computer representation given by the evaluation of f[s,i]. When s is one of 2, 3, 4, or 5, it is clear that each of inv[s,1], ..., inv[s,d[s]] can be obtained by evaluating some linear combination over $\mathbb{Q}$ of f[s,1], ..., f[s,d[s]].

To conduct a similar verification when s is 6, we successively evaluate

```
invariants[6]

Do[ expr[6,i] = ( inv[6,i]
  -Sum[c[6,i,j]*f[6,j],{j,1,d[6]}] ), {i,1,d[6]} ]

Do[y[6,i,k] = Coefficient[ expr[6,i], monM[6,k]],
  {k,1,nM[6]}, {i,1,d[6]}]

sol[6,i_] := Solve[ Table[y[6,i,k] == 0,{k,1,nM[6]}],
  Table[c[6,i,j],{j,1,d[6]}] ]
```

with the selected version. Now, the evaluation of sol[6,1] gives the output

{{c[6,1,1] -> 1, c[6,1,2] -> 0,
   c[6,1,3] -> 0, c[6,1,4] -> 0, c[6,1,5] -> 0}}

and therefore demonstrates that f[6,1] yields inv[6,1]. Similarly, the evaluations of sol[6,2], sol[6,3], sol[6,4], and sol[6,5] show that:

the evaluation of (1/2)f[6,5] yields inv[6,2],
the evaluation of f[6,2] yields inv[6,3],
the evaluation of f[6,4]+(1/8)f[6,5] yields inv[6,4], and
the evaluation of f[6,3]+(1/3)f[6,4]+(1/40)f[6,5] yields inv[6,5].

Consequently, a basis for the vector space $\mathcal{V}_{2,2;6}$ is provided by the five relative invariants $\boldsymbol{F}_{6,1}$, $\boldsymbol{F}_{6,2}$, $\boldsymbol{F}_{6,3}$, $\boldsymbol{F}_{6,4}$, and $\boldsymbol{F}_{6,5}$ of Section 12.3.

The verifications for each s satisfying $7 \leq s \leq 12$ are made in a manner that is completely analogous to the procedure for $s = 6$.

Part 5

Modifications Required for Developments before 1989

CHAPTER 15

Suitable Formulas for Transformations of Homogeneous Linear Differential Equations

15.1. Introduction.

For a search of the literature to find adequate formulas that yield all of the coefficients of homogeneous linear differential equations of any order m that result from changes of the independent variable, MathSciNet is currently not helpful. However, readers of this monograph are aware that such formulas exist because they were developed in [14, 16, 17] and were essential for [19]. The natural questions are: *how do we know that formulas like them were not previously published and why was there very little progress about relative invariants during the years* 1890–1988?

To answer these questions, library usage like that indicated in
http://homepages.uc.edu/~chalklr/Library.pdf
may not be possible today. However, by checking the research articles devoted to relative invariants for monic homogeneous linear differential equations, one finds that, prior to 1989, each such publication used binomial coefficients to express their equations in a form analogous to

$$(15.1) \qquad y^{(m)}(z) + \sum_{i=1}^{m} \binom{m}{i} C_i(z) \, y^{(m-i)}(z) = 0, \quad \text{with } C_0(z) \equiv 1,$$

and failed to provide general transformation formulas for (15.1) corresponding to a change of the independent variable.

15.1.1. Transformations of the first kind for (15.1). When meromorphic functions $C_1(z), C_2(z), \ldots, C_m(z)$ are given on a region Ω of the complex plane and $\rho(z)$ is a given not-identically-zero meromorphic function on Ω, there are unique meromorphic functions $C_1^*(z), C_2^*(z), \ldots, C_m^*(z)$ on Ω such that the substitution

$$(15.2) \qquad y(z) = \rho(z)\, v(z),$$

viewed as a change of the dependent variable from y to v, transforms (15.1) into

$$(15.3) \qquad v^{(m)}(z) + \sum_{i=1}^{m} \binom{m}{i} C_i^*(z) \, v^{(m-i)}(z) = 0, \quad \text{on } \Omega \text{ with } C_0^*(z) \equiv 1.$$

It is easy to establish directly the validity for (15.3) of

$$(15.4) \qquad C_i^*(z) \equiv \sum_{j=0}^{i} \binom{i}{j} \frac{\rho^{(i-j)}(z)}{\rho(z)} \, C_j(z), \quad \text{on } \Omega \text{ when } 0 \le i \le m.$$

However, it is convenient here for us to use (15.12) and (15.13) to see immediately that (15.4) is valid. Note that $C_i^*(z)$ in (15.4) is the same for any $m \ge i$.

15.1.2. Transformations of the second kind for (15.1).
Let meromorphic functions $C_1(z)$, $C_2(z)$, ..., $C_m(z)$ be given on a region Ω and let $z = f(\zeta)$ denote a univalent analytic function on a region Ω^{**} such that $f(\Omega^{**}) = \Omega$. Then, there are unique meromorphic functions $C_1^{**}(\zeta)$, $C_2^{**}(\zeta)$, ..., $C_m^{**}(\zeta)$ on Ω^{**} such that the substitution

$$(15.5) \qquad z = f(\zeta) \quad \text{with} \quad u(\zeta) = y(f(\zeta)),$$

viewed as a change of the independent variable from z to ζ, transforms (15.1) into

$$(15.6) \qquad u^{(m)}(\zeta) + \sum_{i=1}^{m} \binom{m}{i} C_i^{**}(\zeta)\, u^{(m-i)}(\zeta) = 0, \quad \text{on } \Omega^{**} \text{ with } C_0^{**}(\zeta) \equiv 1.$$

However, *no previous publication has presented explicit formulas for the coefficients $C_i^{**}(\zeta)$ of (15.6) when m and i are any integers that satisfy $1 \leq i \leq m$.* We shall establish in Theorem 15.1 on page 159 that those coefficients are given by

$$(15.7) \qquad C_i^{**}(\zeta) \equiv \sum_{j=0}^{i} \beta_{m,i-j,m-i}(\zeta)\, (f'(\zeta))^j\, C_j(f(\zeta)), \quad \text{on } \Omega^{**} \text{ when } 0 \leq i \leq m,$$

where $\beta_{m,r,s}(\zeta)$ is the analytic function defined via $\alpha_{i,j}(\zeta)$ from (15.17)–(15.18) by

$$(15.8) \qquad \beta_{m,r,s}(\zeta) \equiv \left[\frac{\prod_{k=1}^{r}(m-s-k+1)}{\prod_{k=1}^{r}(s+k)} \right] \alpha_{r,s}(\zeta), \quad \text{on } \Omega^{**} \text{ for } r,\, s \geq 0.$$

15.2. Consequences due to an improved notation

For given meromorphic functions $c_1(z)$, ..., $c_m(z)$ on a region Ω, we write a monic mth-order homogeneous linear differential equation in the form

$$(15.9) \qquad y^{(m)}(z) + \sum_{i=1}^{m} c_i(z)\, y^{(m-i)}(z) = 0, \quad \text{on } \Omega \text{ with } c_0(z) \equiv 1.$$

15.2.1. Transformations of the first kind for (15.9).
For any given not-identically-zero meromorphic function $\rho(z)$ on Ω, there are unique meromorphic functions $c_1^*(z)$, ..., $c_m^*(z)$ on Ω such that the substitution

$$(15.10) \qquad y(z) = \rho(z)\, v(z),$$

viewed as a change of the dependent variable from y to v, transforms (15.9) into

$$(15.11) \qquad v^{(m)}(z) + \sum_{i=1}^{m} c_i^*(z)\, v^{(m-i)}(z) = 0, \quad \text{on } \Omega \text{ with } c_0^*(z) \equiv 1.$$

The special case $n = 1$ of (3.4) and (3.5) on page 19 yields

$$(15.12) \qquad c_i^*(z) \equiv \sum_{j=0}^{i} \binom{m-j}{i-j} \frac{\rho^{(i-j)}(z)}{\rho(z)}\, c_j(z), \quad \text{on } \Omega \text{ for } 0 \leq i \leq m.$$

In view of $c_i(z) \equiv \binom{m}{i} C_i(z)$ and $c_i^*(z) \equiv \binom{m}{i} C_i^*(z)$, for $0 \leq i \leq m$, the identity

$$(15.13) \qquad \binom{m-j}{i-j}\binom{m}{j} \Big/ \binom{m}{i} \equiv \binom{i}{j}, \quad \text{for } 0 \leq j \leq i \leq m,$$

shows that each of (15.4) and (15.12) is a consequence of the other.

15.2.2. Transformations of the second kind for (15.9). Let $z = f(\zeta)$ denote the inverse function for a univalent analytic function $\zeta = g(z)$ on Ω. Hence, $z = f(\zeta)$ is a univalent analytic function on $\Omega^{**} = g(\Omega)$ that yields $f(\Omega^{**}) = \Omega$ and satisfies $f'(\zeta) \neq 0$, for each ζ in Ω^{**}. Then, there are unique meromorphic functions $c_1^{**}(\zeta), \ldots, c_m^{**}(\zeta)$ on Ω^{**} such that the substitution

$$\text{(15.14)} \qquad z = f(\zeta) \quad \text{with} \quad u(\zeta) = y(f(\zeta)),$$

viewed as a change of the independent variable from z to ζ, transforms (15.9) into

$$\text{(15.15)} \qquad u^{(m)}(\zeta) + \sum_{i=1}^{m} c_i^{**}(\zeta) \, u^{(m-i)}(\zeta) = 0, \quad \text{on } \Omega^{**} \text{ with } c_0^{**}(\zeta) \equiv 1.$$

The special case $n = 1$ of (3.21), (3.22), (3.23), and (3.24) on page 24 yields

$$\text{(15.16)} \qquad c_i^{**}(\zeta) \equiv \sum_{j=0}^{i} \alpha_{i-j,m-i}(\zeta) \left(f'(\zeta)\right)^j c_j(f(\zeta)), \quad \text{on } \Omega^{**} \text{ for } 0 \leq i \leq m,$$

where the definitions of the analytic functions $\alpha_{i,j}(\zeta)$ on Ω^{**} from (3.23)–(3.24) are

$$\text{(15.17)} \qquad \alpha_{0,j}(\zeta) \equiv 1, \quad \text{on } \Omega^{**} \text{ for any } j,$$

and

$$\text{(15.18)} \qquad \alpha_{i,j}(\zeta) \equiv \sum_{k=1}^{j} \left[\alpha_{i-1,k}^{(1)}(\zeta) - (i+k-1) \, \frac{f''(\zeta)}{f'(\zeta)} \, \alpha_{i-1,k}(\zeta) \right],$$
$$\text{on } \Omega^{**} \text{ for } i \geq 1 \text{ and any } j.$$

In contrast to the development of (15.16) for (15.15) in [**19**, pages 135–137], we shall see next how the notation (15.1) complicated the situation for $C_i^{**}(\zeta)$ in (15.6).

15.3. Previously missing essential formula for older research

THEOREM 15.1. *The coefficients $C_i^{**}(\zeta)$ of (15.6) are given by (15.7).*

PROOF. For $m \geq 1$, $0 \leq i \leq m$, and ζ in Ω^{**}, we rewrite (15.16) to obtain

$$\binom{m}{i} C_i^{**}(\zeta) \equiv \sum_{j=0}^{i} \alpha_{i-j,m-i}(\zeta) \left(f'(\zeta)\right)^j \binom{m}{j} C_j(f(\zeta)).$$

Thus, we have

$$\text{(15.19)} \qquad C_i^{**}(\zeta) \equiv \sum_{j=0}^{i} \gamma_{m,i,j}(\zeta) \left(f'(\zeta)\right)^j C_j(f(\zeta)),$$

where

$$\text{(15.20)} \qquad \gamma_{m,i,j}(\zeta) \equiv \frac{\binom{m}{j}}{\binom{m}{i}} \alpha_{i-j,m-i}(\zeta) \equiv \left[\frac{\prod_{k=1}^{i-j}(i-k+1)}{\prod_{k=1}^{i-j}(m-i+k)} \right] \alpha_{i-j,m-i}(\zeta).$$

With $\beta_{m,r,s}$ defined by (15.8), we see that (15.20) gives $\gamma_{m,i,j}(\zeta) \equiv \beta_{m,i-j,m-i}(\zeta)$ and therefore (15.19) yields (15.7) for (15.6). This completes the proof. $\square$

CHAPTER 16

Computer Algebra with Formulas (15.9)–(15.18)

The research presented in [**19, 20, 21**] was made possible when (15.16) was discovered and systems of computer algebra could then be used to find several key identities through trial-and-error experimentation. Similarly, one can make interesting discoveries or rediscoveries merely by using the formulas for $c_i^*(z)$ and $c_i^{**}(\zeta)$ with a few basic commands in a system of computer algebra. Here, we illustrate how that can be done by selecting a version of *Mathematica* from [**55, 56, 57, 58, 59**] as the system. The names of its commands indicate well what they do.

16.1. Computer representations for $c_i^*(z)$ and $c_i^{**}(\zeta)$

We apply (15.9), (15.12), (15.17), (15.18), and (15.16) with the selected version of Mathematica to conclude that successive notebook evaluations of

```
c[m_,0][z_] := 1

cS[m_,i_][z_] := Sum[Binomial[m-j,i-j]*
   (D[rho[z],{z,i-j}]/rho[z])*c[m,j][z],{j,0,i}]

alpha[0,j_][zeta_] := 1

alpha[i_,j_][zeta_] := ( Sum[alpha[i-1,k]'[zeta]
   -(i-1+k)(f''[zeta]/f'[zeta])*
   alpha[i-1,k][zeta],{k,1,j}] ) /; i >= 1

cSS[m_,i_][zeta_] := Sum[alpha[i-j,m-i][zeta]*
   (f'[zeta])^j*c[m, j][f[zeta]],{j,0,i}]
```

enable *Mathematica* to then give computer representations for $c_i^*(z)$ and $c_i^{**}(\zeta)$, when $i = 0, 1, 2, \ldots$ and m can remain a symbol for any positive integer $\geq i$. For instance, the computer representations for the evaluations of cS[m,1][z] and cSS[m,1][zeta] show that $c_1^*(z)$ and $c_1^{**}(\zeta)$ are respectively given by

$$c_1^*(z) \equiv c_1(z) + m\frac{\rho'(z)}{\rho(z)} \quad \text{and} \quad c_1^{**}(\zeta) \equiv f'(\zeta)\,c_1\bigl(f(\zeta)\bigr) - \binom{m}{2}\frac{f''(\zeta)}{f'(\zeta)}.$$

Also, the computer representation for the evaluation of cS[m,2][z] yields

$$c_2^*(z) \equiv c_2(z) + (m-1)\,c_1(z)\,\frac{\rho'(z)}{\rho(z)} + \binom{m}{2}\frac{\rho''(z)}{\rho(z)}.$$

161

16.2. Applications based on the representations for $c_i^*(z)$ and $c_i^{**}(\zeta)$

EXAMPLE 16.1. With $m \geq 2$ and symbols r_1, r_2 for rational numbers, we set

(16.1) $$\boldsymbol{P}_{m,2} \equiv \boldsymbol{w}_2^{(0)} + r_1\left(\boldsymbol{w}_1^{(0)}\right)^2 + r_2\boldsymbol{w}_1^{(1)}.$$

In regard to the function $P_{m,2}(z)$ on Ω that is obtained by replacing each $\boldsymbol{w}_i^{(j)}$ in $\boldsymbol{P}_{m,2}$ with the corresponding $c_i^{(j)}(z)$ from (15.9), we see that the evaluation of

```
P[z_] := c[m,2][z] + r1*c[m,1][z]^2 + r2*c[m,1]'[z]
```

represents $P_{m,2}(z)$. Also, for the function $P_{m,2}^*(z)$ on Ω that is obtained by replacing each $\boldsymbol{w}_i^{(j)}$ in $\boldsymbol{P}_{m,2}$ with the corresponding $c_i^{*(j)}(z)$ from (15.12), the evaluation of

```
PS[z_] := cS[m,2][z] + r1*cS[m,1][z]^2 + r2*cS[m,1]'[z]
```

represents $P_{m,2}^*(z)$. There are eight terms in the output for the evaluation of

```
dif1[z_] = Expand[ PS[z] - P[z] ]
```

and in those terms the parts not involving m, r1, r2 are equal to the evaluations of

```
b[1] = c[m,1][z]*rho'[z]/rho[z];

b[2] = (rho'[z]/rho[z])^2;

b[3] = rho''[z]/rho[z];
```

while the evaluations of

```
a[1] = Coefficient[dif1[z],b[1]];

a[2] = Coefficient[dif1[z],b[2]];

a[3] = Coefficient[dif1[z],b[3]];
```

then yield the respective coefficients a[1], a[2], a[3] of b[1], b[2], b[3] in dif1[z]. Of course, if r1 = r_1 and r2 = r_2 are specific rational numbers, then we see that: a[1], a[2], a[3] are zero if and only if PS(z) - P(z) is zero and $P_{m,2}^*(z) \equiv P_{m,2}(z)$. After the evaluation of

```
list1 = {a[1]==0, a[2]==0, a[3]==0}
```

as a system of three linear equations in r1 and r2, the evaluation of

```
Solve[list1, {r1,r2}]
```

yields a unique solution that corresponds to

(16.2) $$r_1 \equiv -\frac{(m-1)}{2m} \quad \text{and} \quad r_2 \equiv -\frac{(m-1)}{2}.$$

Thus, when r_1, r_2 for (16.1) are defined by (16.2), we have $P_{m,2}^*(z) \equiv P_{m,2}(z)$ on Ω as a valid identity for any (15.9) on Ω having $m \geq 2$ and any transformation (15.10) of that (15.9) into a corresponding equation (15.11) on Ω.

16.2. APPLICATIONS BASED ON THE REPRESENTATIONS FOR $c_i^*(z)$ AND $c_i^{**}(\zeta)$

EXAMPLE 16.2. With $m \geq 2$ and symbols s_1, s_2 for rational numbers, we set
$$(16.3) \qquad \boldsymbol{Q}_{m,2} \equiv \boldsymbol{w}_2^{(0)} + s_1 \left(\boldsymbol{w}_1^{(0)}\right)^2 + s_2 \boldsymbol{w}_1^{(1)}.$$

In regard to the function $Q_{m,2}(z)$ on Ω that is obtained by replacing each $\boldsymbol{w}_i^{(j)}$ in $\boldsymbol{Q}_{m,2}$ with the corresponding $c_i^{(j)}(z)$ from (15.9), we see that the evaluation of

```
Q[z_] := c[m,2][z] + s1*c[m,1][z]^2 + s2*c[m,1]'[z]
```

represents $Q_{m,2}(z)$. For the function $Q_{m,2}^{**}(\zeta)$ on Ω^{**} that is obtained by replacing each $\boldsymbol{w}_i^{(j)}$ in $\boldsymbol{Q}_{m,2}$ with the corresponding $c_i^{**(j)}(\zeta)$ from (15.16), the evaluation of

```
QSS[zeta_] := ( cSS[m,2][zeta] + s1*cSS[m,1][zeta]^2
              + s2*cSS[m,1]'[zeta] )
```

represents $Q_{m,2}^{**}(\zeta)$. There are twenty terms in the output for the evaluation of

```
dif2[zeta_] = Expand[ QSS[zeta] - (f'[zeta])^2*Q[f[zeta]] ]
```

and in those terms the parts not involving m, s1, s2 are given by the evaluations of

```
b[4] = c[m, 1][f[zeta]] f''[zeta];

b[5] = (f''[zeta]/f'[zeta])^2;

b[6] = f'''[zeta]/f'[zeta];
```

while the evaluations of

```
a[4] = Coefficient[dif2[zeta],b[4]];

a[5] = Coefficient[dif2[zeta],b[5]];

a[6] = Coefficient[dif2[zeta],b[6]];
```

give the coefficients of b[4], b[5], b[6] in dif2[zeta]. Naturally, if s1 $= s_1$ and s2 $= s_2$ are specific rational numbers, then we see that: a[4], a[5], a[6] are zero if and only if QSS[zeta] - (f'[zeta])^2*Q[f[zeta]] is zero and we have the identity $Q_{m,2}^{**}(\zeta) \equiv \bigl(f'(\zeta)\bigr)^2 Q_{m,2}(f(\zeta))$. After the evaluation of

```
list2 = {a[4]==0, a[5]==0, a[6]==0}
```

as a system of three linear equations in s1 and s2, the evaluation of

```
Solve[list2, {s1,s2}]
```

yields a unique solution that corresponds to

$$(16.4) \qquad s_1 \equiv -\frac{(m-2)(3m-1)}{6m(m-1)} \quad \text{and} \quad s_2 \equiv -\frac{m-2}{3}.$$

Thus, for s_1, s_2 in (16.3) defined by (16.4), we have $Q_{m,2}^{**}(\zeta) \equiv \bigl(f'(\zeta)\bigr)^2 Q_{m,2}(f(\zeta))$ on Ω^{**} as a valid identity for any equation (15.9) on Ω having $m \geq 2$ and any transformation (15.14) of that (15.9) into a corresponding equation (15.15) on Ω^{**}.

EXAMPLE 16.3. Here, we use the computer representations for $c_i^*(z)$ and $c_i^{**}(\zeta)$ in Section 16.1 to check that the expression for $\mathcal{I}_{4,1;4}$ in (1.17) on page 4 is printed correctly. We find that the evaluation of

```
Simplify[ ( cS[4,4][z] -(1/4)cS[4,1][z]*cS[4,3][z]
    -(1/2)cS[4,3]'[z] -(9/100)cS[4,2][z]^2
    +(1/5)cS[4,2]''[z] +(13/100)cS[4,1][z]^2*cS[4,2][z]
    +(27/100)cS[4,1]'[z]*cS[4,2][z] +(1/4)cS[4,1][z]*cS[4,2]'[z]
    -(39/1600)cS[4,1][z]^4 -(39/200)cS[4,1][z]^2*cS[4,1]'[z]
    -(33/200)(cS[4,1]'[z])^2 -(3/20)cS[4,1][z]*cS[4,1]''[z]
    -(1/20)cS[4,1]'''[z] )
        - ( c[4,4][z] -(1/4)c[4,1][z]*c[4,3][z]
    -(1/2)c[4,3]'[z] -(9/100)c[4,2][z]^2
    +(1/5)c[4,2]''[z] +(13/100)c[4,1][z]^2*c[4,2][z]
    +(27/100)c[4,1]'[z]*c[4,2][z] +(1/4)c[4,1][z]*c[4,2]'[z]
    -(39/1600)c[4,1][z]^4 -(39/200)c[4,1][z]^2*c[4,1]'[z]
    -(33/200)(c[4,1]'[z])^2 -(3/20)c[4,1][z]*c[4,1]''[z]
    -(1/20)c[4,1]'''[z] ) ]
```

is zero and the evaluation of

```
Simplify[ ( cSS[4,4][zeta]
    -(1/4)cSS[4,1][zeta]*cSS[4,3][zeta]
    -(1/2)cSS[4,3]'[zeta] -(9/100)cSS[4,2][zeta]^2
    +(1/5)cSS[4,2]''[zeta]
    +(13/100)cSS[4,1][zeta]^2*cSS[4,2][zeta]
    +(27/100)cSS[4,1]'[zeta]*cSS[4,2][zeta]
    +(1/4)cSS[4,1][zeta]*cSS[4,2]'[zeta]
    -(39/1600)cSS[4,1][zeta]^4
    -(39/200)cSS[4,1][zeta]^2*cSS[4,1]'[zeta]
    -(33/200)(cSS[4,1]'[zeta])^2
    -(3/20)cSS[4,1][zeta]*cSS[4,1]''[zeta]
    -(1/20)cSS[4,1]'''[zeta] )
        - (f'[zeta])^4( c[4,4][f[zeta]]
    -(1/4)c[4,1][f[zeta]]*c[4,3][f[zeta]]
    -(1/2)c[4,3]'[f[zeta]] -(9/100)c[4,2][f[zeta]]^2
    +(1/5)c[4,2]''[f[zeta]]
    +(13/100)c[4,1][f[zeta]]^2*c[4,2][f[zeta]]
    +(27/100)c[4,1]'[f[zeta]]*c[4,2][f[zeta]]
    +(1/4)c[4,1][f[zeta]]*c[4,2]'[f[zeta]]
    -(39/1600)c[4,1][f[zeta]]^4
    -(39/200)c[4,1][f[zeta]]^2*c[4,1]'[f[zeta]]
    -(33/200)(c[4,1]'[f[zeta]])^2
    -(3/20)c[4,1][f[zeta]]*c[4,1]''[f[zeta]]
    -(1/20)c[4,1]'''[f[zeta]] ) ]
```

is zero. Consequently, $\mathcal{I}_{4,1;4}$ as presented in (1.17) on page 4 is a relative invariant of weight $s = 4$ for the equations (15.9) on page 158 having order $m = 4$.

16.2. APPLICATIONS BASED ON THE REPRESENTATIONS FOR $c_i^*(z)$ AND $c_i^{**}(\zeta)$

EXAMPLE 16.4. With $m \geq 3$ and symbols t_1, t_2, t_3, t_4, t_5 representing rational numbers, we introduce

$$(16.5) \qquad \boldsymbol{I}_{m,3} \equiv \boldsymbol{w}_3 + t_1\,\boldsymbol{w}_1\,\boldsymbol{w}_2 + t_2\,(\boldsymbol{w}_1)^3 + t_3\,\boldsymbol{w}_2^{(1)} + t_4\,\boldsymbol{w}_1\,\boldsymbol{w}_1^{(1)} + t_5\,\boldsymbol{w}_1^{(2)}.$$

For the function $I_{m,3}(z)$ on Ω that is obtained by replacing each $\boldsymbol{w}_i^{(j)}$ in $\boldsymbol{I}_{m,3}$ with the corresponding $c_i^{(j)}(z)$ from (15.9), the evaluation of

```
Inv[z_] := ( c[m,3][z] + t1*c[m,1][z]*c[m,2][z]
           + t2*c[m,1][z]^3 + t3*c[m,2]'[z]
           + t4*c[m,1][z]*c[m,1]'[z] + t5*c[m,1]''[z] )
```

represents $I_{m,3}(z)$. For the function $I_{m,3}^*(z)$ on Ω that is obtained by replacing each $\boldsymbol{w}_i^{(j)}$ in $\boldsymbol{I}_{m,3}$ with the corresponding $c_i^{*(j)}(z)$ from (15.12), the evaluation of

```
InvS[z_] := ( cS[m,3][z] + t1*cS[m,1][z]*cS[m,2][z]
            + t2*cS[m,1][z]^3 + t3*cS[m,2]'[z]
            + t4*cS[m,1][z]*cS[m,1]'[z] + t5*cS[m,1]''[z] )
```

represents $I_{m,3}^*(z)$. For the function $I_{m,3}^{**}(\zeta)$ on Ω^{**} that is obtained by replacing each $\boldsymbol{w}_i^{(j)}$ in $\boldsymbol{I}_{m,3}$ with the corresponding $c_i^{**(j)}(\zeta)$ from (15.16), the evaluation of

```
InvSS[zeta_] := ( cSS[m,3][zeta]
             + t1*cSS[m,1][zeta]*cSS[m,2][zeta] + t2*cSS[m,1][zeta]^3
             + t3*cSS[m,2]'[zeta] + t4*cSS[m,1][zeta]*cSS[m,1]'[zeta]
             + t5*cSS[m,1]''[zeta] )
```

represents $I_{m,3}^{**}(\zeta)$. We note that t_1, t_2, t_3, t_4, t_5 for (16.5) yield

$$(16.6) \qquad I_{m,3}^*(z) \equiv I_{m,3}(z) \text{ on } \Omega, \text{ and } I_{m,3}^{**}(\zeta) \equiv \bigl(f'(\zeta)\bigr)^3 I_{m,3}\bigl(f(\zeta)\bigr), \text{ on } \Omega^{**}.$$

if and only if their representations t1, t2, t3, t4, t5 make the evaluations of

```
diff1[z_] = Expand[ InvS[z] - Inv[z] ]

diff2[zeta_] = Expand[InvSS[zeta]-(f'[zeta])^3*Inv[f[zeta]]]
```

identically zero. Among the thirty-eight terms in the expansion of diff1[z], there are eight parts that do not involve m, t1, t2, t3, t4, t5. Let them be copied individually from the output, pasted into individual input cells, given the names b3[1], b3[2], ..., b3[8], and then evaluated. Among the ninety-three terms in the expansion of diff2[zeta], there are eight parts that do not involve m, t1, t2, t3, t4, t5. Let them be copied from the output, pasted into input cells, given the names b3[9], b3[10], ..., b3[16], and then be evaluated. We evaluate

```
Do[a3[k] = Coefficient[diff1[z], b3[k]], {k,1,8}];

Do[a3[k] = Coefficient[diff2[zeta], b3[k]], {k,9,16}];
```

and then find that the evaluation of

```
Solve[ Table[a3[k] == 0, {k,1,16}], {t1,t2,t3,t4,t5} ]
```

yields a unique solution. As expressed for (16.5), it is given by

(16.7) $$t_1 = -\frac{m-2}{m}, \quad t_2 = \frac{(m-1)(m-2)}{3m^2}, \quad t_3 = -\frac{m-2}{2},$$
$$t_4 = \frac{(m-1)(m-2)}{2m}, \quad \text{and} \quad t_5 = \frac{(m-1)(m-2)}{12}.$$

Thus, (16.6) is satisfied by (16.5) with (16.7) for each equation (15.9) having $m \geq 3$ as well as each transformation (15.10) of (15.9) into a corresponding (15.11) and each transformation (15.14) of (15.9) into a corresponding (15.15). In this regard, see (1.13) of page 3. If the definitions of b3[1], b3[2], ..., b3[16] give difficulty, use the Google browser *Chrome* to visit

http://homepages.uc.edu/~chalklr/Chapter-16.html

and then download the *Mathematica* notebook available there. Details are also given in that notebook for Examples 16.1, 16.2, 16.3, and 16.5.

EXAMPLE 16.5. There are unique rational numbers $u_1, u_2, \ldots, u_{12}$ for

(16.8) $$\boldsymbol{I}_{m,4} \equiv \boldsymbol{w}_4 + u_1 \boldsymbol{w}_1 \boldsymbol{w}_3 + u_2 \boldsymbol{w}_3^{(1)} + u_3 (\boldsymbol{w}_2)^2 + u_4 \boldsymbol{w}_2^{(2)} + u_5 (\boldsymbol{w}_1)^2 \boldsymbol{w}_2$$
$$+ u_6 \boldsymbol{w}_1^{(1)} \boldsymbol{w}_2 + u_7 \boldsymbol{w}_1 \boldsymbol{w}_2^{(1)} + u_8 (\boldsymbol{w}_1)^4 + u_9 (\boldsymbol{w}_1)^2 \boldsymbol{w}_1^{(1)}$$
$$+ u_{10} (\boldsymbol{w}_1^{(1)})^2 + u_{11} \boldsymbol{w}_1 \boldsymbol{w}_1^{(2)} + u_{12} \boldsymbol{w}_1^{(3)}, \quad \text{with } m \geq 4,$$

such that the functions $I_{m,4}(z)$ on Ω, $I_{m,4}^*(z)$ on Ω, and $I_{m,4}^{**}(\zeta)$ on Ω^{**} that are obtained by replacing each $\boldsymbol{w}_i^{(j)}$ in $\boldsymbol{I}_{m,4}$ with the corresponding $c_i^{(j)}(z)$ from (15.9), with the $c_i^{*(j)}(z)$ from (15.11), and with the $c_i^{**(j)}(\zeta)$ from (15.15), satisfy both

$$I_{m,4}^*(z) \equiv I_{m,4}(z) \quad \text{on } \Omega, \quad \text{and} \quad I_{m,4}^{**}(\zeta) \equiv (f'(\zeta))^4 I_{m,4}(f(\zeta)), \quad \text{on } \Omega^{**}.$$

When the technique of Example 4.4 is repeated here, the main difference is that: in place of the copy and paste for Example 4.4 where b3[k] was obtained separately for $1 \leq k \leq 8$ and $9 \leq k \leq 16$, we now use copy and paste to obtain b4[k] separately for $1 \leq k \leq 20$ and for $21 \leq k \leq 40$. Of course, this requires more patience. However, when details similar to those of Example 4.4 are carried out, the coefficients for $\boldsymbol{I}_{m,4}$ in (16.8) are found to be

(16.9) $$u_1 = -\frac{m-3}{m}, \quad u_2 = -\frac{m-3}{2}, \quad u_3 = -\frac{(m-2)(m-3)(5m+7)}{10(m+1)(m)(m-1)},$$
$$u_4 = \frac{(m-2)(m-3)}{10}, \quad u_5 = \frac{(m-2)(m-3)(5m+6)}{5(m+1)m^2}, \quad u_6 = \frac{(m-2)(m-3)(5m+7)}{10(m+1)m},$$
$$u_7 = \frac{(m-2)(m-3)}{2m}, \quad u_8 = -\frac{(m-1)(m-2)(m-3)(5m+6)}{20(m+1)m^3},$$
$$u_9 = -\frac{(m-1)(m-2)(m-3)(5m+6)}{10(m+1)m^2}, \quad u_{10} = -\frac{(m-1)(m-2)(m-3)(2m+3)}{20(m+1)m},$$
$$u_{11} = -\frac{(m-1)(m-2)(m-3)}{10m}, \quad u_{12} = -\frac{(m-1)(m-2)(m-3)}{120}.$$

By setting $m = 4$ in these formulas, we obtain the coefficients for (1.17) on page 4.

Observation. The basic relative invariants $\boldsymbol{\mathcal{I}}_{m,1;s}$ of weight $s \geq 3$ for the equations (15.9) of order $m \geq s$ are given explicitly by the computer program in Section 6.1 on pages 53–54. We note that $\boldsymbol{I}_{m,3}$ in (16.5) with the coefficients of (16.7) is equal to $\boldsymbol{\mathcal{I}}_{m,1;3}$. Also, $\boldsymbol{I}_{m,4}$ in (16.8) with the coefficients of (16.9) is equal to $\boldsymbol{\mathcal{I}}_{m,1;4}$.

CHAPTER 17

Computer Algebra with Formulas (15.1)–(15.8)

Various identities that involve the coefficients of (15.1), (15.3), and (15.6) exist in the mathematical literature. Their derivations are sometimes doubtful. Now, they can be verified directly with computer algebra based on the formulas (15.4) and (15.7) for $C_i^*(z)$ and $C_i^{**}(\zeta)$. An explanation is given next.

17.1. Computer-algebra representations of $C_i^*(z)$ and $C_i^{**}(\zeta)$.

For a version of *Mathematica* from [55, 56, 57, 58, 59] as the system, we apply (15.1), (15.4), (15.17), (15.18), (15.8), and (15.7) to see that the evaluations of

```
Ce[m_,0][z_] := 1

CeS[m_,i_][z_] := Sum[Binomial[i,j]*
   (D[rho[z],{z,i-j}]/rho[z])*Ce[m,j][z],{j,0,i}]

alpha[0,j_][zeta_] := 1

alpha[i_,j_][zeta_] := ( Sum[alpha[i-1,k]'[zeta]
   -(i-1+k)(f''[zeta]/f'[zeta])*
   alpha[i-1,k][zeta],{k,1,j}] ) /; i >= 1

beta[m_,r_,s_][zeta_]:=(Product[(m-s-k+1),{k,1,r}]/
   Product[(s+k),{k,1,r}])*alpha[r,s][zeta]

CeSS[m_,i_][zeta_] := Sum[beta[m,i-j,m-i][zeta]*
   (f'[zeta])^j*Ce[m, j][f[zeta]],{j,0,i}]
```

enable representations for $C_1^*(z), C_2^*(z), \ldots, C_1^{**}(\zeta), C_2^{**}(\zeta), \ldots$ to be obtained as the evaluations of the corresponding

```
CeS[m,1][z], CeS[m,2][z], ..., CeSS[m,1][zeta], CeSS[m,2][zeta], ....
```

For instance, computer representations for the evaluations of `CeS[m,1][z]` and `CeSS[m,1][zeta]` show that $C_1^*(z)$ and $C_1^{**}(\zeta)$ are respectively given by

$$(17.1) \quad C_1^*(z) \equiv C_1(z) + \frac{\rho'(z)}{\rho(z)} \quad \text{and} \quad C_1^{**}(\zeta) \equiv f'(\zeta)\, C_1\big(f(\zeta)\big) - \frac{(m-1)}{2}\frac{f''(\zeta)}{f'(\zeta)}.$$

Also, the computer representation for the evaluation of `CeS[m,2][z]` yields

$$C_2^*(z) \equiv C_2(z) + 2\,C_1(z)\frac{\rho'(z)}{\rho(z)} + \frac{\rho''(z)}{\rho(z)}.$$

17.2. Computer-algebra verifications.

EXAMPLE 17.1. James Cockle was aware in [**22**, page 533, (11)] of 1862 that, for any $m \geq 2$ as well as any (15.1) and substitution (15.2) that transforms (15.1) into (15.3), the coefficients of (15.1) and (15.3) satisfy

$$(17.2) \quad C_2^*(z) - \bigl(C_1^*(z)\bigr)^2 - C_1^{*(1)}(z) \equiv C_2(z) - \bigl(C_1(z)\bigr)^2 - C_1^{(1)}(z).$$

For an independent verification, the evaluation of

```
Expand[ ( CeS[m,2][z] - CeS[m,1][z]^2 - CeS[m,1]'[z] )
       - ( Ce[m,2][z]  - Ce[m,1][z]^2  - Ce[m,1]'[z]  ) ]
```

is zero. By making a change of notation, we see that (17.2) corresponds to the semi-invariant of the first kind $\boldsymbol{P}_{m,2}$ defined by (16.1) and (16.2).

EXAMPLE 17.2. For any (15.1) and substitution (15.5) that transforms (15.1) into (15.6), there is an identity that corresponds to one presented by James Cockle in [**23**, page 446] of 1875. We multiply it by $-\bigl(3(m-1)\bigr)/\bigl(2(m-2)\bigr)$ to obtain

$$(17.3) \quad C_2^{**}(\zeta) - \frac{(m-2)(3m-1)}{3(m-1)^2}\bigl(C_1^{**}(\zeta)\bigr)^2 - \frac{2(m-2)}{3(m-1)}C_1^{**(1)}(\zeta)$$

$$\equiv \bigl(f'(\zeta)\bigr)^2 \left[\begin{array}{l} C_2\bigl(f(\zeta)\bigr) - \dfrac{(m-2)(3m-1)}{3(m-1)^2}C_1\bigl(f(\zeta)\bigr)\bigr)^2 \\ \qquad - \dfrac{2(m-2)}{3(m-1)}C_1^{(1)}\bigl(f(\zeta)\bigr) \end{array} \right].$$

To verify that (17.3) is valid for ζ in Ω^{**} and any integer m that satisfies $m \geq 2$, we check that the evaluation of

```
Together[ ( CeSS[m,2][zeta]
          - ((m-2)(3m-1)/(3(m-1)^2))CeSS[m,1][zeta]^2
          - (2(m-2)/(3(m-1)))CeSS[m,1]'[zeta] )
          - f'[zeta]^2 ( Ce[m,2][f[zeta]]
          - ((m-2)(3m-1)/(3(m-1)^2))Ce[m, 1][f[zeta]]^2
          - (2(m-2)/(3(m-1)))Ce[m,1]'[f[zeta]] ) ]
```

is zero. A change of notation shows that (17.3) corresponds to the semi-invariant of the second kind $\boldsymbol{Q}_{m,2}$ of weight 2 defined by (16.3) and (16.4).

EXAMPLE 17.3. Let m represent an integer that satisfies $m \geq 3$. For any (15.1) and transformation (15.2) of (15.1) into a corresponding (15.3), the research of Georges-Henri Halphen in [**32**, page 127] of 1884 indicates that the coefficients of (15.3) and (15.1) satisfy

$$(17.4) \quad \begin{bmatrix} C_3^*(z) \\ -3\,C_2^*(z)\,C_1^*(z) \\ +2\,\bigl(C_1^*(z)\bigr)^3 \\ -(3/2)\,C_2^{*(1)}(z) \\ +3\,C_1^*(z)\,C_1^{*(1)}(z) \\ +(1/2)\,C_1^{*(2)}(z) \end{bmatrix} \equiv \begin{bmatrix} C_3(z) \\ -3\,C_2(z)\,C_1(z) \\ +2\,\bigl(C_1(z)\bigr)^3 \\ -(3/2)\,C_2^{(1)}(z) \\ +3\,C_1(z)\,C_1^{(1)}(z) \\ +(1/2)\,C_1^{(2)}(z) \end{bmatrix}, \quad \text{for } z \text{ in } \Omega;$$

17.2. COMPUTER-ALGEBRA VERIFICATIONS.

while, for any (15.1) and transformation (15.5) of (15.1) into a corresponding (15.6), the research in [**32**, page 127] indicates that the coefficients of (15.6) and (15.1) are related by

$$(17.5) \quad \begin{bmatrix} C_3^{**}(\zeta) \\ -3\,C_2^{**}(\zeta)\,C_1^{**}(\zeta) \\ +2\,(C_1^{**}(\zeta))^3 \\ -(3/2)\,C_2^{**(1)}(\zeta) \\ +3\,C_1^{**}(\zeta)\,C_1^{**(1)}(\zeta) \\ +(1/2)\,C_1^{**(2)}(\zeta) \end{bmatrix} \equiv (f'(\zeta))^3 \begin{bmatrix} C_3(f(\zeta)) \\ -3\,C_2(f(\zeta))\,C_1(f(\zeta)) \\ +2\,(C_1(f(\zeta)))^3 \\ -(3/2)\,C_2^{(1)}(f(\zeta)) \\ +3\,C_1(f(\zeta))\,C_1^{(1)}(f(\zeta)) \\ +(1/2)\,C_1^{(2)}(f(\zeta)) \end{bmatrix},$$

for ζ in Ω^{**}. To independently verify (17.4), we note that the evaluation of

```
Together[ ( CeS[m, 3][z] - 3 CeS[m, 2][z]*CeS[m, 1][z]
         + 2 CeS[m, 1][z]^3 - (3/2)CeS[m, 2]'[z]
         + 3 CeS[m, 1][z]*CeS[m, 1]'[z] + (1/2)CeS[m, 1]''[z] )
       - ( Ce[m, 3][z] - 3 Ce[m, 2][z]*Ce[m, 1][z]
         + 2 Ce[m, 1][z]^3 - (3/2)Ce[m, 2]'[z]
         + 3 Ce[m, 1][z]*Ce[m, 1]'[z] + (1/2)Ce[m, 1]''[z] ) ]
```

is zero. For an independent verification of (17.5), we find that the evaluation of

```
Together[ (  CeSS[m,3][zeta] - 3CeSS[m,2][zeta]*CeSS[m,1][zeta]
           + 2 CeSS[m,1][zeta]^3 - (3/2)CeSS[m,2]'[zeta]
           + 3 CeSS[m,1][zeta]*CeSS[m,1]'[zeta]
           + (1/2)CeSS[m,1]''[zeta] )
         - (f'[zeta]^3)( Ce[m,3][f[zeta]]
           - 3 Ce[m,2][f[zeta]]*Ce[m,1][f[zeta]]
           + 2 Ce[m,1][f[zeta]]^3 - (3/2)Ce[m,2]'[f[zeta]]
           + 3 Ce[m,1][f[zeta]]*Ce[m,1]'[f[zeta]]
           + (1/2)Ce[m,1]''[f[zeta]] ) ]
```

is zero.

A *Mathematica* notebook that contains evaluations for the input commands of this chapter may be downloaded from
   http://homepages.uc.edu/~chalklr/Chapter-17.html
by using the Google browser *Chrome*.

The fact that the identities (17.4) and (17.5) do not involve the order $m \geq 3$ of the corresponding equations (15.1), (15.3), and (15.6) was thought to be desirable. That situation is reflected in the formulas (18.4) and (18.17). However, the order m is involved in formulas of greater complexity such as (18.18)–(18.21) on page 176.

Formulas (18.17)–(18.21) correspond to the ones that Andrew Forsyth derived for relatve invariants of weights $s = 3, 4, 5, 6, 7$ in [**28**, pages 398–401, (14)–(18)] based on his use of infinitesimal transformations. The previously unavailable (15.7) of page 158 enables a direct verification of Forsyth's formulas to be made based on the technique of this chapter. In this regard, see the verifications in Section 18.5 for Theorem 18.7 on page 177.

CHAPTER 18

Suitable Context for Older Notation

The principal failing of the notation for (15.1), (15.3), and (15.6) that involves binomial coefficients was its role before 1989 in greatly hindering the discovery of suitable formulas for the coefficients of (15.6) corresponding to a change (15.5) of the independent variable. The details about this in Chapter 15 make it clear that the notation involving binomial coefficient should never have been adopted.

However, before we abandon that notation, an explanation should be given to explain why truly remarkable results like (17.4)–(17.5) of Edmund Laguerre and Georges-Henri Halphen were known to only a few mathematicians in 1989. Thus, we provide in Section 18.1 a previously missing precise context about invariants for equations like (15.1). However, to truly honor Laguerre and Halphen, their results should be presented in a form like that of Section 1.3 where binomial coefficients are avoided as a needless distraction.

18.1. Symbolism and terminology

In previous research about invariants for equations written as (15.1), instead of constructing semi-invariants and relative invariants from polynomials upon which algebraic operations can be performed and into which substitutions can be made, the semi-invariants and relative invariants were represented by functions without mention of substitutions. For example, the expression $C_2(z) - (C_1(z))^2 - C_1^{(1)}(z)$ in the right member of (17.2), the expression inside the brackets of the right member for (17.3), and the expression in the right member of (17.4) were described as invariants.

For suitable notation, let $\mathfrak{R}_{m,1}$ be the ring of polynomials over $\mathbb{Q}$ in the variables

(18.1) $\qquad \boldsymbol{W}_i^{(j)}, \quad \text{for } 1 \leq i \leq m \text{ and } j \geq 0;$

set $\boldsymbol{W}_i \equiv \boldsymbol{W}_i^{(0)}$, for $1 \leq i \leq m$; and let $'$ denote the unique derivation for $\mathfrak{R}_{m,1}$ such that $(\boldsymbol{W}_i^{(j)})' \equiv \boldsymbol{W}_i^{(j+1)}$, when $1 \leq i \leq m$ and $j \geq 0$. The constants of $\mathfrak{R}_{m,1}$ (i.e., the elements γ in $\mathfrak{R}_{m,1}$ having $\gamma' = 0$) are the elements of $\mathbb{Q}$. The *weight* of $\boldsymbol{W}_i^{(j)}$ is $i + j$; the *weight* of a nonzero element of $\mathbb{Q}$ is 0; and the *weight* of any nonzero monomial in $\mathfrak{R}_{m,1}$ is the sum of the weights of its factors. A polynomial in $\mathfrak{R}_{m,1}$ is said to be *isobaric* when it is nonzero and the weights of its nonzero terms are equal. The *weight* of an isobaric polynomial is the weight of any nonzero term.

For any polynomial $\widehat{\boldsymbol{R}}$ in $\mathfrak{R}_{m,1}$, let $\widehat{R}(z)$ denote the function on Ω that is obtained by replacing each $\boldsymbol{W}_i^{(j)}$ of $\widehat{\boldsymbol{R}}$ with the corresponding $C_i^{(j)}(z)$ from (15.1), let $\widehat{R}^*(z)$ denote the function on Ω obtained by replacing each $\boldsymbol{W}_i^{(j)}$ of $\widehat{\boldsymbol{R}}$ with the corresponding $C_i^{*(j)}(z)$ from (15.3), and let let $\widehat{R}^{**}(\zeta)$ denote the function on Ω^{**} obtained by replacing each $\boldsymbol{W}_i^{(j)}$ of $\widehat{\boldsymbol{R}}$ with the corresponding $C_i^{**(j)}(\zeta)$ from (15.6).

For instance, in terms of the polynomial
$$\widehat{P}_2 \equiv W_2 - (W_1)^2 - W_1^{(1)}, \tag{18.2}$$
the identity (17.2) of James Cockle for $m \geq 2$ is $\widehat{P}_2^*(z) \equiv \widehat{P}_2(z)$, on Ω. Also, for
$$\widehat{Q}_{m,2} \equiv W_2 - \frac{(m-2)(3m-1)}{3(m-1)^2}(W_1)^2 - \frac{2(m-2)}{3(m-1)}W_1^{(1)}, \tag{18.3}$$
the identity (17.3) of Cockle is given by $\widehat{Q}_{m,2}^{**}(\zeta) \equiv (f'(\zeta))^2 \widehat{Q}_{m,2}(f(\zeta))$, on Ω^{**}. The identities (17.4) and (17.5) of Halphen are represented with the polynomial
$$\widehat{H}_3 \equiv W_3 - 3W_2 W_1 + 2(W_1)^3 - \tfrac{3}{2}W_2^{(1)} + 3W_1 W_1^{(1)} + \tfrac{1}{2}W_1^{(2)}. \tag{18.4}$$
by $\widehat{H}_3^*(z) \equiv \widehat{H}_3(z)$, on Ω, and $\widehat{H}_3^{**}(\zeta) \equiv (f'(\zeta))^3 \widehat{H}_3(f(\zeta))$, on Ω^{**}. In fact, if the variables $W_i^{(j)}$ of (18.1) are introduced so that they are related to the variables $w_i^{(j)}$ for (1.13) by $w_i^{(j)} \equiv \binom{m}{i} W_i^{(j)}$, then $\mathcal{I}_{m,1;3}$ in (1.13) yields $\mathcal{I}_{m,1;3} \equiv \binom{m}{3}\widehat{H}_3$.

DEFINITION 18.1. A polynomial $\widehat{R}$ in $\mathfrak{R}_{m,1}$ is a *Cockle-semi-invariant of the first kind* for equations of the form (15.1) when it is not a constant and yields
$$\widehat{R}^*(z) \equiv \widehat{R}(z), \tag{18.5}$$
for each (15.1) and each transformation (15.2) of (15.1) into a corresponding (15.3).

DEFINITION 18.2. A polynomial $\widehat{R}$ in $\mathfrak{R}_{m,1}$ is a *Cockle-semi-invariant of the second kind* for equations of the form (15.1) when it is not a constant and there is an integer s such that
$$\widehat{R}^{**}(\zeta) \equiv (f'(\zeta))^s \widehat{R}(f(\zeta)), \tag{18.6}$$
for each (15.1) and each transformation (15.5) of (15.1) into a corresponding (15.6).

DEFINITION 18.3. A polynomial $\widehat{R}$ in $\mathfrak{R}_{m,1}$ is a *Laguerre-Halphen relative invariant* for equations of the form (15.1) when it is both a Cockle-semi-invariant of the first kind and a Cockle-semi-invariant of the second kind for such equations.

As examples, we note that $\widehat{P}_2$ in (18.2) is a Cockle-semi-invariant of the first kind; $\widehat{Q}_{m,2}$ in (18.3) is a Cockle-semi-invariant of the second kind; and $\widehat{H}_3$ in (18.4) is a Laguerre-Halphen relative invariant.

18.2. Our viewpoint abut the older Cockle-semi-invariants

To illustrate how the context of Section 18.1 can be employed, we begin by defining $\widehat{F}_i$ and $\widehat{G}_i$ in $\mathfrak{R}_{m,1}$ through
$$\widehat{F}_0 \equiv 1, \quad \widehat{F}_1 \equiv -W_1, \quad \widehat{F}_{i+1} \equiv \widehat{F}_i^{(1)} + \widehat{F}_1 \widehat{F}_i, \quad \text{for } i \geq 1, \tag{18.7}$$
and, with the introduction of $W_0 \equiv 1$,
$$\widehat{G}_i \equiv \sum_{j=0}^{i} \binom{i}{j} \widehat{F}_{i-j} W_j, \quad \text{when } 0 \leq i \leq m. \tag{18.8}$$
We have $\widehat{G}_0 \equiv 1$, $\widehat{G}_1 \equiv 0$, and $\widehat{G}_2 \equiv \widehat{P}_2$ in (18.2). The coefficients of $\widehat{F}_i$ and $\widehat{G}_i$ are polynomials in the variables $W_j^{(k)}$ over $\mathbb{Q}$. They do not involve m.

To obtain $\widehat{G}_i(z)$ on Ω or $\widehat{G}_i^*(z)$ on Ω, we replace each $W_j^{(k)}$ of $\widehat{G}_i$ with the corresponding $C_j^{(k)}(z)$ from (15.1) or the corresponding $C_j^{*(k)}(z)$ from (15.3).

18.2. OUR VIEWPOINT ABUT THE OLDER COCKLE-SEMI-INVARIANTS

THEOREM 18.4. *For $2 \leq i \leq m$ and the equations (15.1), $\widehat{G}_i$ is an isobaric Cockle-semi-invariant of the first kind having weight i.*

PROOF. For $2 \leq i \leq m$, we use (18.8) and (18.7) to see that the coefficient of $\boldsymbol{W}_i$ in $\widehat{\boldsymbol{G}}_i$ is 1 and therefore $\widehat{\boldsymbol{G}}_i$ is not a constant. Since (18.7) shows that $\widehat{\boldsymbol{F}}_i$ is an isobaric polynomial of weight i for $i \geq 0$, we apply (18.8) to conclude, for $2 \leq i \leq m$, that $\widehat{\boldsymbol{G}}_i$ is an isobaric polynomial of weight i.

For $0 \leq i \leq m$ and a transformation (15.2) of (15.1) into a corresponding (15.3), we employ (18.8), (15.4), and the identity

$$\binom{i}{j}\binom{j}{k} \equiv \binom{i}{k}\binom{i-k}{j-k}, \quad \text{for } 0 \leq k \leq j \leq i,$$

to obtain

(18.9) $\quad \widehat{G}_i^*(z) \equiv \sum_{j=0}^{i} \binom{i}{j} \widehat{F}_{i-j}^*(z)\, C_j^*(z)$

$\equiv \sum_{j=0}^{i} \binom{i}{j} \widehat{F}_{i-j}^*(z) \sum_{k=0}^{j} \binom{j}{k} \frac{\rho^{(j-k)}(z)}{\rho(z)} C_k(z)$

$\equiv \sum_{k=0}^{i} \binom{i}{k} C_k(z) \sum_{j=k}^{i} \binom{i-k}{j-k} \frac{\rho^{(j-k)}(z)}{\rho(z)} \widehat{F}_{i-j}^*(z)$

$\equiv \sum_{k=0}^{i} \binom{i}{k} C_k(z) \sum_{\nu=0}^{i-k} \binom{i-k}{\nu} \frac{\rho^{(\nu)}(z)}{\rho(z)} \widehat{F}_{i-k-\nu}^*(z)$

$\equiv \sum_{k=0}^{i} \binom{i}{k} C_k(z)\, S_{i-k}(z), \quad \text{on } \Omega,$

where

(18.10) $\quad S_\mu(z) \equiv \sum_{\nu=0}^{\mu} \binom{\mu}{\nu} \frac{\rho^{(\nu)}(z)}{\rho(z)} \widehat{F}_{\mu-\nu}^*(z), \quad \text{on } \Omega \text{ for } \mu \geq 0.$

We note that (18.10), (18.7), and (17.1) yield $S_0(z) \equiv \widehat{F}_0(z)$ and

$$S_1(z) \equiv \widehat{F}_1^*(z) + \frac{\rho^{(1)}(z)}{\rho(z)} \equiv -C_1^*(z) + \frac{\rho^{(1)}(z)}{\rho(z)} \equiv -C_1(z) \equiv \widehat{F}_1(z).$$

Let μ be an integer satisfying $\mu \geq 1$ such that $S_\mu(z) \equiv \widehat{F}_\mu(z)$. Then, we use (18.7), $\widehat{F}_\mu(z) \equiv S_\mu(z)$, $\widehat{F}_1(z) \equiv \rho^{(1)}(z)/\rho(z) + \widehat{F}_1^*(z)$, and (18.10) to verify that

$\widehat{F}_{\mu+1}(z) \equiv \widehat{F}_\mu^{(1)}(z) + \widehat{F}_1(z)\, \widehat{F}_\mu(z)$

$\equiv S_\mu^{(1)}(z) + \frac{\rho^{(1)}(z)}{\rho(z)} S_\mu(z) + \widehat{F}_1^*(z)\, S_\mu(z)$

$\equiv \sum_{\nu=0}^{\mu} \binom{\mu}{\nu} \frac{\rho^{(\nu+1)}(z)}{\rho(z)} \widehat{F}_{\mu-\nu}^*(z) - \frac{\rho^{(1)}(z)}{\rho(z)} S_\mu(z) + \frac{\rho^{(1)}(z)}{\rho(z)} S_\mu(z)$

$+ \left(\sum_{\nu=0}^{\mu} \binom{\mu}{\nu} \frac{\rho^{(\nu)}(z)}{\rho(z)} \widehat{F}_{\mu-\nu}^{*(1)}(z) + \sum_{\nu=0}^{\mu} \binom{\mu}{\nu} \frac{\rho^{(\nu)}(z)}{\rho(z)} \widehat{F}_1^*(z)\, \widehat{F}_{\mu-\nu}^*(z) \right)$

and, since the relation $\widehat{\boldsymbol{F}}_{i+1} \equiv \widehat{\boldsymbol{F}}_i^{(1)} + \widehat{\boldsymbol{F}}_1 \widehat{\boldsymbol{F}}_i$ in (18.7) is also valid for $i = 0$,

$$\widehat{F}_{\mu+1}(z) \equiv \sum_{\nu=1}^{\mu+1} \binom{\mu}{\nu-1} \frac{\rho^{(\nu)}(z)}{\rho(z)} \widehat{F}^*_{\mu+1-\nu}(z) + \sum_{\nu=0}^{\mu} \binom{\mu}{\nu} \frac{\rho^{(\nu)}(z)}{\rho(z)} \widehat{F}^*_{\mu+1-\nu}(z)$$

$$\equiv \sum_{\nu=0}^{\mu+1} \binom{\mu+1}{\nu} \frac{\rho^{(\nu)}(z)}{\rho(z)} \widehat{F}^*_{\mu+1-\nu}(z) \equiv S_{\mu+1}(z), \quad \text{on } \Omega.$$

Thus, $S_\mu(z) \equiv \widehat{F}_\mu(z)$ is valid on Ω for $\mu \geq 0$. We replace $S_{i-k}(z)$ in (18.9) with $\widehat{F}_{i-k}(z)$ and compare the result with (18.8) to see, for any (15.1) and (15.2), that

$$\widehat{G}^*_i(z) \equiv \sum_{k=0}^{i} \binom{i}{k} C_k(z) \widehat{F}_{i-k}(z) \equiv \widehat{G}_i(z), \quad \text{on } \Omega \text{ when } 0 \leq i \leq m.$$

Hence, for $2 \leq i \leq m$, $\widehat{G}_i$ is an isobaric Cockle-semi-invariant of the first kind having weight i. This completes the proof. □

This proof of Theorem 18.4 illustrates how the context of Section 18.1 can be applied. To connect it with the argument in Subsection 18.3.2 as the only one available for earlier researchers, we have the following result.

THEOREM 18.5. *For an equation* (15.1) *on Ω having order $m \geq 2$, repeated as*

$$y^{(m)}(z) + \sum_{i=1}^{m} \binom{m}{i} C_i(z) y^{(m-i)}(z) = 0,$$

suppose that $\rho_1(z) \not\equiv 0$ is a meromorphic function on a subregion $\mathcal{U}_1$ of Ω such that the substitution $y(z) = \rho_1(z) t(z)$ transforms the restriction to $\mathcal{U}_1$ of (15.1) *into*

$$(18.11) \qquad t^{(m)}(z) + \sum_{i=2}^{m} \binom{m}{i} d_i(z) t^{(m-i)}(z) = 0, \quad \text{on } \mathcal{U}_1 \text{ with } d_1(z) \equiv 0.$$

Then, for $1 \leq i \leq m$, $d_i(z)$ of (18.11) *is given by $d_i(z) \equiv \widehat{G}_i(z)$, on $\mathcal{U}_1$.*

PROOF. For the indicated transformation of (15.1) on $\mathcal{U}_1$ into (18.11) on $\mathcal{U}_1$, we find that (15.4) and (18.7) yield

$$(18.12) \quad d_1(z) \equiv \frac{\rho_1^{(1)}(z)}{\rho_1(z)} + C_1(z) \equiv 0 \quad \text{and} \quad \widehat{F}_1(z) \equiv -C_1(z) \equiv \frac{\rho_1^{(1)}(z)}{\rho_1(z)}, \quad \text{on } \mathcal{U}_1.$$

We use (18.7) and (18.12) to see that the formula

$$(18.13) \qquad \widehat{F}_k(z) \equiv \frac{\rho_1^{(k)}(z)}{\rho_1(z)}, \quad \text{on } \mathcal{U}_1,$$

is true for $k = 0$ and $k = 1$. In terms of any positive integer k for which (18.13) is valid, we observe that (18.7) and (18.13) yield

$$\widehat{F}_{k+1}(z) \equiv \widehat{F}_k^{(1)}(z) + \widehat{F}_1(z) \widehat{F}_k(z)$$

$$\equiv \frac{\rho_1^{(k+1)}(z)}{\rho_1(z)} - \frac{\rho_1^{(k)}(z) \rho_1^{(1)}(z)}{(\rho_1(z))^2} + \frac{\rho_1^{(1)}(z)}{\rho_1(z)} \frac{\rho_1^{(k)}(z)}{\rho_1(z)} \equiv \frac{\rho_1^{(k+1)}(z)}{\rho_1(z)}.$$

Thus, (18.13) is true for $k \geq 0$. Using (15.4), (18.13), and (18.8), we notice that the substitution $y(z) = \rho_1(z) v(z)$ transforms the restriction to $\mathcal{U}_1$ of (15.1) into the equation (18.11) on $\mathcal{U}_1$ having

$$d_i(z) \equiv \sum_{j=0}^{i} \binom{i}{j} \frac{\rho_1^{(i-j)}(z)}{\rho_1(z)} C_j(z) \equiv \sum_{j=0}^{i} \binom{i}{j} \widehat{F}_{i-j}(z) C_j(z) \equiv \widehat{G}_i(z),$$

for $2 \leq i \leq m$. This completes the proof. □

18.3. Original introduction of $\widehat{G}_i(z)$, $\widehat{G}_i^*(z)$, and $\widehat{G}_i^*(z) \equiv \widehat{G}_i(z)$

Let (15.1) be an equation of order $m \geq 2$ on Ω and let $y(z) = \rho(z)\,v(z)$ be a substitution as described for (15.2) that transforms (15.1) on Ω into (15.3) on Ω. Here, we are to ignore the content of Sections 18.1 and 18.2 in order to view the way that Georges-Henri Halphen in [**32**] and Andrew Forsyth in [**28**] introduced the Cockle-semi-invariants of the first kind that were essential for their constructions.

18.3.1. Halphen canonical form for (15.1).
Let $\rho_1(z)$ be a meromorphic function on a subregion $\mathcal{U}_1$ of Ω such that $\rho_1^{(1)}(z) + C_1(z)\,\rho_1(z) \equiv 0$ on $\mathcal{U}_1$ and $\rho_1(z) \not\equiv 0$. Then, we use (15.4) to see that the substitution $y(z) = \rho_1(z)\,t(z)$ transforms the restriction to $\mathcal{U}_1$ of (15.1) into the equation on $\mathcal{U}_1$ given by

$$(18.14) \qquad t^{(m)}(z) + \sum_{i=2}^{m} \binom{m}{i} \widehat{G}_i(z)\, t^{(m-i)}(z) = 0, \quad \text{with } \widehat{G}_1(z) \equiv 0,$$

where explicit expressions for $\widehat{G}_2(z), \ldots, \widehat{G}_m(z)$ are obtained by substituting

$$\frac{\rho_1^{(1)}(z)}{\rho_1(z)} \equiv -C_1(z), \quad \frac{\rho_1^{(2)}(z)}{\rho_1(z)} \equiv -C_1^{(1)}(z) + \left(\frac{\rho_1^{(1)}(z)}{\rho_1(z)}\right)^2 \equiv -C_1^{(1)}(z) + \left(C_1(z)\right)^2,$$

and

$$\frac{\rho_1^{(k+1)}(z)}{\rho_1(z)} \equiv \left(\frac{\rho_1^{(k)}(z)}{\rho_1(z)}\right)' + \frac{\rho_1^{(k)}(z)}{\rho_1(z)}\frac{\rho_1^{(1)}(z)}{\rho_1(z)}, \quad \text{for } k \geq 2,$$

into (15.4) to obtain $\widehat{G}_1(z) \equiv 0$, $\widehat{G}_2(z) \equiv C_2(z) - \left(C_1(z)\right)^2 - C_1^{(1)}(z)$, ... But, the coefficients of (18.14) are defined on all of Ω and they are uniquely specified by the given (15.1). In this way, apart from the selection of the variable t, the equation (15.1) on Ω uniquely determines (18.14) on Ω as its *Halphen canonical form*.

18.3.2. Deduction of $\widehat{G}_i^*(z) \equiv \widehat{G}_i(z)$.
Just as (15.1) on Ω specifies (18.14) on Ω as its Halphen canonical form, the equation (15.3) on Ω specifies

$$(18.15) \qquad t^{(m)}(z) + \sum_{i=2}^{m} \binom{m}{i} \widehat{G}_i^*(z)\, t^{(m-i)}(z) = 0, \quad \text{with } \widehat{G}_1^*(z) \equiv 0,$$

on Ω as its Halphen canonical form where $\widehat{G}_2^*(z) \equiv C_2^*(z) - \left(C_1^*(z)\right)^2 - C_1^{*(1)}(z)$, and so on. For $1 \leq i \leq m$, $\widehat{G}_i^*(z)$ was regarded as obtained from $\widehat{G}_i(z)$ by replacing in $\widehat{G}_i(z)$ each $C_j^{(k)}(z)$ from (15.1) with the corresponding $C_j^{*(k)}(z)$ from (15.3).

With reference to $\mathcal{U}_1$ for the local transformation $y(z) = \rho_1(z)\,t(z)$ of (15.1) on $\mathcal{U}_1$ into (18.14) on $\mathcal{U}_1$, we observe that the substitution $v(z) = (1/\rho(z))y(z)$ transforms the restriction to $\mathcal{U}_1$ of (15.3) into the restriction to $\mathcal{U}_1$ of (15.1). Hence, the substitution $v(z) = (\rho_1(z)/\rho(z))t(z)$ transforms the restriction to $\mathcal{U}_1$ of (15.3) into the restriction to $\mathcal{U}_1$ of (18.14). Consequently, both (18.14) and (18.15) are Halphen canonical forms for (15.3). *This requires $\widehat{G}_i^*(z) \equiv \widehat{G}_i(z)$, for $2 \leq i \leq m$.*

Since Georges-Henri Halphen in [**32**] of 1884, Andrew Forsyth in [**28**] of 1888, and other researchers did not employ polynomials into which substitutions from (15.1) or (15.3) or (15.6) could be performed, the function $\widehat{G}_i(z)$, for $2 \leq i \leq m$, was referred to as *an isobaric semi-invariant of the first kind having weight i.*

Sections 18.1 and 18.2 provide clarification. See Theorems 18.4 and 18.5.

18.4. Results of Forsyth in the context for Sections 18.1 and 18.2

Andrew Forsyth employed infinitesimal transformations in [**28**, pages 398–401] of 1888 to derive a necessary structure for Laguerre-Halphen relative invariants of weights $s = 3, 4, 5, 6, 7$ for equations (15.1) of order $m \geq s$. The deduction for the weight $s = 3$ in [**28**, page 398, (14)] corresponds to the notation

$$(18.16) \quad \widehat{\Theta}_3(z) \equiv \widehat{G}_3(z) - (3/2)\widehat{G}_2^{(1)}(z), \quad \text{for equations (15.1) of order } m \geq 3.$$

The right member of (18.16) is equal to the right member of (17.4) and was known to Georges-Henri Halphen in [**32**] of 1884. That it does not involve m was noted on page 169. In that regard, the notation of Sections 18.1 and 18.2 yields the identity

$$(18.17) \quad \widehat{\Theta}_3 \equiv \widehat{G}_3 - (3/2)\widehat{G}_2^{(1)} \equiv \widehat{H}_3,$$

where $\widehat{H}_3$ is given by (18.4).

With respect to the notation of Section 18.1 on page 171, we use (18.8) to see that the polynomials of interest that correspond to the formulas for $\widehat{\Theta}_4(z)$, $\widehat{\Theta}_5(z)$, $\widehat{\Theta}_6(z)$, and $\widehat{\Theta}_7(z)$ in [**28**, pages 399–401, (15), (16),(17),(18)] are given by

$$(18.18) \quad \widehat{\Theta}_4 \equiv \widehat{G}_4 - 2\widehat{G}_3^{(1)} + (6/5)\widehat{G}_2^{(2)} - \frac{3(5m+7)}{5(m+1)}\left(\widehat{G}_2\right)^2,$$

for equations (15.1) of order $m \geq 4$,

$$(18.19) \quad \widehat{\Theta}_5 \equiv \widehat{G}_5 - \frac{5}{2}\widehat{G}_4^{(1)} + \frac{15}{7}\widehat{G}_3^{(2)} - \frac{5}{7}\widehat{G}_2^{(3)} - \frac{10(7m+13)}{7(m+1)}\widehat{G}_2\widehat{\Theta}_3,$$

for equations (15.1) of order $m \geq 5$,

$$(18.20) \quad \widehat{\Theta}_6 \equiv \widehat{G}_6 - 3\widehat{G}_5^{(1)} + (10/3)\widehat{G}_4^{(2)} - (5/3)\widehat{G}_3^{(3)} + (5/14)\widehat{G}_2^{(4)}$$
$$+ \frac{30(7m^2+28m+25)}{7(m+1)^2}\left(\widehat{G}_2\right)^3 + \frac{5(7m+8)}{14(m+1)}\left(\widehat{G}_2^{(1)}\right)^2$$
$$- 5\frac{3m+7}{m+1}\widehat{G}_2\left(\widehat{G}_4 - 2\widehat{G}_3^{(1)} + \frac{2(14m+31)}{7(3m+7)}\widehat{G}_2^{(2)}\right),$$

for equations (15.1) of order $m \geq 6$,

and

$$(18.21) \quad \widehat{\Theta}_7 \equiv \widehat{G}_7 - \frac{7}{2}\widehat{G}_6^{(1)} + \frac{105}{22}\widehat{G}_5^{(2)} - \frac{35}{11}\widehat{G}_4^{(3)} + \frac{35}{33}\widehat{G}_3^{(4)} - \frac{7}{44}\widehat{G}_2^{(5)}$$
$$- \frac{7\widehat{G}_2}{11(m+1)}\left[\begin{array}{l}(3/2)(11m+31)\left(2\widehat{G}_5 - 5\widehat{G}_4^{(1)}\right)\\ + 5(15m+41)\widehat{G}_3^{(2)} - 15(2m+5)\widehat{G}_2^{(3)}\end{array}\right]$$
$$- \frac{7(3m+4)}{11(m+1)}\left[3\widehat{G}_2^{(2)}\left(\widehat{G}_3 + \widehat{G}_2^{(1)}\right) - 5\widehat{G}_2^{(1)}\widehat{G}_3^{(1)}\right]$$
$$+ \frac{21(55m^2+288m+329)}{11(m+1)^2}\left(\widehat{G}_2\right)^2\widehat{\Theta}_3,$$

for equations (15.1) of order $m \geq 7$,

except that: the denominator $11(m+1)^2$ appearing in the last fraction of (18.21) is a correction for the denominator $22(m+1)^2$ that [**28**, page 401, (18)] would give. For details about that misprint, see [**19**, page 79].

The restriction to infinitesimal transformations is removed in Theorem 18.7.

18.4. RESULTS OF FORSYTH IN THE CONTEXT FOR SECTIONS 18.1 AND 18.2

18.4.1. Deduction for $\widehat{G}_i^{(k)}$. For any polynomial $\widehat{P}$ in $\mathfrak{R}_{m,1}$ of page 171 and the derivation ′ for $\mathfrak{R}_{m,1}$, we write $\widehat{P}^{(0)} \equiv \widehat{P}$, $\widehat{P}^{(1)} \equiv (\widehat{P}^{(0)})'$, $\widehat{P}^{(2)} \equiv (\widehat{P}^{(1)})'$, ... Thus, for any $k \geq 0$, we use the notation $\widehat{P}^{(k)}$ for the polynomial in $\mathfrak{R}_{m,1}$ obtained from $\widehat{P}$ by repeatedly applying k times the derivation ′ defined for $\mathfrak{R}_{m,1}$.

We recall that $\widehat{P}(z)$ designates the function on Ω obtained from $\widehat{P}$ by replacing each $W_i^{(j)}$ in $\widehat{P}$ with the corresponding $C_i^{(j)}(z)$ from (15.1). However, due to properties of the derivation ′ for $\mathfrak{R}_{m,1}$, we see that the function $(\widehat{P}^{(k)})(z)$ on Ω that is obtained from $\widehat{P}^{(k)}$ by replacing each $W_i^{(j)}$ in $\widehat{P}^{(k)}$ with the corresponding $C_i^{(j)}(z)$ from (15.1) is equal to the kth derivative with respect to z of $\widehat{P}(z)$. Similarly, the function $(\widehat{P}^{(k)})^*(z)$ on Ω obtained from $\widehat{P}^{(k)}$ by replacing each $W_i^{(j)}$ in $\widehat{P}^{(k)}$ with the corresponding $C_i^{*(j)}(z)$ from (15.3) is equal to the kth derivative with respect to z of the function $\widehat{P}^*(z)$ obtained from $\widehat{P}$ by replacing each $W_i^{(j)}$ in $\widehat{P}$ with the corresponding $C_i^{*(j)}(z)$ from (15.3).

PROPOSITION 18.6. *For $2 \leq i \leq m$ and $k \geq 0$, $\widehat{G}_i^{(k)}$ is an isobaric Cockle-semi-invariant of the first kind having weight $i+k$.*

PROOF. For $2 \leq i \leq m$ and $k \geq 0$, we use (18.8) to see that the coefficient of $W_i^{(k)}$ in $\widehat{G}_i^{(k)}$ is 1 and $\widehat{G}_i^{(k)}$ is an isobaric polynomial of weight $i+k$. Theorem 18.4 on page 173 yields $\widehat{G}_i^*(z) \equiv \widehat{G}_i(z)$, on Ω, from which we deduce

$$(18.22) \qquad (\widehat{G}_i^{(k)})^*(z) \equiv \frac{d^k}{dz^k} \widehat{G}_i^*(z) \equiv \frac{d^k}{dz^k} \widehat{G}_i(z) \equiv (\widehat{G}_i^{(k)})(z), \quad \text{on } \Omega.$$

We compare (18.22) with (18.6) of Definition 18.1 on page 172 to complete the proof. □

18.4.2. Properties of $\widehat{\Theta}_3$, $\widehat{\Theta}_4$, $\widehat{\Theta}_5$, $\widehat{\Theta}_6$, and $\widehat{\Theta}_7$. We know that $\widehat{\Theta}_3$ in (18.17) is a Laguerre-Halphen relative invariant for equations (15.1) of order $m \geq 3$ because $\widehat{H}_3$ in (18.4) on page 172 has that property with respect to Definition 18.3.

THEOREM 18.7. *For $3 \leq s \leq 7$, $\widehat{\Theta}_s$ is a Laguerre-Halphen relative invariant of weight s for the equations (15.1) of order $m \geq s$.*

PROOF. Let s satisfy $3 \leq s \leq 7$. We use (18.17)–(18.21) to see that the coefficient of W_s in $\widehat{\Theta}_s$ is equal to the coefficient 1 of W_s in $\widehat{G}_s$. Since Definition 18.1 on page 172 shows that a nonzero sum of Cockle-semi-invariants of the first kind is a Cockle-semi-invariant of the first kind, we apply Proposition 18.6 to conclude that $\widehat{\Theta}_s$ is an isobaric Cockle-semi-invariant of the first kind having weight s.

We establish in Section 18.5 that: when an equation (15.1) on Ω is transformed by $z = f(\zeta)$ of (15.5) into a corresponding equation (15.6) on Ω^{**}, the identity

$$(18.23) \qquad \widehat{\Theta}_s^{**}(\zeta) - (f'(\zeta))^s \widehat{\Theta}_s(f(\zeta)) \equiv 0, \quad \text{on } \Omega^{**} \text{ for } 3 \leq s \leq 7,$$

is valid, where $\widehat{\Theta}_s(z)$ on Ω and $\widehat{\Theta}_s^{**}(\zeta)$ on Ω^{**} are obtained by replacing each $W_i^{(j)}$ in $\widehat{\Theta}_s$ with the corresponding $C_i^{(j)}(z)$ from (15.1) and $C_i^{**(j)}(\zeta)$ from (15.6). In view of (18.23) and Definitions 18.2–18.3, we conclude that $\widehat{\Theta}_s$ is a Laguerre-Halphen relative invariant of weight s. □

18.5. Computer-algebra verification of (18.23)

After selecting a version of *Mathematica* from [55, 56, 57, 58, 59] as the system, we recall from page 167 that the evaluations of

```
Ce[m_,0][z_] := 1

alpha[0,j_][zeta_] := 1

alpha[i_,j_][zeta_] := ( Sum[alpha[i-1,k]'[zeta]
  -(i-1+k)(f''[zeta]/f'[zeta])*
  alpha[i-1,k][zeta],{k,1,j}] ) /; i >= 1

beta[m_,r_,s_][zeta_]:=(Product[(m-s-k+1),{k,1,r}]/
  Product[(s+k),{k,1,r}])*alpha[r,s][zeta]

CeSS[m_,i_][zeta_] := Sum[beta[m,i-j,m-i][zeta]*
  (f'[zeta])^j*Ce[m, j][f[zeta]],{j,0,i}]
```

provide representations for the coefficients $C_1^{**}(\zeta)$, $C_2^{**}(\zeta)$, ... of (15.6). We use (18.7) and (18.8) to see that the evaluations of

```
F[0][z_] := 1

F[1][z_] := - Ce[m,1][z]

F[i_][z_] := F[i-1]'[z]+F[1][z]*F[i-1][z] /; i >= 2

G[i_][z_] := Sum[Binomial[i,j]*F[i-j][z]*Ce[m,j][z],{j,0,i}]

FSS[0][zeta_] := 1

FSS[1][zeta_] := - CeSS[m,1][zeta]

FSS[i_][zeta_] :=
  FSS[i-1]'[zeta]+FSS[1][zeta]*FSS[i-1][zeta] /; i >= 2

GSS[i_][zeta_] :=
  Sum[Binomial[i,j]*FSS[i-j][zeta]*CeSS[m,j][zeta],{j,0,i}]
```

provide representations for the function $\widehat{G}_i(z)$ on Ω and $\widehat{G}_i^{**}(\zeta)$ on Ω^{**} obtained from $\widehat{G}_i$ by replacing each W_j in $\widehat{G}_i$ with the corresponding $C_j(z)$ from (15.1) or $C_j^{**}(\zeta)$ from (15.6). In view of (18.17)–(18.21), we observe that the evaluations of

```
Theta[3][z_] := ( G[3][z]-(3/2)G[2]'[z] )

ThetaSS[3][zeta_] := ( GSS[3][zeta]-(3/2)GSS[2]'[zeta] )

Theta[4][z_] := ( G[4][z]-2G[3]'[z]+(6/5)G[2]''[z]
  -(3/5)((5m+7)/(m+1))G[2][z]^2 )
```

18.5. COMPUTER-ALGEBRA VERIFICATION OF (18.23)

```
ThetaSS[4][zeta_] := ( GSS[4][zeta]-2GSS[3]'[zeta]
  +(6/5)GSS[2]''[zeta]-(3/5)((5m+7)/(m+1))GSS[2][zeta]^2 )

Theta[5][z_] := ( G[5][z]-(5/2)G[4]'[z]+(15/7)G[3]''[z]
  -(5/7)G[2]'''[z]-(10/7)((7m+13)/(m+1))G[2][z]*Theta[3][z] )

ThetaSS[5][zeta_] := ( GSS[5][zeta]-(5/2)GSS[4]'[zeta]
  +(15/7)GSS[3]''[zeta]-(5/7)GSS[2]'''[zeta]
  -(10/7)((7m+13)/(m+1))GSS[2][zeta]*ThetaSS[3][zeta] )

Theta[6][z_] := ( G[6][z]-3G[5]'[z]+(10/3)G[4]''[z]
  -(5/3)G[3]'''[z]+(5/14)G[2]''''[z]
  +(30/7)G[2][z]^3*((7m^2+28m+25)/(m+1)^2)
  +(5/14)((7m+8)/(m+1))G[2]'[z]^2
  -5((3m+7)/(m+1))G[2][z]*
      (G[4][z]-2G[3]'[z]+(2/7)((14m+31)/(3m+7))G[2]''[z]) )

ThetaSS[6][zeta_] := ( GSS[6][zeta]-3GSS[5]'[zeta]
  +(10/3)GSS[4]''[zeta]-(5/3)GSS[3]'''[zeta]
  +(5/14)GSS[2]''''[zeta]
  +(30/7)GSS[2][zeta]^3*((7m^2+28m+25)/(m+1)^2)
  +(5/14)((7m+8)/(m+1))GSS[2]'[zeta]^2
  -5((3m+7)/(m+1))GSS[2][zeta]*(GSS[4][zeta]
     -2GSS[3]'[zeta]+(2/7)((14m+31)/(3m+7))GSS[2]''[zeta]) )

Theta[7][z_] := ( G[7][z]-(7/2)G[6]'[z]+(105/22)G[5]''[z]
  -(35/11)G[4]'''[z]+(35/33)G[3]''''[z]-(7/44)G[2]'''''[z]
  -(7/11)(G[2][z]/(m+1))*( (3/2)(11m+31)(2G[5][z]-5G[4]'[z])
     +5(15m+41)G[3]''[z]-15(2m+5)G[2]'''[z] )
  -(7/11)((3m+4)/(m+1))*( 3G[2]''[z]( G[3][z]+G[2]'[z] )
     - 5G[2]'[z]*G[3]'[z] )
  +G[2][z]^2*Theta[3][z]*((1155m^2+6048m+6909)/(11(m+1)^2)) )

ThetaSS[7][zeta_] := ( GSS[7][zeta]-(7/2)GSS[6]'[zeta]
  +(105/22)GSS[5]''[zeta]-(35/11)GSS[4]'''[zeta]
  +(35/33)GSS[3]''''[zeta]-(7/44)GSS[2]'''''[zeta]
  -(7/11)(GSS[2][zeta]/(m+1))*( (3/2)(11m+31)(2GSS[5][zeta]
     -5GSS[4]'[zeta])+5(15m+41)GSS[3]''[zeta]
     -15(2m+5)GSS[2]'''[zeta] )
  - (7/11)((3m+4)/(m+1))( 3GSS[2]''[zeta]*
     (GSS[3][zeta]+GSS[2]'[zeta])-5GSS[2]'[zeta]*GSS[3]'[zeta])
  +GSS[2][zeta]^2*ThetaSS[3][zeta]*
     (21(55m^2+288m+329)/(11(m+1)^2)) )
```

represent the functions $\widehat{\Theta}_s(z)$ on Ω and $\widehat{\Theta}_s^{**}(\zeta)$ on Ω^{**}, for $3 \leq s \leq 7$, that appear

180   18. SUITABLE CONTEXT FOR OLDER NOTATION

in (18.23) on page 177. Since the evaluation for each of

```
FullSimplify[ ThetaSS[3][zeta] - f'[zeta]^3*Theta[3][f[zeta]] ]

FullSimplify[ ThetaSS[4][zeta] - f'[zeta]^4*Theta[4][f[zeta]] ]

FullSimplify[ ThetaSS[5][zeta] - f'[zeta]^5*Theta[5][f[zeta]] ]

FullSimplify[ ThetaSS[6][zeta] - f'[zeta]^6*Theta[6][f[zeta]] ]

FullSimplify[ ThetaSS[7][zeta] - f'[zeta]^7*Theta[7][f[zeta]] ]
```

is zero, we conclude that (18.23) is valid. A *Mathematica* notebook containing the preceding evaluations can be downloaded from
    http://homepages.uc.edu/~chalklr/Chapter-18.html
with the Google browser *Chrome*. It illustrates well the technique of Chapter 17.

18.6. Several observations

A different argument to verify Theorem 18.7 was employed for [**19**, page 79]. There, after finding explicit formulas for all of the basic relative invariants of the equations (15.9), we used computer algebra with the substitution $w_i^{(j)} = \binom{m}{i} W_i^{(j)}$ in the basic relative invariants $\mathcal{I}_{m,1;3}, \ldots, \mathcal{I}_{m,1;7}$ to verify that

$$(18.24) \qquad \mathcal{I}_{m,1;s} \equiv \binom{m}{s} \widehat{\Theta}_s, \quad \text{for } 3 \leq s \leq 7 \text{ and } m \geq s,$$

where $\widehat{\Theta}_3, \ldots, \widehat{\Theta}_7$ are given by (18.17)–(18.21). The properties of $\widehat{\Theta}_s$ as a Laguerre-Halphen relative invariant for the equations (15.1) then follow from properties of $\mathcal{I}_{m,1;s}$ as a relative invariant for the equations (15.9).

The formulas for $\widehat{\Theta}_3(z), \widehat{\Theta}_4(z), \widehat{\Theta}_5(z), \widehat{\Theta}_6(z)$, and $\widehat{\Theta}_7(z)$ in [**28**, pages 398–401] and their rewritten versions appearing in [**8**, page 235] did not lead to general results. Francesc Brioschi introduced errors when he rewrote $\widehat{\Theta}_7(z)$ for [**8**, page 235] of 1891 and those errors were copied in the expression for $\widehat{\Theta}_7(z)$ that Ludwig Schlesinger included in [**47**, page 196] of 1897.

18.7. Computer-algebra verification of (18.24)

We continue with the *Mathematica* notebook that was begun on page 178 and includes the sixteen commands of page 178, the seven commands of page 179, and the five commands above. At this point, the evaluations of `Theta[3][z]`, `Theta[4][z]`, `Theta[5][z]`, `Theta[6][z]`, `Theta[7][z]` are representations for the functions $\widehat{\Theta}_3(z), \widehat{\Theta}_4(z), \widehat{\Theta}_5(z), \widehat{\Theta}_6(z)$, and $\widehat{\Theta}_7(z)$ on Ω that are obtained by replacing each $W_j^{(k)}$ in $\widehat{\Theta}_3, \widehat{\Theta}_4, \widehat{\Theta}_5, \widehat{\Theta}_6$, and $\widehat{\Theta}_7$ of (18.17)–(18.21) with $C_j^{(k)}(z)$ from (15.1). Next, we evaluate

```
Ce[m_,i_][z_] = W[i][z]
```

and recognize that the evaluations of `Theta[3][z]`, `Theta[4][z]`, `Theta[5][z]`, `Theta[6][z]`, `Theta[7][z]` now represent the polynomials $\widehat{\Theta}_3, \widehat{\Theta}_4, \widehat{\Theta}_5, \widehat{\Theta}_6$, and $\widehat{\Theta}_7$ of (18.17)–(18.21) in the variables $W_j^{(k)}$ over $\mathbb{Q}$.

18.7. COMPUTER-ALGEBRA VERIFICATION OF (18.24)

With respect to the equations (15.9) on page 158, in order for the evaluations of basicInv[m,1,3][z], basicInv[m,1,4][z], basicInv[m,1,5][z] as well as basicInv[m,1,6][z] and basicInv[m,1,7][z], to represent the basic relative invariants $\mathcal{I}_{m,1;3}$, $\mathcal{I}_{m,1;4}$, $\mathcal{I}_{m,1;5}$, $\mathcal{I}_{m,1;6}$, and $\mathcal{I}_{m,1;7}$ as polynomials over $\mathbb{Q}$ in the variables $w_i^{(j)}$, for $1 \le i \le m$ and $j \ge 0$, we evaluate the input commands

```
a[m_,1][z_] := (1/Binomial[m+1,3])( w[2][z]
  -((m-1)/2)w[1]'[z]-((m-1)/(2m))w[1][z]^2 )

d[m_,1][z_] := (1/(m(m-1)))w[1][z]

K[m_,1,i_,j_][z_] := 0 /; i <= -1

K[m_,1,0,j_][z_] := 1

K[m_,1,i_,j_][z_] :=
  ( Sum[( D[K[m,1,i-1,k][z],z]
  -(m-1)*d[m,1][z]*K[m,1,i-1,k][z]
  +(m+2-i-k)(2-i-k)a[m,1][z]*
  K[m,1,i-2,k][z]),{k,j+1,m}] ) /; i >= 1

w[0][z_] = 1;    X[k_][z_] := w[k][z]

L[m_,1,i_][z_] :=
  Sum[ K[m,1,i-j,j][z]*X[j][z], {j, 0, i}]

M[m_,1,e1_,i_][z_] :=
  FunctionExpand[Binomial[m-i,e1-i]]*
  Product[(e1-r), {r,1,e1-i}]*L[m,1,i][z]

A[e1_,i_] := -1/(e1+i-1) /; i >= 1

B[e1_,i_] := (e1-i)/(e1+i-2) /; i >= 1

inv[m_,1,e1_,0][z_] := 0

inv[m_,1,e1_,1][z_] := 0

inv[m_,1,e1_,i_][z_] := ( M[m,1,e1,i][z]
  +A[e1,i-1]*D[ inv[m,1,e1,i-1][z], z]
  +B[e1,i-1]*a[m,1][z]*
  inv[m,1,e1,i-2][z] ) /; i >= 2

basicInv[m_,1,e1_][z_] := inv[m,1,e1,e1][z]
```

from pages 53–54 of Section 6.1. For the relation $w_j^{(k)} = \binom{m}{j} W_j^{(k)}$, we evaluate

```
binomial[m_,i_] := Product[m-k,{k,0,i-1}]/i!

w[i_][z_] := binomial[m,i]*W[i][z]
```

where the first of these two input commands enables $\binom{m}{i}$ to be evaluated for any nonnegative integer i even when m is merely a symbol. Since the evaluations for each of the input commands

```
FullSimplify[ basicInv[m,1,3][z] - binomial[m,3]*Theta[3][z] ]

FullSimplify[ basicInv[m,1,4][z] - binomial[m,4]*Theta[4][z] ]

FullSimplify[ basicInv[m,1,5][z] - binomial[m,5]*Theta[5][z] ]

FullSimplify[ basicInv[m,1,6][z] - binomial[m,6]*Theta[6][z] ]

FullSimplify[ basicInv[m,1,7][z] - binomial[m,7]*Theta[7][z] ]
```

is zero, we conclude that (18.24) on page 180 is valid. The *Mathematica* notebook that is downloadable from
> http://homepages.uc.edu/~chalklr/Chapter-18.html

with the Google browser *Chrome* contains evaluations for the input statements of this chapter.

18.8. Brief summary

The subject needed a simpler notation, precise definitions, explicit formulas of a general character for the coefficients of equations resulting from a change of the independent variable, and a symmetrical development with respect to the two types of semi-invariants. Instead, after the research of Andrew Forsyth in [28] of 1888, the subject was identified with the performance of infinitesimal transformations. For example, see [7] of 1899 and [53] of 1906. Biographies of Georges-Henri Halphen reveal the attitudes that prevailed by incorrectly implying the subject of invariants for differential equations was merely a detail in the theory of continuous groups. Also, since Halphen's research about invariants did not fit into that context, it should have been praised rather than claimed by biographers to be "no longer in the mainstream." Thus, because the subject had become so thoroughly muddled, it needed the complete redevelopment that we began in 1989.

There are numerous areas of mathematics where considerable effort would be required for a neophyte to understand the contributions made by experts or to fit those contributions into an interesting historical perspective.

In contrast, the subject of relative invariants has a long history. Moreover, it should now be intelligible to anyone knowledgeable about the differential calculus and the concept of a polynomial in algebra.

We are pleased to have advanced this remarkable area of mathematics.

Bibliography

1. P. Appell, *Sur les équations différentielles algébriques et homogènes par rapport à la fonction inconnue et à ses dérivées*, C. R. Acad. Sci. Paris **104** (1887), 1776–1779.
2. _____, *Sur une classe d'équations réductibles aux équations linéaires*, C. R. Acad. Sci. Paris **107** (1888), 776–778.
3. _____, *Équations différentielles homogènes du second ordre à coefficients constants*, Ann. Fac. Sci. Toulouse Math. (1) **3** (1889), K1–K12.
4. _____, *Sur les invariants de quelques équations différentielles*, J. Math. Pures Appl. (4) **5** (1889), 361–423.
5. N. Bourbaki, *Éléments de Mathématique, Livre II, Algèbre, Chapitres IV et V*, Hermann, Paris, 1950.
6. _____, *Elements of Mathematics, Algebra II, Chapters 4–7*, Springer-Verlag, Berlin, 1989.
7. C. L. Bouton, *Invariants of the general linear differential equation and their relation to the theory of continuous groups*, Amer. J. Math. **21** (1899), 25–84.
8. F. Brioschi, *Les invariants des équations différentielles linéaires*, Acta Math. **14** (1891), 233–248.
9. D. Caligo, *Sopra una classe di equazioni differenziali non lineari*, Mem. Accad. Sci. Torino Cl. Sci. Fis. Mat. Nat. (3) **1** (1952), 1–24.
10. _____, *Sulla integrazione delle equazioni differenziali del secondo ordine a riferimento razionale*, Rend. Mat. Appl. (5) **11** (1952), 299–314.
11. R. Campbell, *Les Intégrales Eulériennes et Leurs Applications: Étude approfondie de la fonction gamma*, Dunod, Paris, 1966.
12. R. Chalkley, *On the second-order homogeneous quadratic differential equation*, Math. Ann. **141** (1960), 87–98.
13. _____, *New contributions to the related work of Paul Appell, Lazarus Fuchs, Georg Hamel, and Paul Painlevé on nonlinear differential equations whose solutions are free of movable branch points*, J. Differential Equations **68** (1987), 72–117.
14. _____, *Relative invariants for homogeneous linear differential equations*, J. Differential Equations **80** (1989), 107–153.
15. _____, *The differential equation $Q = 0$ in which Q is a quadratic form in y'', y', y having meromorphic coefficients*, Proc. Amer. Math. Soc. **116** (1992), 427–435.
16. _____, *A formula giving the known relative invariants for homogeneous linear differential equations*, J. Differential Equations **100** (1992), 379–404.
17. _____, *Semi-invariants and relative invariants for homogeneous linear differential equations*, J. Math. Anal. Appl. **176** (1993), 49–75.
18. _____, *A persymmetric determinant*, J. Math. Anal. Appl. **176** (1994), 107–117.
19. _____, *Basic Global Relative Invariants for Homogeneous Linear Differential Equations*, no. 744, Memoirs Amer. Math. Soc., Providence, 2002, 1–204.
20. _____, *Basic Global Relative Invariants for Nonlinear Differential Equations*, no. 888, Memoirs Amer. Math. Soc., Providence, 2007, 1–365.
21. _____, *Relative Invariants from 1879 Onward: Their Evolution for Differential Equations*, Lumina Press, Plantation, Florida, 2014, 1–145 + xviii.
22. J. Cockle, *Correlations of analysis*, The London, Edinburgh, and Dublin Philosophical Magazine and Journal of Science (4) **24** (1862), 531–534.
23. _____, *On a differential criticoid*, Philos. Mag. (4) **50** (1875), 440–446.
24. C. M. Cosgrove, *New family of exact stationary axisymmetric gravitational fields generalizing the Tomimatsu-Sato solutions*, J. Phys. A **10** (1977), 1481–1524.

25. _____, *A new formulation of the field equations for the stationary axisymmetric vacuum gravitational field I. general theory*, J. Phys. A **11** (1978), 365–382.
26. D. R. Curtiss, *On the invariants of a homogeneous quadratic differential equation of the second order*, Amer. J. Math. (1903), 365–382.
27. B. M. Dubrov, *Generalized Wilczynski invaiants for nonlinear ordinary differential equations*, Symmetries and overdetermined systems of partial differential equations (M. Eastwood, ed.), IMA Vol. Math. Appl., vol. 144, Proceedings of the IMA summer program, Minneapolis, MN, July 17 - August 4, 2006, Springer, New York, 2008, pp. 25–40.
28. A. R. Forsyth, *Invariants, covariants and quotient derivatives associated with linear differential equations*, Philosophical Transactions of the Royal Society of London **179** (1888), 377–489.
29. N. V. Grigorenko, *Web-based review in* Zentralblatt MATH. Zbl 1006.34084 of [19], European Mathematical Society, FIZ Karlsruhe & Springer-Verlag, online 2002 to time of this writing.
30. _____, *Web-based review in* Zentralblatt MATH. Zbl 1136.34001 of [20], European Mathematical Society, FIZ Karlsruhe & Springer-Verlag, online 2007 to time of this writing.
31. G.-H. Halphen, *Sur les invariants des équations différentielles linéaires du quatrième ordre*, Acta Math. **3** (1883), 325–380.
32. _____, *Mémoire sur la Réduction des Équations Différentielles Linéaires aux Formes Intégrables*, Mémoires présentés par divers savants à l'Académe des Sciences de l'Institut de France (2) **28** (1884), 1–301.
33. C. Hermit, H. Poincaré, and E. Rouché (eds.), *Oeuvres de Laguerre*, vol. 1, pp. 420–427, Gauthier-Villars, Paris, 1898.
34. O. Hölder, *Ueber die Eigenschaft der Gammafunction keiner algebraischen Differentialgleichung zu genügen*, Math. Ann. **28** (1887), 1–13.
35. C. Jordan, H. Poincaré, É. Picard, and E. Vessiot (eds.), *Oeuvres de G.-H. Halphen*, vol. 3, pp. 1–260 and 463–514, Gauthier-Villars, Paris, 1921.
36. E. R. Kolchin, *Differential Algebra and Algebraic Groups*, Academic Press, New York, 1973.
37. E. Laguerre, *Sur les équations différentielles linéaires du troisième ordre*, C. R. Acad. Sci. Paris **88** (1879), 116–119.
38. _____, *Sur quelques invariants des équations différentielles linéaires*, C. R. Acad. Sci. Paris **88** (1879), 224–227.
39. S. Lang, *Algebra*, Addison Wesley, New York, 1984.
40. G. Metzler, *Invariants and equations associated with the general linear differential equation*, Ph.D. thesis, John Hopkins University, Baltimore, 1891.
41. H. Morikawa, *On differential invariants of holomorphic projective curves*, Nagoya Math. J. **77** (1980), 75–87.
42. _____, *Some analytic and geometric applications of the invariant-theoretic methods*, Nagoya Math. J. **80** (1980), 1–47.
43. J. C. Ndogmo, *A method for the equivalence group and its infinitesimal generators*, J. Phys. A, Math. Theor. **41** (2008), no. 10.
44. _____, *Generating relative and absolute invariants of linear differential equations*, Int. Math. Forum **4** (2009), no. 17-20, 873–886.
45. F. Neuman, *Global properties of linear ordinary differential equations*, Kluwer, Dordrecht/Boston/London, 1991.
46. J. R. Ritt, *Differential Algebra*, Amer. Math. Soc. Colloq. Publ., vol. 33, Amer. Math. Soc., New York, 1950.
47. L. Schlesinger, *Handbuch der Theorie der linearen Differentialgleichungen*, vol. 2, Teubner, Leipzig, 1897.
48. Y. Se-ashi, *A geometric construction of Laguerre-Forsyth's canonical forms of linear ordinary differential equations*, Adv. Studies Pure Math. **22** (1993), 265–297.
49. P. R. Vein and P. Dale, *Determinants, their derivatives and nonlinear differential equations*, J. Math. Anal. Appl. **74** (1980), 599–634.
50. _____, *Determinants and Their Application in Mathematical Physics*, Applied Mathematical Sciences, vol. 134, Springer, New York, 1999, page 149, Theorem 4.53.
51. G. Wallenberg, *Ueber nichtlinear homogene Differentialgleichungen zweiter Ordnung*, J. Reine Angew. Math. **119** (1898), 87–113.
52. B. Weisfeiler, *Comments on differential invariants*, Infinite Dimensional Groups with Applications (V. Kac, ed.), vol. 4, Math. Sci. Res. Inst. Publ., 1985, pp. 355–370.

53. E. J. Wilczynski, *Projective Differential Geometry of Curves and Ruled Surfaces*, Teubner, Leipzig, 1906, Reprinted by Chelsea, New York, 1961.
54. Wolfram, *Mathematica*, Version 3.0, Wolfram Research Inc., Champaign, Illinois, 1996.
55. _____, *Mathematica*, Version 7.0.1, Wolfram Research Inc., Champaign, Illinois, 2009.
56. _____, *Mathematica*, Version 8.0.1, Wolfram Research Inc., Champaign, Illinois, 2011.
57. _____, *Mathematica*, Version 9.0.1, Wolfram Research Inc., Champaign, Illinois, 2013.
58. _____, *Mathematica*, Version 10.1, Wolfram Research Inc., Champaign, Illinois, 2015.
59. _____, *Mathematica*, Version 11.2, Wolfram Research Inc., Champaign, Illinois, 2017.

Index

$c_{i_1, i_2, \ldots, i_n}(z)$, coefficient of (4.1), 19, 24, 29
$c^*_{i_1, i_2, \ldots, i_n}(z)$, coefficient of (4.4), 19, 29
$c^{**}_{i_1, i_2, \ldots, i_n}(\zeta)$, coefficient of (4.7), 24, 29
m, order of (4.1), (4.4), and (4.7), 29
n, degree of (4.1), (4.4), and (4.7), 29
Ω, domain of (4.1) and (4.4), 29
Ω^{**}, domain of (4.7), 29
$y(z) = \rho(z)\, v(z)$, on Ω, 29
$z = f(\zeta)$, on $\Omega^{**} = g(\Omega)$, 29
$u(\zeta) \equiv (y \circ f)(\zeta)$, on Ω^{**}, 29
$\mathcal{C}_{m,n}$, class of equations, 27, 29
$w^{(k)}_{i_1, i_2, \ldots, i_n}$, variables, 30, 48
$\mathcal{R}_{m,n}$, polynomial ring over $\mathbb{Q}$, 30, 48
$\widehat{\mathcal{R}}_{m,n}$, polynomial ring over $\mathbb{C}$, 48
$\mathbb{Q}$, field of rational numbers, 3, 30, 48
$\mathbb{C}$, field of complex numbers, 48, 66
$\mathcal{Q}_{m,n}$, quotient field for $\mathcal{R}_{m,n}$, 72
$\mathfrak{F}_\Omega$, field of functions on Ω, 65
$'$, derivation for $\mathcal{R}_{m,n}$, 30
$'$, derivation for $\widehat{\mathcal{R}}_{m,n}$, 48
$F, P, Q, \ldots$ in $\mathcal{R}_{m,n}$, 30
$F(z), P(z), Q(z), \ldots$ on Ω, 30
$F^*(z), P^*(z), Q^*(z), \ldots$ on Ω, 30
$F^{**}(\zeta), P^{**}(\zeta), Q^{**}(\zeta), \ldots$ on Ω^{**}, 30
$C_{p,q,r}(\boldsymbol{P}, \boldsymbol{Q})$, differential polynomial specified by $p, q, r, \boldsymbol{P}$, and $\boldsymbol{Q}$, 36
$a_{m,n}$, 32, 36, 42, 50
$b_{m,n}$, 34, 46, 107
$d_{m,n}$, 32, 34, 44, 108
$e_{m,n}$, 116
$w^{(k)}_{i_1, i_2, \ldots, i_n}$, 30, 48
$A_{p,q,r,s,t}$, 36
$\boldsymbol{B}_{1,1}, \boldsymbol{B}_{1,2}, \boldsymbol{B}_{2,2}$, 66
$\boldsymbol{B}_{h,i,j}$, 36
$C_{p,q,0}(\boldsymbol{P}, \boldsymbol{Q})$, 36, 59
$C_{p,q,1}(\boldsymbol{P}, \boldsymbol{Q})$, 36, 59, 71
$C_{p,q,2}(\boldsymbol{P}, \boldsymbol{Q})$, 37, 60
$C_{p,q,3}(\boldsymbol{P}, \boldsymbol{Q})$, 37, 60
$C_{p,q,4}(\boldsymbol{P}, \boldsymbol{Q})$, 61
$C_{p,q,r}(\boldsymbol{P}, \boldsymbol{Q})$, 36
$\boldsymbol{D}_2$, 65, 66, 70

$\boldsymbol{E}_6$, 66, 69
$\boldsymbol{E}_7$, 66, 69
$\boldsymbol{F}_{m,n}$, 37
$\boldsymbol{F}_{p,q,r,s,t}$, 111
$\boldsymbol{F}_{s,i}$, 130–132
$\boldsymbol{G}_{m,n}$, 41
$\boldsymbol{G}_{p,q,r,s,t}$, 108
$\boldsymbol{H}_7$, 66
$\boldsymbol{H}_{h,i,j}$, 108
$\boldsymbol{H}_{m,n}$, 45
$\boldsymbol{I}_{m,n;\,e_1,\ldots,e_n;\,h_1,\ldots,h_{n-1},i}$, 33
$\boldsymbol{J}_{n;m,\,e_1,\ldots,e_n,\,h_1,\ldots,h_{n-1},i}$, 35
$\boldsymbol{K}_{h,i,j}$, 109
$\boldsymbol{K}_{m,n;\,i,j}$, 32
$\boldsymbol{L}_{h,i,j}$, 112
$\boldsymbol{L}_{m,n;\,i_1,\ldots,i_n}$, 32
$\boldsymbol{L}_{s,i}$, 38, 39
$\boldsymbol{M}_{h,i,j,\nu}$, 116
$\boldsymbol{M}_{m,n;\,e_1,\ldots,e_n;\,h_1,\ldots,h_{n-1},i}$, 33
$\boldsymbol{M}_{s,i}$, 127, 128
$\boldsymbol{N}_{p,q,r,s,t,k}$, 121
$\boldsymbol{N}_{s,i}$, 129
$\boldsymbol{R}$, 77, 107
$\boldsymbol{S}$, 108
$\boldsymbol{U}_{p,t,\mu,k}$, 113
$\boldsymbol{U}_{m,n;\,i,j}$, 34
$\boldsymbol{V}_{q,r,s,t,\mu,\nu,k}$, 113
$\boldsymbol{V}_{m,n;\,i_1,\ldots,i_n}$, 34
$\boldsymbol{W}_{m,n;\,e_1,\ldots,e_n,\,h_1,\ldots,h_{n-1},h_n}$, 35
$\boldsymbol{X}_{j_1, j_2, \ldots, j_n}$, 30
$\boldsymbol{Z}_{p,q,r,s,t,\mu,\nu}$, 111
$\mathcal{C}_{p,q,4}(\boldsymbol{P}, \boldsymbol{Q})$, 61
$\mathcal{I}_{m,n;\,e_1,\ldots,e_n}$, 32
$\mathcal{J}_{m,n;\,e_1,\ldots,e_n}$, 34
$\mathcal{P}_k$, 108
$\mathcal{Q}_k$, 108
$\mathcal{T}_j$, 110
$a^{**}_{m,n}(\zeta), a^{**}_{m,n}(g(z))$, 83
$c_{i_1, i_2, \ldots, i_n}(z)$, 19, 24, 29
$c^*_{i_1, i_2, \ldots, i_n}(z)$, 19, 29
$c^{**}_{i_1, i_2, \ldots, i_n}(\zeta)$, 24, 29
$d_{m,n}(s)$, 38
$f(\zeta), g(z), u(\zeta), v(z), y \circ f$, 29

INDEX

$A_{e_1,...,e_n,i}$, 33, 35
$B_{e_1,...,e_n,i}$, 33, 35
$E_{h,i,j}(z)$, 81
$F_1(z), F_2(z)$, 20
$F_{h,i,j,\nu}(z)$, 87
$G_1(z), G_2(\zeta)$, 25
$G_{p,q,r,s,t,k}(z)$, 92
$H_{m,n;e_1,...,e_n;h_1,...,h_n}$, 32, 34
$H_{p,q,r,t,k,\nu}$, 93
V, 3
$X_{p,q,r,s,t,\mu,\nu}(\zeta)$, 80
$Y_{p,t,v,k}(z)$, 82
$Z_{q,r,s,t,u,v,k}(z)$, 82
$\mathcal{A}_{j_1,j_2,...,j_n}^{i_1,i_2,...,i_n}(z)$, 19
$\mathcal{A}_{p,q,i}$, 99
$\mathcal{B}_{j_1,j_2,...,j_n}^{i_1,i_2,...,i_n}(\zeta)$, 24
$\mathcal{B}_{p,q,i}$, 100
$\mathcal{D}(z)$, 82
$\mathcal{F}_{j_1,j_2,...,j_n}^{i_1,i_2,...,i_n}(z)$, 20
$\mathcal{G}_{j_1,j_2,...,j_n}^{i_1,i_2,...,i_n}(\zeta)$, 25
$\mathcal{H}$, 27
$\mathcal{H}^*$, 27
$\mathcal{H}^{**}$, 27
$\mathcal{L}_{j,n}$, 47
$\mathcal{M}_{m,n}$, 48
$\mathcal{N}_{m,n}$, 47, 48
$\mathcal{O}(M)$, 101
$\mathcal{Q}_{m,n}$, 72
$\mathcal{R}_{m,n}$, 30
$\widehat{\mathcal{R}}_{m,n}$, 48
$\mathcal{V}_{m,n;s}$, 38
$\mathfrak{w}_1, \mathfrak{w}_2$, 42, 45
$\mathfrak{c}_i(z)$, 42, 45
$\mathfrak{c}_i^*(z)$, 42
$\mathfrak{c}_i^{**}(\zeta)$, 45
$\mathfrak{A}_{p,q,r,s,t}(\zeta)$, 79
$\mathfrak{A}_{p,q,r,s,t}(g(z))$, 81
$\mathfrak{C}_{p,q,r,\mu}$, 36, 108
$\mathfrak{F}_\Omega$, 65
$\mathfrak{J}$, 27
$\mathfrak{P}_\nu$, 89
$\mathfrak{S}_\nu$, 48
$\alpha_{i,j}(\zeta)$, 24
$\alpha_{p,q,i}$, 99
$\beta_{p,q,i}$, 100
$\gamma_{h,i,j}$, 99
π, 19, 24, 25, 29
$\rho(z)$, 29
σ, 30
$\phi_{h,i,j}(z)$, 77
$\Gamma_{h,i,j}$, 99
$\Theta_{p,q,r,t,k,\nu}$, 122
Ω, 29
Ω^{**}, 29
Com[r,P,Q], 140, 152
poly[p,q,r,P,Q][z], 60

Paul Émile Appell, xi, 1, 5, 9, 40, 65, 66
Appell's study of equations in $\mathcal{C}_{2,2}$ when
   a nontrivial fatorization exists, 65
   there is no nontrivial factorization and
     his condition (7.4) is satisfied, 65, 66

basic polynomial
   definition of, 31
   index of, 31
basic relative invariant
   $\mathcal{I}_{3,1;3}$ in $\mathcal{R}_{3,1}$ for $\mathcal{C}_{3,1}$, 2, 38
   $\mathcal{I}_{4,1;3}$ in $\mathcal{R}_{4,1}$ for $\mathcal{C}_{4,1}$, 4, 127
   $\mathcal{I}_{4,1;4}$ in $\mathcal{R}_{4,1}$ for $\mathcal{C}_{4,1}$, 4, 127
   $\mathcal{I}_{5,1;3}$ in $\mathcal{R}_{5,1}$ for $\mathcal{C}_{5,1}$, 128
   $\mathcal{I}_{5,1;4}$ in $\mathcal{R}_{5,1}$ for $\mathcal{C}_{5,1}$, 128
   $\mathcal{I}_{5,1;5}$ in $\mathcal{R}_{5,1}$ for $\mathcal{C}_{5,1}$, 128
   $\mathcal{I}_{2,2;1,1}$ in $\mathcal{R}_{2,2}$ for $\mathcal{C}_{2,2}$, 67, 130
   $\mathcal{I}_{2,2;1,2}$ in $\mathcal{R}_{2,2}$ for $\mathcal{C}_{2,2}$, 67, 130
   $\mathcal{I}_{2,2;2,2}$ in $\mathcal{R}_{2,2}$ for $\mathcal{C}_{2,2}$, 67, 130
   $\mathcal{I}_{m,1;3}$ in $\mathcal{R}_{m,1}$ for $\mathcal{C}_{m,1}$, 3
   $\mathcal{I}_{m,2;1,1}$ in $\mathcal{R}_{m,2}$ for $\mathcal{C}_{m,2}$, 57
   $\mathcal{I}_{m,2;1,2}$ in $\mathcal{R}_{m,2}$ for $\mathcal{C}_{m,2}$, 57
   $\mathcal{I}_{m,2;2,2}$ in $\mathcal{R}_{m,2}$ for $\mathcal{C}_{m,2}$, 58
   $\mathcal{I}_{m,n;e_1,...,e_n}$ in $\mathcal{R}_{m,n}$ for $\mathcal{C}_{m,n}$, 32–35
   $\mathcal{J}_{m,n;e_1,...,e_n}$ in $\mathcal{R}_{m,n}$ for $\mathcal{C}_{m,n}$, 34, 35
   definition of, 31
   index of, 31
basic relative invariants in $\mathcal{R}_{m,n}$ for $\mathcal{C}_{m,n}$
   explicit formulas for all of them, 32–35
   number of, 47
Francesco Brioschi, 9, 180

James Cockle, 8, 168, 171
constant of $\mathcal{R}_{m,n}$, 30

definition of
   $\mathcal{C}_{m,n}$, 27, 29
   $\mathcal{R}_{m,n}$, 30
   $\widehat{\mathcal{R}}_{m,n}$, 48
   basic polynomial, 31
   basic relative invariant, 31
   constant, 30
   differential polynomials, 30
   differential-polynomial combination, 30
   index of basic polynomial, 31
   index of basic relative invariant, 31
   relative invariant, 31
   semi-invariant of the first kind, 8, 31
   semi-invariant of the second kind, 8, 31
   the derivation ' for $\mathcal{R}_{m,n}$, 30
   the derivation ' for $\widehat{\mathcal{R}}_{m,n}$, 48
   the order-sum of a monomial, 101
differential polynomial $C_{p,q,r}(P, Q)$
   applications, 39, 69, 70, 127–132
   definition of, 36
   expansion for, 59–61
   for machine representations, 139, 151
   identities involving, 62, 74

INDEX

Andrew Russell Forsyth, 1, 4, 5, 9, 37, 169, 175, 176, 182

Georges-Henri Halphen, xi, 1, 3–5, 168, 171, 172, 175, 176, 182
hindrances for research before 1989
   counter-productive notation, 1, 157, 158, 171
   distracting infinitesimal transformations, 182
   misleading generalizations, 182
   missing precise definitions, 171, 172
   undiscovered key formulas
      for changes of independent variable, 21, 22, 157, 158, 171

identities
   for $\boldsymbol{C}_{p,q,1}(\boldsymbol{P}, \boldsymbol{Q})$, 74
   for $\boldsymbol{C}_{p,q,2}(\boldsymbol{P}, \boldsymbol{Q})$, 62
   $\boldsymbol{b}_{m,n} \equiv \boldsymbol{a}_{m,n} + \boldsymbol{d}_{m,n}^{(1)} + (\boldsymbol{d}_{m,n})^2$, 108
index of basic polynomial, 31
index of basic relative invariant, 31
invariant
   as a polynomial
      relative invariant, 31
      semi-invariant of the first kind, 31
      semi-invariant of the second kind, 31
   as a quotient of polynomials
      absolute invariant, 39, 40
      rational relative invariant, 40
      rational semi-invariant of first kind, 40
      rational semi-invariant of second kind, 40

Edmund Nicolas Laguerre, xi, 1, 2, 171
Laguerre-Forsyth canonical form, 5, 9
linear invariants of Forsyth, 5, 9

Mathematica
   program for
      $\boldsymbol{C}_{p,q,r}(\boldsymbol{P}, \boldsymbol{Q})$, 59–61, 139, 140, 151, 152
      $\boldsymbol{\mathcal{I}}_{m,1;\,e_1}$, 53–56
      $\boldsymbol{\mathcal{I}}_{m,2;\,e_1,e_2}$, 56–59
      $\boldsymbol{\mathcal{J}}_{m,1;\,e_1}$, 54–56
      $\boldsymbol{\mathcal{J}}_{m,2;\,e_1,e_2}$, 58, 59
      relative invariants in $\mathcal{R}_{m,1}$ of weight s, 134–138
      relative invariants in $\mathcal{R}_{m,2}$ of weight s, 146–150
      to illustrate concepts clearly, 161–163, 165, 166
   Version 3.0, 63, 143
   Versions 7.0.1, 8.0.1, 9.0.1, 10.1, 11.2, 10, 53, 54, 59, 63, 133, 135–143, 145, 147, 149, 151–153, 161, 167

František Neuman, 9
   global versus local, 9

Jules Henri Poincaré, xi

reference to work of
   Paul Émile Appell, xi, 5, 9, 65, 66
   Nicolas Bourbaki, 30, 48, 72
   Charles Leonard Bouton, 8, 9, 182
   Francesco Brioschi, 9, 180
   Domenico Caligo, 9
   Robert Campbell, 48
   Roger Chalkley, xi, 1, 3, 5–7, 9, 31–33, 35, 37, 40, 42, 61, 63, 65, 66, 70, 107, 132, 143, 157, 161, 176, 180
   James Cockle, 8, 168
   Christopher M. Cosgrove, 9
   David Raymond Curtiss, 9
   Paul Dale, 9
   Boris Mikhailovich Dubrov, 9
   Andrew Russell Forsyth, 4, 5, 9, 37, 169, 174, 175, 180, 182
   Georges-Henri Halphen, xi, 3, 4, 8, 42, 168, 169, 174–176
   Ludwig Otto Hölder, 48
   Ellis Robert Kolchin, 30, 48
   Edmund Nicolas Laguerre, xi, 1, 2, 8
   Serge Lang, 73
   George F Metzler, 9
   Hisasi Morikawa, 9
   Jean-Claude Ndogmo, 9
   František Neuman, 9
   Joseph Fels Ritt, 30, 48
   Ludwig Schlesinger, 9, 180
   Yutaka Se-ashi, 9
   P. Robert Vein, 9
   Georg Jacob Wallenberg, 9
   Borus Weisfeiler, 9
   Ernest Julius Wilczynski, 9, 182
   Wolfram Research Inc., 53–63, 67–70, 133–143, 145–153, 161, 167, 178
relative invariants
   all of weight $s \leq 13$ for $\mathcal{C}_{3,1}$, 38, 39
   all of weight $s \leq 12$ for $\mathcal{C}_{4,1}$, 127, 128
   all of weight $s \leq 12$ for $\mathcal{C}_{5,1}$, 128, 129
   all of weight $s \leq 12$ for $\mathcal{C}_{2,2}$, 130–132
   basic ones denoted by $\boldsymbol{\mathcal{I}}_2, \boldsymbol{\mathcal{I}}_3, \boldsymbol{\mathcal{I}}_4, \boldsymbol{\mathcal{I}}_5$
      depend on context where employed, 127–132
   defined, 31
   denoted by $\boldsymbol{D}_2$, $\boldsymbol{E}_6$, and $\boldsymbol{E}_7$ for $\mathcal{C}_{2,2}$
      defined, 66
      to illustrate Theorem 4.10, 69, 70
      to recognize the equations that satisfy Appell's condition of solvability, 66
   specified via $\boldsymbol{C}_{p,q,r}(\boldsymbol{P}, \boldsymbol{Q})$
      in $\mathcal{R}_{3,1}$ for $\mathcal{C}_{3,1}$, $\boldsymbol{L}_{s,i}$, 39, 140, 141
      in $\mathcal{R}_{4,1}$ for $\mathcal{C}_{4,1}$, $\boldsymbol{M}_{s,i}$, 127, 128, 141, 142
      in $\mathcal{R}_{5,1}$ for $\mathcal{C}_{5,1}$, $\boldsymbol{N}_{s,i}$, 129, 142
      in $\mathcal{R}_{2,2}$ for $\mathcal{C}_{2,2}$, $\boldsymbol{F}_{s,i}$, 130–132, 153

representation of $C_{p,q,r}(P, Q)$ for $r \geq 2$
   as a differential-polynomial combination
      of P, Q, and $a_{m,n}$, 59–61
   in terms of semi-invariants
      of the first kind, 36
      of the second kind, 107, 108, 123, 124
representations via computer algebra
   of $\boldsymbol{I}_{m,1;\,e_1}$ for $3 \leq e_1 \leq m$, 53–56
   of $\boldsymbol{J}_{m,1;\,e_1}$ for $3 \leq e_1 \leq m$, 54–56
   of $\boldsymbol{I}_{m,2;\,e_1,e_2}$ for $m \geq 2$
      and $max\{1, e_1\} \leq e_2 \leq m$, 56–59
   of $\boldsymbol{J}_{m,2;\,e_1,e_2}$ for $m \geq 2$
      and $max\{1, e_1\} \leq e_2 \leq m$, 58, 59
   of relative invariants having weight s
      for $\mathcal{C}_{m,1}$ with $m \geq 3$, 133–138
      for $\mathcal{C}_{m,2}$ with $m \geq 2$, 145–151
   of $C_{p,q,r}(P, Q)$ in terms of given
      machine representations for P and Q,
      139, 140, 151, 152

Ludwig Schlesinger, 9, 180
semi-invariant of the first kind, 8, 31, 41, 42
   $a_{m,n}$, 32, 36, 42, 108
semi-invariant of the second kind, 8, 31,
   42–47
   $b_{m,n}$, 34, 47, 108
set $\mathcal{C}_{m,n}$ of differential equations, 27
substitutions in differential polynomials, 30

Terminology for pre-1989 equations
   involving binomial coefficients
   Cockle-semi-invariant
      of the first kind, 172
      of the second kind, 172
   Laguerre-Halphen relative invriant, 172
transformation
   of first kind, $y(z) = \rho(z)\,v(z)$, 1, 3, 5, 29,
      31, 157, 158
   of second kind, $z = f(\zeta)$, 2, 3, 5, 24, 29,
      31, 158, 159

vector space $\mathcal{V}_{m,n;\,s}$ for $\mathcal{C}_{m,n}$, 38, 127
   basis element $\boldsymbol{E}_{s,i}$ for $\mathcal{V}_{m,n;\,s}$, 38
   basis element $\boldsymbol{F}_{s,i}$ for $\mathcal{V}_{2,2;\,s}$, 130
   basis element $\boldsymbol{L}_{s,i}$ for $\mathcal{V}_{3,1;\,s}$, 38
   basis element $\boldsymbol{M}_{s,i}$ for $\mathcal{V}_{4,1;\,s}$, 127
   basis element $\boldsymbol{N}_{s,i}$ for $\mathcal{V}_{5,1;\,s}$, 129

www.ingramcontent.com/pod-product-compliance
Lightning Source LLC
Chambersburg PA
CBHW062214220526
45471CB00009B/3195